ENHANCED RECOVERY
OF RESIDUAL AND HEAVY OILS

ENHANCED RECOVERY
OF RESIDUAL
AND HEAVY OILS

Second Edition

Edited by M.M. Schumacher

NOYES DATA CORPORATION

Park Ridge, New Jersey, U.S.A.

1980

Published in the United States of America by
Noyes Data Corporation
Noyes Building, Park Ridge, New Jersey 07656

Library of Congress Cataloging in Publication Data

Schumacher, M M
    Enhanced recovery of residual and heavy oils.

    (Energy technology review ; no. 59) (Chemical technology
review ; no. 174)
    Edition for 1978 published under title: Enhanced oil
recovery.
    Bibliography: p.
    Includes index.
    1. Secondary recovery of oil. I. Title. II. Se-
ries.
TN871.S34 1980   622'.3382   80-19812
ISBN 0-8155-0816-6

FOREWORD

The recovery of residual and heavy oils, from difficultly accessible reserves, has become increasingly attractive as the world price of oil has risen. Because of the need for additional methods to achieve self-sufficiency in energy production, this book, an updated edition of our 1978 title, *Enhanced Oil Recovery—Secondary and Tertiary Methods,* has been expanded to include more material on recovery methods for high viscosity (heavy) oil, information on possible methods of mining petroleum, and a discussion of carbon dioxide sources. In addition, case studies of various secondary and tertiary processes for recovery of residual oil have been presented, and discussions of the chemicals and methods employed and environmental aspects have been brought up-to-date for this volume.

The material included here will be of interest to chemical and petroleum engineers and anyone doing research relating to improved yields from residual or heavy crude reserves.

The data in this book are based on federally funded studies. Because the information is taken from many sources, it is possible that certain portions of this book may disagree or conflict with other parts of the book. This is especially true of monetary values and opinions of future potential. We chose to include these different points of view, however, in order to make the book more valuable to the reader. Cost figures provided are those given in the cited report, the date of which is always given in the bibliography. When the dates of the cost figures are given, we have included them in this book.

Advanced composition and production methods developed by Noyes Data are employed to bring these durably bound books to you in a minimum of time. Special techniques are used to close the gap between "manuscript" and "completed book." Technological progress is so rapid that time-honored, conventional typesetting, binding and shipping methods are no longer suitable. We have by-passed the delays in the conventional book publishing cycle and provide the user with an effective and convenient means of reviewing up-to-date information in depth.

The table of contents is organized in such a way as to serve as a subject index and provides easy access to the information contained in this book. Each chapter is followed by a list of references giving further details on these timely topics. The Sources Utilized section at the end of the volume lists the government reports upon which the book is based.

Some of the illustrations in this book may be less clear than could be desired; however, they are reproduced from the best material available to us.

CONTENTS AND SUBJECT INDEX

INTRODUCTION

When *Enhanced Oil Recovery—Secondary and Tertiary Methods* appeared two years ago, petroleum was in limited supply. Few people, however, anticipated that the shortage would reach the crisis stage it has in the two short years that have elapsed.

Technology is available to mitigate this crisis, but the application of such technology, as described in this text, in the best way possible to meet our needs remains a challenge. With the Organization of Petroleum Exporting Countries raising the price of crude oil periodically, it is expedient that the United States and its major allies implement these techniques to the fullest extent to reduce dependency on imported oil.

Oil-consuming nations have little choice if they wish to achieve energy self-sufficiency. They must explore every possibility. In addition to locating and exploiting new oil fields, including the promising ones offshore, they must extract as much residual oil as possible from already existing fields by applying more thorough recovery procedures.

Our petroleum engineers must also look to mining as a means of producing heretofore "unrecoverable" petroleum. By merging petroleum and mining technology they should be able to add considerably to our oil supply.

During the first half of the twentieth century, when new discoveries of oil were made repeatedly, there was little incentive to produce petroleum efficiently. Much was wasted when it was left behind in the reservoir. In most petroleum fields less than one-third of the original oil in place was recovered using only the original reservoir pressure to drive the oil into the producing well.

As petroleum has grown more precious, the petroleum industry has effected curbs to stop such waste by initiating measures to insure more exacting, more complete recovery. Furthermore, because many of the major domestic fields have passed into an advanced stage of decline, sophisticated techniques have become necessary to recover the oil still remaining in the field after primary production. These techniques have become known as enhanced oil recovery.

Enhanced oil recovery, then, is a collection of general methods, each with its own unique capability to drive the most oil from a particular reservoir. Each has been investigated rather thoroughly on a theoretical and laboratory basis, and to a lesser extent, in the field.

The early chapters of this book detail the various enhanced recovery methods for petroleum. Two chapters on mining follow. Selection of the appropriate enhanced recovery method, description of chemicals used, the effects of enhanced recovery on the environment, and the status of field testing are topics then treated at length. Considerable attention is given to recovery of heavy oil. Production of this petroleum, which has been slight because of the high viscosities involved, would add greatly to the nation's oil reserves. To further expound on tertiary techniques, the text includes reports on specific projects and research. It concludes with a discussion on the future of enhanced oil recovery.

Until recently, the costliness of EOR discouraged its widespread use. So long as the price of crude oil was relatively low, little was done to develop this technology. Now that the economic climate has changed, however, enhanced oil recovery no longer remains a system to be saved for a later date. The time for its extensive use has arrived.

AUGMENTING PETROLEUM PRODUCTION

The information in this chapter is based on:

An Investigation of Primary Factors Affecting Federal Participation in R&D Pertaining to the Accelerated Production of Crude Oil, prepared by J.M. Sharp, of Gulf Universities Research Consortium, for National Science Foundation, under Contract NSF-C-942 (GURC Report 140), September 15, 1974.

A Survey of Field Tests of Enhanced Recovery Methods for Crude Oil, prepared by J.M. Sharp, of Gulf Universities Research Consortium, for National Science Foundation under Contract NSF-C-942 (GURC Report 140-S), December 27, 1974.

Assessment of Enhanced Recovery Technology as a Means for Increasing Total Crude Oil Recovery in Texas, NSF Report RA-N-74-251, prepared by E.A. Lohse, of Texas Governor's Energy Advisory Council, for National Science Foundation, January 6, 1975.

The Estimated Recovery Potential of Conventional Source Domestic Crude Oil, prepared by J.W. Devanney III, R. Ciliano and R.J. Stewart, of Mathematica, Inc., for U.S. Environmental Protection Agency, under Contract 68-01-2445, May 1975.

Review of Secondary and Tertiary Recovery of Crude Oil, FEA Report G-75/482, prepared by Lewin and Associates, Inc., for Federal Energy Administration, June 1975.

The Potential and Economics of Enhanced Oil Recovery, FEA Report B-76/221, prepared by Lewin and Associates, Inc., for Federal Energy Administration, April 1976.

Potential Environmental Consequences of Tertiary Oil Recovery, prepared by C. Braxton, R. Stephens, C. Muller, J. White, J. Post, J. Norton, M. Goldberg, and P. Stevenson, of Energy Resources Co., Inc., for U.S. Environmental Protection Agency, under Contract 68-01-1912, July 1976.

Tertiary Oil Recovery: Potential Application and Constraints,
PNL Report RAP-25, prepared by C.A. Geffen, of Battelle's
Pacific Northwest Laboratory, for U.S. Department of Energy,
June 1978.

Mining for Petroleum: Feasibility Study, BuMines Report OFR
56-79, prepared by J.S. Hutchins, E. Bond and D.M. Bass, of
Energy Development Consultants, Inc., for U.S. Bureau of Mines,
July 1978.

*Oil Mining. A Technical and Economic Feasibility Study of Oil
Production by Mining Methods,* BuMines Report 55-79, prepared
by A. Edey, B.A. Kennedy and L.A. Readdy, of Golder Associates,
Inc., for U.S. Bureau of Mines, October 1978.

PETROLEUM—DEMAND AND SUPPLY

The modern state of development of industry is characterized by consumption
of enormous quantities of petroleum. It is not used simply for the production
of various fuels and lubricants. With each passing year more and more petroleum is used for manufacturing synthetic rubber, synthetic fibers, plastics, drugs
and thousands of other products.

While demand for petroleum products continues to rise in this country's economy, domestic petroleum production in the United States is in steady decline.
Proved reserves are shrinking as production outstrips new discoveries and expansions. Imports have provided about one-half of total demand, with substantial
amounts of these imports coming from OPEC nations. Alternative energy sources
are under development but will not be available in the near-term. In light of
these circumstances, pressures to decrease near-term dependency are high.

One promising step toward reduced reliance on imports is to improve the efficiency of domestic production through the use of advanced recovery techniques.
After conventional production is depleted, substantial quantities of oil remain
in existing, developed oil fields. Until recently, most of this oil was regarded as
impossible or uneconomic to recover.

New developments in technology and the rise in world oil prices, however, give
promise that substantial portions of this otherwise neglected oil can be recovered.
These new technical developments fall under the broad heading of enhanced oil
recovery (EOR).

Other possible ways of recovering a portion of the immense petroleum reserve
lying within the United States involve mining. There is a staggering proportion
of liquid petroleum reserves which are not being recovered from their host rocks.
Innovative mining methods could recover oil that might otherwise be lost.

Today's petroleum technology indicates that only about 30 to 35% of the original oil in place is being produced. 65 to 70% is left in the ground, economically unrecoverable by present technology. The next technological breakthrough
probably will be to merge petroleum engineering and mining to recover most of
the remaining oil from known accumulations.

PRIMARY PRODUCTION OF PETROLEUM

Primary production, as the term suggests, is the first method of producing oil from a well. When discovered, a crude oil reservoir contains a mixture of water, oil and gas in the small pore spaces (holes) in the reservoir rock. Initially, the reservoir holds these fluids under considerable pressure, caused by the hydrostatic pressure of the ground water. At this pressure a large part of the gas is dissolved in the oil. These two fluids, the initial (connate) water and the gas in solution, combine to provide the driving force for moving the oil into the well where it is pushed by the underlying pressure or is lifted by pumps to the surface.

If the pressure on the fluid in the reservoir (reservoir energy) is great enough, the oil will flow into the well and up to the surface. In this case, no pumping equipment is required. If the reservoir energy is not sufficient to force the oil to the surface then the well must be pumped. In either case, nothing is added to the reservoir to increase or maintain the reservoir energy nor to sweep the oil toward the well. The rate of production from a flowing well tends to decline as the natural reservoir energy is expended. When a flowing well is no longer producing at an efficient rate, a pump is installed.

Producing naturally, a field may yield 20 to 30% of the original oil in place (OOIP). In fact, an average of 24% of the estimated original oil in place has been produced from the U.S. as of December 31, 1974 (1).

As oil production continues, the reservoir pressure declines unless a fluid such as water enters the reservoir to replace the produced oil. Only a few oil reservoirs are fortunate enough to have a contiguous aquifer which is able to supply water as fast as the oil is normally produced. Thus, nearly all oil reservoirs experience decreasing reservoir pressures throughout their production history.

While naturally flowing wells are sought after by all producers, only about 70% of the more than 500,000 wells in the U.S. were naturally flowing at the end of 1977. Primary oil production amounted to about half of the U.S. production in 1976. The National Petroleum Council estimates that less than 10% will come from primary production from known fields by the year 2000.

Reservoir pressure decline adversely affects oil production in two ways. First, it diminishes the force which drives oil into the well bore. Second, and more important, a decline in reservoir pressure soon causes some of the gas held in solution to be released as discrete gas bubbles in the pore spaces of the reservoir rock. Such a discrete gas phase impedes the flow of oil toward the well while increasing the flow of gas.

ENHANCED OIL RECOVERY

Pressure Maintenance

A traditional step for increasing oil recovery is to inject gas or water into (or near to) an oil reservoir for the purpose of delaying the pressure decline during oil production, a technique called pressure maintenance. A well-executed pressure maintenance program can substantially increase the amount of economically recoverable oil over that to be expected with no pressure maintenance. Without either fluid injection or an active natural waterdrive (from a suitable contiguous aquifer), oil recovery falls to where further production is no longer economically feasible.

Secondary Recovery

Even after a decline in pressure has caused the oil recovery rate to become uneconomic, oil production can again be increased through a concentrated injection of fluid into the oil reservoir. Historically, these techniques have been called secondary recovery because the fluid injection results in a second crop of oil from the reservoir.

In actual practice, it is often difficult to distinguish between pressure maintenance and secondary recovery. This gray area results from the present custom of starting the gas injection or waterflood before oil rates approach the lower production limits.

Tertiary Recovery

The presence of substantial quantities of oil in the reservoir, even after a successful waterflood, has prompted interest and experimentation in tertiary recovery methods to recover a third crop of oil. These methods make use of substances added before or during fluid injection to increase the recovery efficiency of the injected water or gas. These more efficient methods may be applied immediately without utilizing the older secondary recovery methods; thus, the name tertiary recovery may not be truly applicable.

Definition of Enhanced Oil Recovery

Because of the confusion which often arises in attempting to distinguish between pressure maintenance, secondary recovery, and tertiary recovery projects, it is simpler to use an overall term such as enhanced oil recovery (EOR) to encompass the variety of methods and techniques which increase the recovery of oil above that which would be obtained through primary recovery. This term replaces the older terminology because its methods increasingly are being applied to reservoirs early in their production history rather than after secondary methods have reached their economic limits.

It is important to understand that there are no universally acceptable definitions of secondary and tertiary recovery. Some consider certain techniques to fall within a particular category, while others feel that the terms secondary and tertiary apply only to timing or production sequence. Both secondary and tertiary recovery shall be considered as being "enhanced" and representing a production technique which permits the recovery of a higher percentage of the original oil in place than would have been possible using only primary recovery methods.

The definition of enhanced recovery methods as used by the oil and gas regulatory bodies of the 30 states comprising the Interstate Oil Compact is production of crude oil from reservoirs through actions taken to increase primary reservoir drive via pressure maintenance, waterflooding, gas injection, miscible fluid displacement, microemulsion flooding, in situ combustion, cyclic steam injection, steam flooding, CO_2 injection, polymer flooding, caustic injection or related methods.

Scope: In this text, secondary recovery methods will encompass immiscible gas injection and waterflood or combinations of both. The major tertiary methods will include hydrocarbon miscible, CO_2 and inert gas miscible, micellar miscible, steam drive (cyclic and full flood), and in situ combustion (fireflood).

Essentially the secondary methods are aimed strictly at maintaining reservoir pressure at levels adequate to prevent production cut off beyond the point at which the reservoir's inherent driving forces cease to be effective. Given that beyond a certain point 60% or more of the original oil may still be in place for a typical reservoir undergoing pressure maintenance, there has been considerable interest in methods which go beyond secondary recovery. These methods attempt not only to apply pressure to the oil but also to alter the forces which hold the oil in place. The major tertiary methods fall into two basic categories:

(1) Miscible methods aimed at reducing the surface tension forces between the oil and the driving fluid; and

(2) Thermal methods which are aimed at reducing the viscosity of the oil by heating as well as in some cases changing the chemical character of the oil.

As displayed in Figure 1.1, about 50% of the total domestic crude oil production in 1973 was already derived from fields partially or fully under some form of enhanced recovery.

Figure 1.1: Crude Production by Type of Recovery Technique

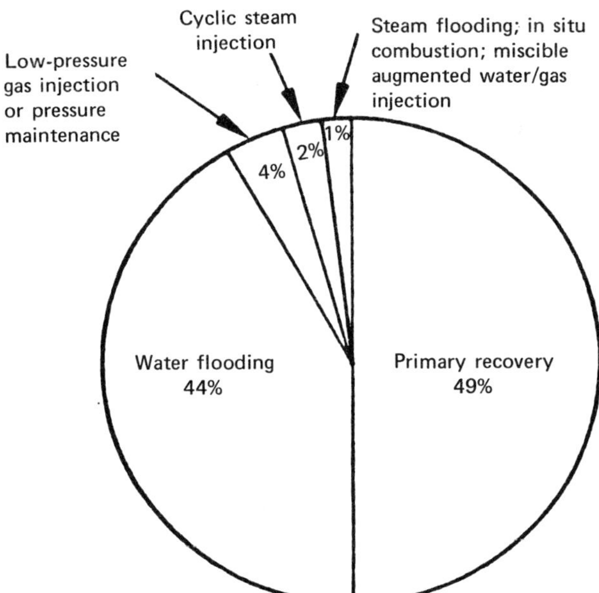

Source: FEA/G-75/482

It is apparent that the great preponderance of enhanced recovery production comes from techniques generally referred to as secondary recovery—waterflooding and pressure maintenance. Traditional estimates state that approximately 8% of the total U.S. original oil in place is recoverable using secondary recovery techniques. Improved recovery planning and higher crude oil prices may improve the recovery rate from secondary recovery to 13% of the original oil in place.

Even with these improvements, the efficiency of recovery from existing domestic oil fields remains low, from 33 to 38% of the original oil in place. Since nearly two-thirds of this recoverable oil has been already consumed, without new discoveries or improvements in the rates of recovery, only a decade or two of reserves may be available.

As displayed in Figure 1.2, however, a carefully designed and executed total program of enhanced recovery can add considerably to our reserves. Importantly, this would come from already discovered oil fields and be realized by the more efficient use of "exotic" enhanced recovery techniques such as miscible augmented water/gas injection, steam flooding and in situ combustion.

**Figure 1.2: Estimates of Recoverable Oil as a Percentage of
Original Oil in Place**

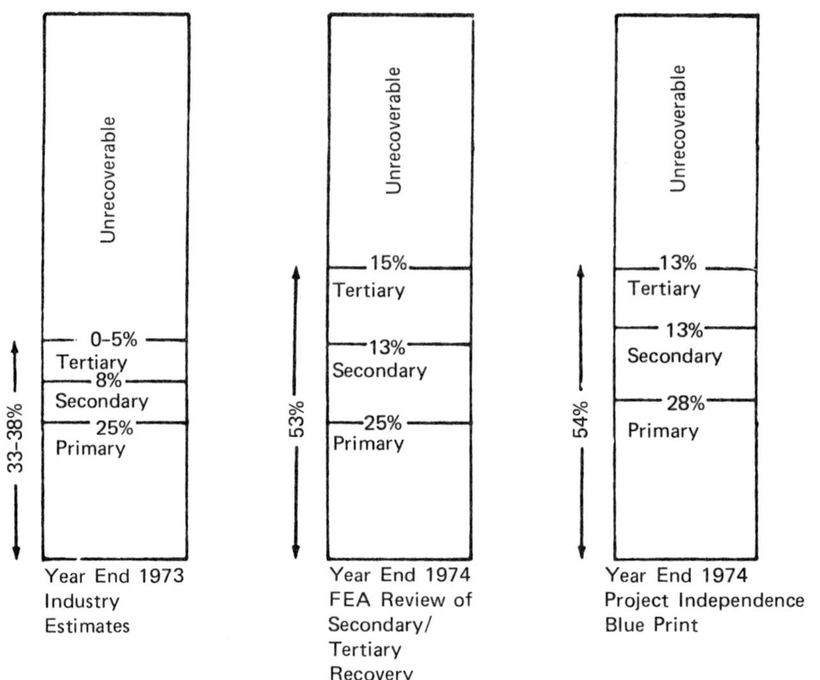

Source: FEA/G-75/482

THE PROMISE AND POTENTIAL OF EOR

The portion of the remaining oil that will be recovered through enhanced oil recovery and the rate of production are the subjects of considerable controversy. Indeed, the specific question of how much oil remains to be found onshore in the lower 48 states is a matter of heated debate. The U.S. Geological Survey (USGS) has officially estimated undiscovered recoverable reserves at between 110 and 214 billion barrels (2). The National Petroleum Council (NPC) correspondingly estimates this range as 53 to 70 billion barrels (3). A Mobil Oil Corporation analysis placed recoverables at approximately 13 billion barrels. A prior study by M.K. Hubbert, a USGS geologist, estimated only 9 billion barrels recoverable. Yet-to-be-discovered oil is not the main thrust of this effort, however. On the other hand, proved reserves of crude oil in the United States as of January 1, 1976, according to API estimates, amounted to 32.7 billion barrels.

Figure 1.3 displays the magnitude of the oil supply that might be available for EOR. The numbers on the chart report oil data for U.S. onshore, lower 48 states. (Oil in the lower 48 states is set as the initial target because the tapping of the EOR potential of Alaska and in offshore areas will require further research and study.)

Figure 1.3: Production, Reserves, and Residual Oil in Place—U.S. Onshore, Except Alaska

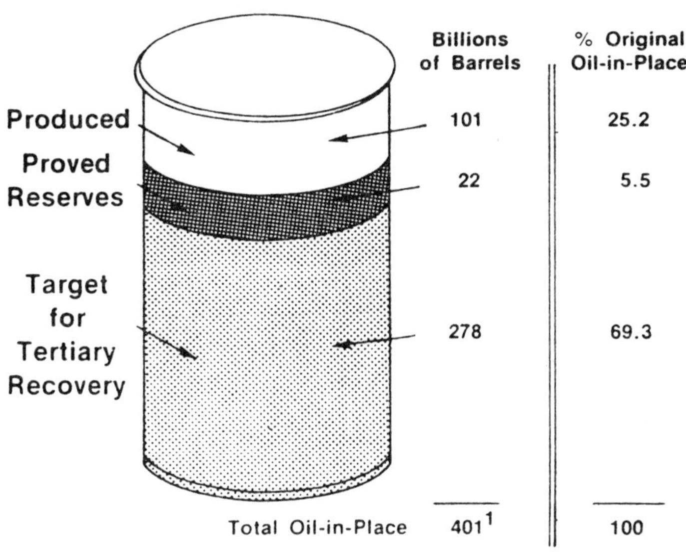

[1] The Comparable Figures for the Total U.S., Including OCS are (as of 1/1/76):

- Produced—106 Billion Barrels
- Proved Reserves—34 Billion Barrels
- Remaining Oil in Place—300 Billion Barrels
- Total Oil in Place—440 Billion Barrels

Source: FEA/B-76/221

As can be seen, the ultimate recovery from primary and secondary operations will yield only slightly more than 30% of the total original oil in place. This will leave almost 70% as the target for tertiary recovery.

The portion of the remainder that will be recovered and the rate of that production are highly uncertain. Predictions by six firms and estimates put forth in four studies differ markedly (Table 1.1). Estimates of potential EOR reserves range from 15 billion barrels to 110 billion barrels. The magnitude of the difference in these opinions as to how much of the remaining oil will be recovered reveals the basic challenge in analyzing the potential of tertiary recovery; no consensus exists as to the amount or timing of the additional oil recovery.

Though the optimistic outlook of the early 1970's toward EOR clouded somewhat, there is little cause for the pessimism reflected in certain of the projections noted in Table 1.1. For example, based on the *Oil and Gas Journal's* survey of enhanced oil recovery (4), it is found that five EOR techniques produced 240,000 barrels per day and, counting only the small number of 1976 efforts and the pending changes in thermal recovery (changing from steam soak to steam drive), EOR production by these methods could reasonably be expected to double. Contributions of the five enhanced oil recovery techniques to this production are shown in Table 1.2.

Table 1.1: Estimates of EOR Potential

	Potential EOR Reserves (billions of barrels)	. .Daily Production. . MM bbl	by Date
Oil companies			
1	15	—	—
2	—	1.0	2005
3	25	0.5	1985
4	18	0.75	1985
5	110	1.0	1985
6	25	1.0–2.0	1990
Other			
FEA/PIR*			
Business as usual ($11)	57	1.8	1985
Accelerated development	57	2.3	1985
FEA**			
($12–$13)	65	—	—
FEA/energy outlook***			
($12)	—	0.9	1985
EPA†	15–25	—	—

Project Independence Report, Federal Energy Administration, November 1974.
**Review of Secondary and Tertiary Recovery of Crude Oil,* Lewin & Associates, Inc. for FEA, June 1975.
***1976 National Energy Outlook,* Federal Energy Administration.
†*The Estimated Recovery Potential of Conventional Source Domestic Crude Oil,* Mathematica, Inc. for the U.S. Environmental Protection Agency, May 1975.

Source: FEA/B-76/221

Table 1.2: Summary of Ongoing/Planned EOR Projects

EOR Technique	...Ongoing...		Planned/Conversions		....Total....	
	acres	bbl/day	acres	bbl/day	acres	bbl/day
(1) Steam	13,658	106,876	27,228	214,925	40,885	321,801
(2) In situ combustion	3,984	7,581	1,164	2,212	5,148	9,793
(3) CO$_2$/hydrocarbon miscible	81,925	113,500	7,100	10,000	89,025	123,500
(4) Surfactant/polymer	758	530	189	132	947	662
(5) Polymer/caustic	13,045	7,442	8,142	4,843	21,187	12,285
Total bbl/day from EOR		235,929		232,112		468,041

Source: FEA/B-76/221

GURC Forecast

As reported by the Gulf Universities Research Consortium, a late 1973 consensus
of about 20 oil producing companies experienced in enhanced oil recovery technology projected a highly probable addition to U.S. proved reserves by 1985 of
18 to 36 billion barrels of crude oil as a result of applying EOR methods to
known domestic fields in the lower 48 states (5). Additionally, the consensus
projected a potential (a) EOR production in 1985 of 1.0 to 1.5 million barrels
per day, and (b) an EOR cumulative production by 1985 of 1.5 to 2.0 billion
barrels, and (c) EOR production of 100,000 to 200,000 barrels per day beginning about 1980.

The additional reserves projection represents an improvement in ultimate recovery of original oil in place of from 4.5 to 9% as compared with the average for
domestic fields of 32 to 33% of the approximately 400 billion barrels of oil discovered thus far. While a few EOR-knowledgeable companies consider this projection to be somewhat optimistic, the majority consider it to be quite realistic.

However, it is particularly important to emphasize that the companies' 1985 production projections specified that a large compressed-time-scale field test program
would have to be initiated in 1974 if this level of production were to be achieved.
A final point of emphasis regarding the 1973 consensus was the companies' firm
position that the required field test program would not be carried out by industry using private risk capital alone because of the unpredictable risk of EOR applications coupled with the very high front-end costs and the long time subsequent to this investment before a positive cash flow could result. It was generally agreed that federal cost-sharing or suitable tax incentives would be required
in order that these early 1980's production potentials be realized. Additionally,
it was stated clearly that this situation would hold regardless of any conceivable
increases in the wellhead price of domestic crude oil.

Economic feasibility field tests of EOR methods require 4 to 7 years. The tran-
sition from test to production requires the acquisition of large quantities of materials, which introduces an additional 2 to 4 years' delay unless large advance com-
mitments can be made. With the recommended federal/industry cooperative program, 1985 is about when major EOR production should begin. Hence, EOR
should make a much greater impact during 1985 to 1995.

Tables 1.3 and 1.4 summarize production and reserves projections developed by
GURC for the AEC. The average "almost assured" forecast of 18.5 billion barrels is the most reliable number of the national industrial consensus. Realization
of either average forecast of 18.5 or 36.3 billion barrels would constitute a major contribution to the nation's future supply of petroleum energy and products.
These values represent 6 to 12% recovery of remaining oil in place, in addition
to the 53-year (1920 through 1972) national historical average recovery factor
of 29.9% of original oil in place. The 18.5 billion barrels would increase the total average recovery of original oil in place to: 29.9% + (0.06 x 67.4%) = 33.94%
(34%). Similarly, the 36.3 billion barrels would increase the total average recovery of original oil in place to 37.98% (38%).

The highest "almost assured" forecast of 60.0 billion barrels of remaining oil in
place represents the opinions of some of the nation's expert petroleum engineers,
even though they comprise a minority opinion in the company consensus.

Should the highest forecasts (60 to 100 billion barrels) prove to be correct, additional oil production by EOR technology would begin to approximate the total
amount that has been produced from U.S. domestic resources through 1972
(99.91 billion barrels).

Table 1.3: 1985 EOR Domestic Production Projections

Cumulative production by 1985:	
"Almost assured" estimate	1.8×10^9 bbl
"Realistic" estimate	3.4×10^9 bbl
"Reasonable" estimate	5.9×10^9 bbl
Annual production in 1985:	
Average projection	423×10^6 bbl
Lowest projection	100×10^6 bbl
Highest projection	800×10^6 bbl

Source: NSF-C-942

Table 1.4: Projections of EOR–Added Domestic Reserves by 1985

"Almost assured" projections:	
Average	18.5×10^9 bbl
Lowest	5.0×10^9 bbl
Highest	60.0×10^9 bbl
"Reasonable" projections:	
Average	36.3×10^9 bbl
Lowest	15.0×10^9 bbl
Highest	100.0×10^9 bbl

Source: NSF-C-942

Effect of Price on EOR Development

Until the price increases in 1973-1974, EOR technology was principally dormant as a viable production method in that laboratory and theoretical work and the limited small-scale field testing which characterized most of industry's EOR efforts rarely indicated the possibility of profitable application. The costs projected on the basis of laboratory and field tests indicated that payout was either impossible, unlikely, or not competitive with the return on investment to be realized from a much more predictable and much less expensive (though possibly less efficient) secondary recovery method. Now that prices have escalated, the economic projections are more likely to be favorable. Pricing actions by OPEC have created market conditions in which it is prudent to seriously consider the use of these high-cost, high-recovery techniques.

EOR development is thus highly sensitive to the price of oil. Given that the greater proportion of tertiary production had been classified as old oil and priced at $5.25 or less, one should not be surprised at the slow rate of EOR adoption.

The successful application of tertiary recovery methods is highly dependent not only on economic, but also on regulatory and technological factors. The uncertainty of future governmental policies in these areas has resulted in a wide range of predictions for the potential of enhanced oil recovery, with estimates of the amount of oil that enhanced techniques could recover ranging from 7 to 76 billion barrels of oil, at prices ranging from $10 to $25 per barrel.

A basic requirement for the successful implementation of tertiary oil recovery is an oil price that will make these projects economically viable. The price level must be sufficient to cover all project costs and include an adequate operator profit. In general, tertiary recovery techniques add at least $2 to $10 to the cost of each additional barrel of oil produced; it has been estimated that some of the chemical recovery methods would have production costs of $20 to $25 per barrel of oil (6).

50 to 80% of the incremental cost of tertiary production is for front-load items such as facilities and injection fluids. Total production costs also include normal well operating costs, the costs of lifting and disposing of water left from previous waterflooding operations and the time value of investment money due to the time delay in income from the project (7). A capital requirement of $2 billion for each 100,000 bbl/day produced by tertiary methods and a manpower need of one engineer for each 1,000 bbl/day of production has been predicted (8).

Such costs, of course, become less prohibitive with each increase in oil prices. As seen in Table 1.5, an increase in oil prices will most likely lead to a more widespread use of the advanced recovery technologies (9)(10)(11). Ranges given in the table represent differences between low- and high-process performance cases.

While the impact of tertiary recovery is subject to many variables, we can nevertheless conclude that enhanced recovery methods:

    (a)   Do offer significant potential for augmenting future crude supplies from conventional sources;

    (b)   Contribute to the national priority goals of energy conservation and efficient resource exploitation;

(c)   Impact favorably upon the movement towards reduced domestic dependence on foreign imports with its attendant security and balance of payments implications; and

(d)   Offer an immediate alternative to other energy development options whose technological, environmental and financial risks are sufficiently undefined so as to create considerable apprehension.

Table 1.5: Potential Oil Recovery

| | Potential Production Rate | | |
| | . . . . (MM bbl/day). . . . . | | |
	1980	1990	2000
Thermal methods			
OTA,* 1977 $/bbl			
$11.62	0.2-0.3	0.4-0.5	0.3-0.4
$13.75	0.2-0.3	0.5-0.6	0.4-0.5
$22	0.3-0.4	0.8-1.1	0.5-0.7
NPC,* 1976 $/bbl			
$10	0.2	0.3-0.4	0.2-0.3
$15	0.2-0.3	0.6-0.8	0.2-0.3
$25	0.3-0.5	0.9-1.1	0.4-0.5
ERDA,* 1976 $/bbl			
$11.63	0.16	0.17	—
CO_2 miscible methods			
OTA, 1977 $/bbl**			
$11.62	0.1	0.05-0.3	0.6-1.6
$13.75	0.1	0.1-0.6	1.1-2.8
$22	0.1	0.5-0.8	3.3-5.6
NPC, 1976 $/bbl			
$10	—	0.1	0.5
$15	—	0.2-0.3	0.9-1.0
$25	—	0.6-0.7	1.1-1.2
ERDA, 1976 $/bbl			
$11.63	—	0.4	0.4
Chemical methods			
OTA, 1977 $/bbl			
$11.62	—	0.1-0.3	0.2-0.9
$13.75	†	0.1-0.5	0.2-1.9
$22	†	0.5-0.8	1.3-2.5
NPC, 1976 $/bbl***			
$10	†	†	†
$15	†	0.3	0.2
$25	†	1.0	1.2
ERDA, 1976 $/bbl			
$11.63	†	†	†

*OTA = Office of Technology Assessment; NPC = National Petroleum Council and ERDA = Energy Research and Development Administration.
**Includes both onshore and offshore reservoirs.
***Micellar-polymer only.
†Less than 0.05 mm bbl/day production.

Source: PNL-RAP-25

REFERENCES

(1) American Petroleum Institute, "Reserves of Crude Oil, Natural Gas Liquids and Natural Gas in the U.S. and Canada" (May 29, 1975).

(2) Theobald, P., Schweinfurth, S., and Duncan, D., "Energy Resources of the United States," Geological Survey Circular 650, Washington (1972).

(3) National Petroleum Council, *U.S. Energy Outlooks,* Washington (1972).

(4) *Oil and Gas Journal* (April 5, 1976).

(5) Crump, J.R., and Sharp, J.M., *Planning Criteria Relative to a National RDT and E Program Directed to the Enhanced Recovery of Crude Oil and Natural Gas,* Gulf Universities Research Consortium Report 130, Galveston, Texas, Report ORO-4507-1 (Appendix E of PB-248 951) (November 30, 1973).

(6) Simpson, J.J., "Financing Enhanced Recovery," *Journal of Petroleum Technology,* pp 771–776 (July 1977).

(7) Geffen, T.M., "Oil Production to Expect from Known Technology," *Oil and Gas Journal,* pp 66–76 (May 7, 1973).

(8) "Yield from Exotic Recovery Seen Slow, Costly," *Oil and Gas Journal,* p 80 (March 29, 1976).

(9) Ham, J.D., Hochheiser, H.W., and Guthrie, H.D., "Long-Range Supply and Logistics Projections for Enhanced Oil Recovery," *Long-Range Oil and Natural Gas Supply, Demand and Logistics Projections,* Energy Research and Development Administration, pp 155–161.

(10) *Enhanced Oil Recovery Potential in the United States,* Office of Technology Assessment (January 1978).

(11) *Enhanced Oil Recovery,* National Petroleum Council (December 1976).

SECONDARY RECOVERY OF CRUDE OIL

The information in this chapter is based on:

"Oil Recovery by Nuclear Explosion," *Nauka i Zhizn* (U.S.S.R.), No. 2, 1973, by L. Shadrin, translation JPRS 59039, prepared by U.S. Joint Publications Research Service, May 17, 1973.

The Estimated Recovery Potential of Conventional Source Domestic Crude Oil, prepared by J.W. Devanney, III, R. Ciliano and R.J. Stewart, of Mathematics, Inc., for U.S. Environmental Protection Agency, under Contract 68-01-2445, May 1975.

Review of Secondary and Tertiary Recovery of Crude Oil, FEA Report G-75/482, prepared by Lewin and Associates, Inc., for Federal Energy Administration, June 1975.

"Treatment of Injection Waters," *Rev. Inst. Fr. Petr.* 20: No. 2, 280–290, 1965, by M. Peinado, translation ERDA-tr-163, prepared by U.S. Joint Publications Research Service for Energy Research and Development Administration, 1976.

The Potential and Economics of Enhanced Oil Recovery, FEA Report B-76/221, prepared by Lewin and Associates, Inc., for Federal Energy Administration, April 1976.

Potential Environmental Consequences of Tertiary Oil Recovery, prepared by C. Braxton, R. Stephens, C. Muller, J. White, J. Post, J. Norton, M. Goldberg and P. Stevenson, of Energy Resources Co., Inc., for U.S. Environmental Protection Agency, under Contract 68-01-1912, July 1976.

When first brought into production, most oil fields have sufficient natural forces to push their oil out of the oil-bearing rock and into the new wells. During this primary production phase, the advance of underground water beneath the oil may push it toward the well, or natural gas dissolved in the crude may expand to force oil out of the rock much as champagne flows from an uncorked bottle. These natural forces are soon depleted, however, and the oil production rate will drop sharply if secondary oil recovery techniques are not applied.

In the secondary recovery phase the oil field operator aids the natural pressure forces by injecting gas or flooding the oil reservoir with water. In appropriate cases, these measures may permit production of an additional 30 to 40% of the original oil in the reservoir.

INITIAL PRODUCTION OF OIL

The oil industry first develops oil-gas deposits which, as geologists say, are indigenous to the most recoverable parts of productive formations—the archparts of the folds, with comparatively high porosity and sufficient permeability, for example, with the capacity to be penetrated by liquids and gases. Rocks with such properties are called reservoirs. Ore reservoirs are sandstones, limestones, dolomites and other permeable rocks, which are isolated from the basic rock mass by such impermeable rocks as clay or shale. The oil deposit is bounded below, as a rule, by water-saturated rocks, and quite often gas accumulates above the deposit, forming a gas cap.

Formation Energy

When a drill hole enters the oil deposit it does not only open up the petroleum pocket, but also uncaps a natural storehouse of energy. The source of this energy is the pressure of the water surrounding (or underlying) the oil formation, the pressure of the gas liberated by the oil which expands as formation pressure decreases and, finally, the force of gravity. All these components comprise what the oil people call formation energy. The oil is forced from the formation into the borehole under the influence of formation energy. As the initial reserves of energy are exhausted the formation pressure decreases and the yield of the well decreases correspondingly, the volume of petroleum given up by it per unit of time decreasing.

By the time the formation energy is exhausted only part of the oil stored in the formation can be extracted. Complete recovery is characterized by the oil recovery coefficient, which is the ratio of the volume of recovered oil to the volume initially contained in the deposit. When only natural formation energy is utilized this coefficient does not usually exceed 30%.

The problem of increasing the petroleum recovery coefficient is an extremely important one from the standpoint of national economy. Only a few percent increase in petroleum recovery yields a considerable profit. Indeed, the cost of additional petroleum recovery at existing fields is immeasurably less than the cost involved in prospecting and opening up new deposits. An increase in the petroleum recovery coefficient of only one percent at oil fields of the Soviet Union, for example, is equivalent to the discovery and operation of a large oil deposit of an annual yield of about four million tons.

PRESSURE MAINTENANCE

Special technological methods are used for more complete recovery of oil. A common method employed today is artificial maintenance of formation pressure.

This traditional step for increasing oil recovery involves the injection of fluid into (or near) an oil reservoir for the purpose of delaying the pressure decline during oil production—a technique called pressure maintenance. A well-executed pressure maintenance program can substantially increase the amount of economically recoverable oil over that to be expected with no pressure maintenance.

Considering all fields, both with and without a natural water drive and/or pressure maintenance, the overall average recovery from primary production is between 20 to 30% of the original oil in place by the time the pressure, and consequently the oil production rate, falls to where further production is no longer economically feasible. Without either fluid injection or an active natural water drive (from a suitable contiguous aquifer), oil recovery is considerably less, usually restricted to a range of 5 to 20% of the oil originally in place.

WATERFLOODING

After a decline in pressure from the water drive or pressure maintenance, production can be increased through a technique called waterflooding, which is the injection of water through injection wells to push crude oil toward producing wells.

Water is pumped into the productive stratum through pressure boreholes in a volume equal to (or greater than) the volume of oil extracted. As a result the formation energy in the deposit is kept at the optimum level. The fountain time of the well is prolonged, which substantially reduces the volume of drilling operations and the cost of the oil.

History of Waterflooding

This first and now most widely used secondary technique had an accidental beginning in the 1870s in the Pithole City area of western Pennsylvania. A leak from an adjacent water-bearing geologic structure ruined production in the affected well, but increased the rate of production in nearby wells. This led to increased ultimate recovery from the field as a whole.

By the early 1900s circle pattern waterflooding, expanding in concentric circles, was used rather extensively, albeit illegally (for fear of contaminating the oil) in the Bradford area of Pennsylvania.

It was not until 1921 that waterflooding became legal. The practice eventually spread from Pennsylvania to Ontario, 1913; California, 1917; Oklahoma, 1931; and Texas, 1936. Expansion after 1940 was much more rapid, as restrictive legislation subsided and the Texas Railroad Commission allowed higher production rates for waterflood projects.

Extent of Use

By 1973 waterflooding had become one of the major contributors to our nation's oil recovery with about one-half of domestic oil being produced from reservoirs partly or completely under waterflooding. In the U.S.S.R. the use of such artificial flooding of strata enabled the Russians to produce an extra one billion tons of petroleum in the ten years prior to 1973. They employed this method at 170 fields, which accounted for 70% of their total petroleum production.

The practice has become so extensive that in 1970 in Texas there were 2,821 active water injection projects covering more than 1,000 fields (1). In Louisiana there were 347 water injection projects active in 1972 (2). In Illinois, 70 to 80% of all production is on waterflood (3).

Economic Aspects

Waterflood has been quite economical at prices below those prevailing before the 1972 to 1973 jump, that is, $3.00 per barrel or less. Even in high transport-cost-to-market areas such as Alaska, considerable water injection has been practiced. The Cook Inlet fields have been on water almost since first production and now have 60 injection wells. The Grand Ile 16 field (Gulf of Mexico) in 1973 had 56 producing and 20 water injection wells in use for some time, a practice considered typical for the larger mature fields in the Gulf.

Even in foreign countries, when the price of oil was $2.00/bbl and less, considerable waterflood was taking place. In Canada in 1966, over 100 separate fields were under waterflood, including the four largest. Full-scale waterflooding of the 400,000-acre Pembina Cardium Field in Alberta was initiated in 1956, three years after discovery. Some water injection has even been practiced in Venezuela. In 1964, when the price (in 1974 dollars) of oil f.o.b. Venezuela must have been less than $2.00, there were 25 plants in operation, injecting 550,000 bpd into 29 reservoirs. Even in the Middle East, there has been some waterflood. Iraq has been injecting water into the Kirkuk Field since 1971, when the 1974 dollar price of oil was no more than $1.50. Abquaia Field in Saudi Arabia has been on water since 1957, and the Ghawar Field was injecting 363,000 bpd in 1966.

We may not see much additional oil from conventional waterflooding due to the price rise because its economic viability at $3.00/bbl appears so substantial. This is most unfortunate, because waterflooding is a rather efficient means of increasing recovery as the examples in Table 2.1 indicate. For a typical gas depletion field with oil above 30° API, waterflood will increase recovery from the range 15 to 20% to 30 to 45%.

Table 2.1: Increased Recovery via Waterflood—Illustrative Effects for Selected Fields

Field	Characteristics	Primary	Waterflood*
		Recovery (%)	
Eldorado	sandstone, D = 650 ft, H = 10-30%; k = 150-1,500 md; 37° API; 4 cp at 70°F	21	43
Jay		17	48
Pembina Cardium	cretaceous sand, thin; D = 5,100 ft; low, irregular permeability	15	~30
Swan Hills	Devonian reef; downdip peripheral	17	44
Weyburn		14	31
West Burk Burnett	gunsite sand	15	32

*Total, not incremental, recovery. Thus, in the Eldorado field, waterflood recovered an additional 22% (43-21) of the OOIP after primary.

Source: EPA 68-01-2445

Geffen (4) places the average cost of waterflood recovery at 30¢ to 50¢/bbl as
past practice. Mobil in the S.E. Breckenridge unit of the Stephens Regular Field
charged the other interests: $60 per injection well per month in maintenance +
$0.0048/bbl water injected + 0.000305 x psi x KWHC (psi is injection pressure
in psi and KWHC is $/kWh). This contract was signed in 1973. The Dickie unit
in the same field estimated a cost of $405,000 to obtain an additional 630,000
barrels, while the Jones unit estimated a cost of $492,000 to obtain an additional
1.2 MM bbl.

The initial cost of a rather large project in the Seminole Field involving 65 injection wells with a capacity of 195,000 barrels of water per day (bwd) was put at
$2.9 million on average. This project hoped to recover an additional 115 million
barrels of oil.

In short, the data substantiate Geffen's claim that even the rather extensive waterflooding which has taken place in the past rarely cost over 50¢/bbl on average,
and sometimes a good deal less. Some recent reports, however, on new projects
indicate somewhat higher costs. A five-spot waterflood replacing gas injection
in the Alvord Field is reported to cost $1.5 million to get 2.4 MM bbl additional
oil at a rate of 600 to 1,000 bpd. At a still higher cost, Mosbacker, a Houston
independent, claims a cost of $10 million to obtain 3 to 4 million additional
barrels. The Jay Field waterflood, operating under very difficult conditions (high
depth, pressure, temperature, and sulfur content) cost $19.5 million to inject
200,000 bpd, but this investment is expected to increase production by 217
MM bbl.

For illustrating the economics of waterflood enhanced recovery, a hypothetical
relatively small field having an estimated 2,600,000 bbl of original oil in place
was studied by Lewin and Associates for the FEA (FEA/G-75/482). The crude
oil was light, 37° API, and low in sulfur content (sweet). The reservoir depth
was medium at an average of 4,500 ft, located in a sandstone formation with
average permeability of 120 md. Solution gas/partial water had provided the
natural primary recovery drive.

> 520,000 barrels (20% of oil in place) were projected as recover-
> able through primary production,
>
> 430,000 additional barrels (16% of oil in place) would be re-
> coverable by using a traditional waterflood recovery technique,
>
> Investment costs would total $500,000, consisting of drilling
> costs for three injection wells at $65,000 per well, and $305,000
> for production equipment; the drilling costs would be expended
> with the production equipment capitalized and depreciated on a
> unit of production basis for tax calculations,
>
> Operating costs for the additional oil would average $1.21/bbl,
>
> Production taxes would average 8% of gross price; the field would
> be under one-eighth royalty, and
>
> The project would experience a small increase in production after
> one year, reaching a peak rate of production in the third year.

Figure 2.1:   Sample Production Curves for Waterflooding

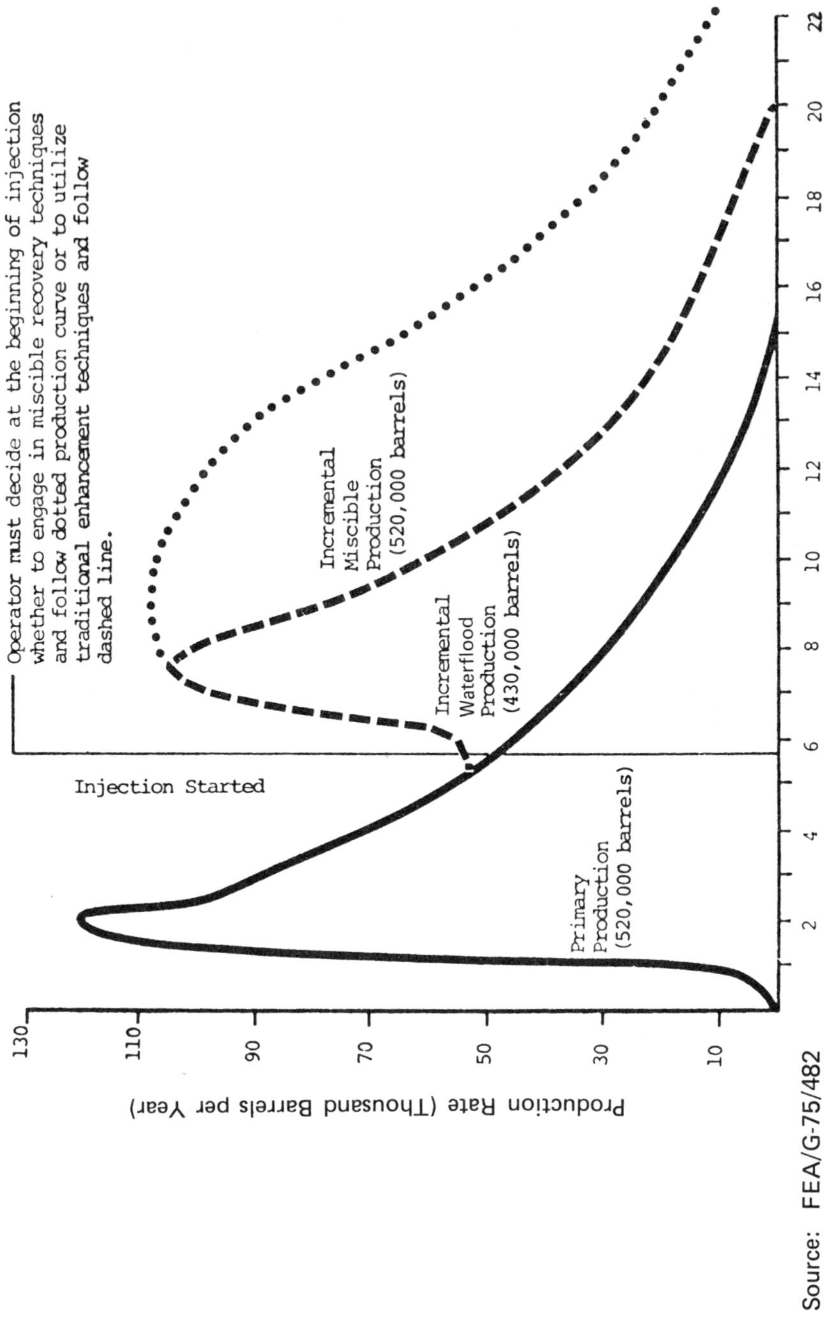

Source: FEA/G-75/482

An economic rate of return (ROR) calculation shows that the project would achieve a 20% rate of return on investment, before interest and income taxes, and 16% after taxes at the controlled old oil price of $5.25/bbl. (A $13.00 market price for the additional production oil would provide the project a 54% after-tax rate of return.) Using a 10% discounted cash flow rate (after income taxes), the traditional waterflood would be profitable at the controlled $5.25 price. The 10% after-tax rate of return would cover the cost of capital, any losses on other fields arising from the higher level of risk in enhanced production projects, research and development costs, and profit for the investor.

Figure 2.1 provides the production curve for the waterflood project described here.

While price increases will probably not induce significant increased recovery via traditional waterflooding, the method will continue as a major recovery technique. Indeed because of its long-standing use, its benefits and shortcomings are better known than those for any of the other enhanced recovery processes.

Advantages and Disadvantages

In waterflooding, treated water is injected into the oil-bearing part of the reservoir to force oil to flow toward the production wells. Water, because of its high density, relatively efficient displacement characteristics and its nearly incompressible nature, can raise the reservoir pressure quickly. This is an important factor in rapidly restoring the oil productivity of older wells in advanced stages of depletion. Most waterfloods are designed so that increased oil production in older wells occurs within a period of six months to a year.

Water, however, has two major problems that impede its efficiency. First, it does not flush all of the oil from the pore spaces as it moves through the reservoir rock. Because water and oil do not mix, 25 to 50% of the oil is left behind in the form of small droplets held within the larger pores. This problem may be prevented and thus additional oil recovered by the use of additives to reduce or eliminate the nonmixing nature of water and oil (as discussed in the chapter on tertiary recovery).

The second limitation is that the advancing water front generally bypasses significant portions of the reservoir due to difficult well placements or unexpected geological configurations. This lack of a perfect sweep efficiency is responsible for leaving behind additional crude oil in areas not reached by the waterflood. Altogether, 50 to 70% of the original oil in place remains after waterflooding.

Little is known about the success rate for this type of oil stimulation, but it is true that only 70% of the waterfloods initiated after pilot tests prove to be successful in recovering a significant amount of oil.

As waterflood operations have amply demonstrated, the sedimentary rocks in which oil reservoirs are found are not homogeneous nor is their configuration necessarily conducive to uniform fluid flow. On the contrary, the natural processes whereby sand, organic material, carbonaceous materials produced by marine animals, clays, etc., were deposited were highly variable. The rocks formed from such deposits are, therefore, highly variable in mineralogy, chemical content and microstructure. Subsequent tectonic and diagenetic processes created further inhomogeneities and numerous discontinuities.

These structural variations and defects are not discernible by remote sensing methods; they can only be inferred from waterflood histories, from in-field core samples, and such geophysical, well logging and outcrop geological information as may be available for a specific reservoir.

Hence, the fluid flow patterns between injection and production wells are not predictable with reasonable accuracy, even after many years of waterflooding from the same injection/production well configurations.

Treatment of Injection Waters for Secondary Recovery

Since most of the failures of waterflooding are attributed to a lack of knowledge of the hydrology of the underground reservoirs, it is important to consider water treatment.

In the secondary recovery of oil from a deposit by the injection of water, the water compatibility factor plays a primary role. Numerous obstructions of pipes and injection and production wells, as well as numerous cases of deterioration due to corrosion of well equipment and collection systems, have actually been observed during the injection of incompatible water.

Such water may permit the materials that they contain in suspension to be deposited in the rocks through which they pass, or may cause swelling of the clays and the formation of insoluble mineral precipitates, such as alkaline-earth carbonates and sulfates (gypsum, barite, calcite, etc.) and gels of metal hydroxides. Gases such as oxygen and the bacteria that they carry are ultimately responsible for corrosion of the material. The result is very costly reconditioning of the wells, replacement of the pipes and, in the most serious cases, loss of the wells.

The object of an investigation by M. Peinado (ERDA-tr-163) was to briefly review the different causes of injection water incompatibility and the means for solving this problem, with particular emphasis on the chemical aspects of the injection of water in briny wells for washing the production tubing.

Injection Water: *Origin* — The injection water used in secondary recovery may be taken from various sources. Rivers and lakes often play a contributing part. Often wastewater from neighboring communities is involved, generally water from wells specially drilled for this purpose is used. Finally, use is also made of the so-called reservoir water, which accompanies the crude petroleum in the producing reservoir. Due to the very diversity of their origins, these categories of water have very different compositions. They vary especially in the nature and the concentrations of the salts that they contain in solution. Thus two categories of water are distinguished:

> Fresh Water — This comes from rivers or lakes and shallow beds and only contains small amounts of salt (a few grams per liter). Fresh water has the particular disadvantage that it contains quantities of oxygen in solution and solid materials in suspension (fine sand, clay, plant and animal substances) which are not negligible.

> Salt Water — This comes either from deep layers close to the layer of oil or from regions near the coasts and contains large quantites of salt in solution (several dozen grams per liter). This type of water can also contain corrosive gas, such as hydrogen sulfide and carbon dioxide and bacteria of the sulfate-reducing type, whose role in corrosion will be discussed later.

Table 2.2

In solution	Solids	Minerals	Anions	Chlorides: (Cl^-); bromides: (Br^-); fluorides: (F^-) Sulfates: (SO_4)$^=$; carbonates: (CO_3)$^=$; bicarbonates: (CO_3H)$^-$
			Cations	Sodium: (Na^+); potassium: (K^+); calcium: (Ca^{++}); barium: (Ba^{++}); magnesium: (Mg^{++}); iron: (Fe^{++}); manganese: (Mn^{++})
		Organic	plants and animals (plankton)	
	Gases	Oxygen: (O_2); nitrogen: (N_2); hydrogen sulfide (H_2S); carbon dioxide (CO_2)		
In suspension	Minerals	Ferric hydroxide: $Fe(OH)_3$ Silica: (SiO_2) and sand Clay and mud Alkaline earth sulfates and carbonates		
	Organic	Algae, bacteria (plant and animal materials) Oil		

Source: ERDA-tr-163

Composition — The substances dissolved or in suspension that may be found in the injection water are listed in Table 2.2.

Physical Treatment of Solid Materials in Suspension: Water scheduled to be injected in a formation whose porosity and permeability are often low must be freed of all the particles in suspension. This operation is achieved easily by decantation, followed or not by filtration.

Determination of the Presence of Materials in Suspension — Turbidity: The particles in suspension are visible either to the naked eye or under an ordinary microscope, but it is easier to evaluate their size by means of turbidimetry. It is a question of measuring the turbidity, or the cloudiness, caused by the particles, a measurement which takes advantage of the capacity of cloudy water to absorb or diffuse light.

Alluviating Power: Filtrating membranes of the Millipore type are used, for which the thickness and size of the pores are perfectly defined. The test consists of measuring the flow as a function of the volume accumulated and, when the results are given in semilogarithmic coordinates, the slope of the line corresponding to filtration is a characteristic of the alluviating power of the water. For a pore dimension of 0.45 μ, a slope ranging between 0 and 0.25 designates a very high quality water. Between 0.25 and 0.50, the slope represents a water of medium quality and higher than 1.25, a water of very poor quality.

Elimination of the Matter in Suspension — One proceeds by decantation in large basins. The sedimentation rate of the particles and especially that of the colloidal particles (ranging between 10 Å and 1 μ), is accelerated by coagulation by means of flocculating ions. The sulfate anions are more effective than the chloride anions for flocculating the positively charged colloids while the aluminum and ferric cations are more effective than the calcium and magnesium cations, which in turn, are more effective than the sodium cations in flocculating the negatively charged colloids, which is why aluminum sulfate is commonly used.

Then, if it is considered necessary, the decanted water is exposed to filtration through filters with sand, anthracite, or diatomites, either under pressure or by means of gravity.

Sloat (5) emphasizes the need to eliminate the oil by coagulation when the oil comes from the reservoir due to its capacity to prompt the development of bacteria generating corrosion and sediments.

Physical and Chemical Treatment of Soluble Gases: The gases that are usually found in injection waters and which are the most harmful from the corrosion point of view are hydrogen sulfide, carbon dioxide, and oxygen. The first two gases are eliminated first by aeration and then the oxygen is eliminated by reduction. This operation does not present any particular problem.

Determination of the Presence of Gases and Their Analyses — Hydrogen sulfide: The presence of hydrogen sulfide is easily detected by the odor and the blackening of a paper soaked with silver acetate. The analysis may be carried out by iodometry or by colorimetry (formation of methylene blue).

Carbon Dioxide: The presence of carbon dioxide in water is detected by the fact that this water forms a precipitate with limewater and by the fact that this precipitate disappears in an acid medium. The determination is made by means of a sodium carbonate solution in the presence of phenolphthalein (colored indicator).

Oxygen: Oxygen is recognized and analyzed by means of a reducer, sulfite or salt of hydrazine, or even better by means of a mixture of manganese sulfate, soda, and potassium iodide. The iodine freed is titrated by means of sodium thiosulfate.

Elimination of the Dissolved Gases — First a physical treatment is applied, aeration of the water to be injected. This process consists either of making the water drip into the atmosphere or of blowing air under pressure into the water. This air expels the hydrogen sulfide and the carbon dioxide. It is also possible to eliminate all the gases by subjecting the water to a relatively intense vacuum.

In the first case, the water still contains oxygen which is generally and very quickly eliminated chemically by addition of sodium sulfite, which proceeds to the state of sodium sulfate. Meanwhile, the ferrous and manganous ions proceed to the state of ferric and manganic ions and precipitate. They are separated by filtration.

Chemical Treatment of Water-Soluble Salts: As already noted, the composition of a body of water is closely connected with the nature of the rocks crossed. Waters which are associated with formations producing oil contain highly variable quantities of salts. The following ions are most commonly found here: sodium (Na^+) and potassium (K^+), magnesium (Mg^{++}) and calcium (Ca^{++}), iron (Fe^{++}), chloride (Cl^-), sulfate ($SO_4^=$), sulfide ($S^=$), and bicarbonate (CO_3H^-) and, more rarely the barium (Ba^{++}) and strontium (Sr^{++}) ions.

The chemical stability of a formation water is also dependent on the temperature, the pressure, and the composition of the gases which come into contact with it. The variation of these factors as it flows modifies the equilibrium of the water and may involve precipitations or the placing of materials in solution.

Precipitation of Insoluble Salts — Compatible waters are waters which can be mixed without producing any undesirable chemical reaction between the constituents dissolved in the different waters (5). By undesirable chemical reactions is meant those reactions capable of causing the formation of insoluble substances which will be deposited and will reduce the permeability of the formation or will plug up the injection wells.

In practice, the injected water is mixed in situ with the interstitial water along a more or less large area where insoluble compounds form when the two waters are incompatible. The volume of water of the mixing area increases with the distance to the axis of the injection well, and in the rather general case of a viscosity ratio for the waters of one, the increase in the mixing area is proportional to the square root of the distance to the axis of the injection well.

When reactions occur between chemical constituents of incompatible waters, all the reacting chemical species are not precipitated. Only those species precipitate whose concentration exceeds the product of thermodynamic solubility and

which therefore produce a saturated solution. It is naturally necessary to take
into account the fact that the solubility of a compound is influenced by the
other ions in the water, the temperature, and sometimes the pressure.

Large excess quantities of ions with respect to the saturation concentration will
cause large quantities of precipitates to be deposited. The most common deposits resulting from incompatibility of the waters are gypsum ($CaSO_4 \cdot 2H_2O$), anhydrite ($CaSO_4$), aragonite ($CaCO_3$), calcite ($CaCO_3$), barite ($BaSO_4$), troilite (FeS),
siderite ($FeCO_3$), and hematite ($Fe_2O_3 \cdot nH_2O$).

Precipitation may take place quickly if the excess quantity of ions is large, and
a certain waiting period called the induction period is required if the excess quantity of ions is small. This period represents the time necessary for the crystalline
seeds to be produced. Once these seeds have been formed, other molecules of
the compound will adhere to the seeds to form a crystal and finally precipitate.
It is interesting to note that the excess quantity of ions necessary to initiate the
immediate precipitation of salts which are slightly soluble, such as $CaCO_3$ and
$BaSO_4$, is much less than that required for the immediate precipitation of a more
soluble salt such as $CaSO_4$.

Swelling of the Clays — The phenomenon of clay swelling in situ during the injection of fresh water has long been known. It may result in a considerable reduction of the permeability of the layer which, in the case of water injection
for secondary recovery, leads to failure of the process.

F.O. Jones (6), working in the laboratory, was able to control the swelling of
sensitive clays by using water containing small proportions of divalent cations
such as Ca^{++} and Mg^{++}. When the water concentration of Ca^{++} and Mg^{++} is only
on the order of one-tenth the total salinity of this water (or even less), the clays
do not become swollen. However, it is necessary to avoid contrasts in salinity
and to initiate injection with a water having about half the salinity as the water
of the formation and to drop step-by-step to the fresh water containing small
quantities of Ca^{++} and Mg^{++}.

The majority of the formation waters contain sufficient Ca^{++} and Mg^{++} but the
formations with water containing only sodium salts are clay formations of the
exchangeable sodium type. The fresh injection water having small quantities of
Ca^{++} and Mg^{++} consequently loses these divalent ions which are bonded to the
clay and are enriched by Na^+ ions. As soon as this water moves farther away,
it damages the bed. Thus, when the formation waters do not contain sufficient
Ca^{++} and Mg^{++}, it becomes necessary to use high-salinity injection waters with a
small concentration of divalent cations.

It must also be mentioned that after injecting a water rich in Na^+ ions, care must
be taken not to inject fresh water without divalent ions.

Water Compatibility Tests in the Laboratory: For the many reasons just reviewed,
it is desirable to carry out water compatibility tests before proceeding with an injection operation for secondary recovery.

Based on analyses of waters capable of mixing with each other, it is possible to
calculate the composition of the water resulting from the mixture and subsequently

to predict the nature of the salt or salts capable of being deposited. It is also possible to determine whether there will be corrosion of the material or not.

A knowledge of the nature of the clays in situ makes it possible in turn to evaluate the possibility of reducing the permeability of the formation by swelling of these clays.

Only tests involving circulation through soil samples removed from the very bed which will be affected by the water injection, however, would indicate the degree of warping possible and the treatments to be applied in order to eliminate this warping.

On the other hand, it has been shown many times that water considered to be incompatible with the interstitial water or with clays in situ was capable of being injected in a sample without filling it, even after a long period of circulation. Nevertheless, this phenomenon often disappears when the injection water is seeded with a few crystals of salts capable of precipitating, or when the thermodynamic conditions of the system are varied, in particular the injection flows (salinity contrasts).

Chemical Treatment of the Waters to Make Them Compatible: The most common precipitates, i.e., those resulting from mixing some waters containing $SO_4^=$ or HCO_3^- anions and other waters containing Ca^{++}, Ba^{++}, and Fe^{++} cations were considered.

It should be pointed out that even in the particular case where the injection water comes from the very reservoir to be treated, incompatible waters may be present, since it is necessary to inject a different water supply if only to compensate the volume of oil produced in the reservoir.

The problem therefore consists of treating the water or the mixture of injection waters to eliminate the ions incompatible with the ions of the interstitial water or to conceal the action of the incompatible ions present in the injection water or in the interstitial water.

Elimination of the Incompatible Ions from the Injection Water — One proceeds by precipitation and decantation or filtration. If it is a question of incompatible anions, they are eliminated by means of barium chloride. If it is a question of alkaline earth cations, they are eliminated by means of sodium sulfate or sodium carbonate. This treatment is one of the easiest.

Sequestration of Alkaline Earth and Ferrous Ions — Sequestration is the phenomenon whereby the precipitation of a cation by an anion is prevented by the formation of a soluble complex compound between the cation and the sequestering anion added. Since it is a question of a competitive reaction of the cation with the precipitating anion on the one hand and the sequestering anion on the other hand, the results will depend on the following factors:

      (a)   reaching equilibrium;
      (b)   nature and concentration of the cation;
      (c)   nature and concentration of the precipitating anion;
      (d)   nature and concentration of the sequestering anion;

    (e)   presence of other metal ions;
    (f)   total ionic charge;
    (g)   pH of the solution; and
    (h)   temperature.

The sequestering agents of the alkaline earth cations (7) most widely used are the sodium polyphosphates and the sodium salt of ethylenediaminetetraacetic acid.

As far as the ferrous cation is concerned it can be transformed into the ferric cation and precipitated out by oxidation (aeration). It can also be sequestered by means of sodium glucoheptonate, or better, by means of citric acid (8). The success of such a treatment is obviously influenced by a strict observation of the factors mentioned previously.

Biological Treatment: Before injecting water in a reservoir for the purpose of secondary oil recovery, it is necessary to make sure by biological analysis that it does not contain any bacteria. Such bacteria must be destroyed.

Analysis — By means of a microscopic examination of the residue obtained after filtration of the water on a Millipore filter, it is possible to determine whether the water contains algae or protozoa, ferruginous or sulfate-reducing bacteria, or fungi. Then, using an adequate culture medium, a complete list of the germs is made and a count of the sulfate-reducing bacteria taken. The latter microorganisms are considered the most harmful.

Algae and Protozoa (Risk of Filling Up by Alluvion): These microorganisms can only proliferate in sunlight and in air. They are eliminated by means of an algicide of the copper sulfate type (2 g/m^3), or by means of a quaternary ammonium salt (5 g/m^3).

Ferruginous Bacteria (Risk of Filling Up by Alluvion): These are aerobic and develop by transforming soluble ferrous salts into insoluble ferric hydroxide which agglutinates with the cellular mass. They are easily eliminated with chlorine or formaldehyde.

Sulfate-Reducing Bacteria (Risk of Corrosion): These are anaerobic but can survive in the presence of oxygen. On the other hand, the presence of water with a high salt content (30% NaCl) may prevent their development. They destroy the sulfate ions by freeing hydrogen sulfide and in this way become very harmful for metal surfaces, which they corrode.

The amines of fatty acid salts, the quaternary ammonium salts, and formaldehyde have been proven effective in eliminating them. In general, the biological treatment is simple; but, there is never any certainty as to its long-range effectiveness (habituation of the bacteria to the bactericides).

Conclusion: The problems posed by water injection in the secondary recovery of crude oil are well known and in general well resolved if the natural permeability of the reservoir or corrosion of the material is considered. Accordingly, the injection water will be subjected to a physical treatment, a chemical treatment, or a biological treatment. None of the treatments, however, aims to improve the injection capacity, i.e., the flow of water injected and consequently the flow of oil produced.

To overcome the forces holding the oil in the rock and to improve the efficiency of waterflooding, the injection of chemicals has been recommended. Such techniques will be discussed in the chapter on Tertiary Recovery of Crude Oil—Miscible Techniques.

IMMISCIBLE GAS INJECTION

The secondary recovery technique of injection of natural gas has been employed since 1900. This low-pressure injection of gas is used to maintain reservoir pressure to prevent production cut-off and thereby increase the rate of production.

In general, gas that is immiscible with the reservoir oil is not as efficient as water. The lower viscosity of the gas causes an increase in the bypassing of the oil, both within the individual pore spaces and in large sections of the reservoir. Recent price increases, especially for intrastate natural gas, have made this technique much less attractive. A number of such projects have been curtailed, and it is expected that there will be few undertaken in the future.

For many years, gas reinjection methods had been widely utilized. The practice of regulating gas well below its energy equivalent price resulted in an economic situation where it paid the operator to reinject associated gas even when the increase in oil recovery was even rather marginal. Under oil/gas price ratios prevailing in the United States prior to the price jumps, it paid an operator to reinject rather large proportions (60 to 80%) of produced gas over a wide range of gas drive reservoir parameters. This was especially true if the operator expected gas prices to rise at rates higher than his cost of capital.

Efficiency

A major problem with gas reinjection is its inefficiency. For typical gas solution drive fields, reinjection will only increase oil recovery from the neighborhood of 15 to 20% of the original oil in place to perhaps 25%, still leaving the great bulk of the oil in the ground. This is due largely to the unfavorable mobility ratio. (The mobility of a fluid in a reservoir is equal to the ratio of the permeability to viscosity of that fluid. The mobility ratio, then, is equal to the ratio of the mobility of the displacing fluid to the fluid being displaced.)

Since waterflood is usually much more efficient, it will generally dominate gas reinjection when prices support the more expensive waterflood, which has already happened in the United States. For example, in 1970 it was reported that only 208 of the more than 3,000 injection and pressure maintenance projects in Texas used gas as the injection fluid.

Gas Reinjection in Conjunction with Waterflood

Occasionally, gas reinjection can be used in conjunction with waterflood as has been done by Creole in Lake Maracaibo, where gas is injected into the cap with water down dip. The gas prevents the oil from pushing into the gas cap, thus wetting additional rock; however, this technique is useful only in certain rather specialized cases. Even then, the differences with and without gas injection are not striking.

Gas Injection Sequential to Waterflood

Gas injection can also be used sequentially in conjunction with waterflood. In water wet reservoirs, capillary forces will tend to pull the water through the narrow passages, bypassing the wider passages. The opposite will tend to happen with gas. Thus, in some cases, extra production can be obtained by cycling gas and water injection.

There has been one sizable project of this type, an Amoco project in the Gillock South field in Galveston, Texas, approved in 1965. This project has since been superseded by a CO_2 flood approved in 1971. One of the problems of gas injection relative to water injection is the comparatively higher cost of compression as opposed to pumping. Again, for the Gillock South field, of the total cost of the secondary recovery of $4 million, $1.8 million was the initial cost of gas injection facilities and $1.7 million was the operating cost of the gas systems. In return, the project yielded an additional 7 to 8% of OOIP (8.9 million barrels) at a marginal cost of roughly 50¢ per barrel.

Future Prospects

In summary, future prospects for gas injection are sharply limited. Gas injection will continue to be used for pressure maintenance purposes and to prevent oil from migrating into gas caps, but there is no reason to believe that gas injection can make a sizable dent in the recovery of oil.

REFERENCES

(1)　Railroad Commission of Texas, "A Survey of Secondary Recovery and Pressure Maintenance Operations in Texas to 1970," Bulletin 70, Austin, Texas.

(2)　Louisiana Department of Conservation, "Secondary Recovery and Pressure Maintenance Operations in Louisiana," 1972 Report, Baton Rouge, Louisiana.

(3)　"Rollback Threatens New Illinois World," *Oil and Gas Journal,* (February 18, 1974).

(4)　Geffen, T., "Oil Production to Expect from Known Technology," *Oil and Gas Journal,* (May 7, 1973).

(5)　Sloat, B., "Injection Water Treatment Starts in the Producing Wells," SPE Paper 875, Sixth Biennial Secondary Recovery Symposium, Wichita Falls, Texas, (May 4-5, 1964).

(6)　Jones, F.O., "Influence of Chemical Composition of Water on Clay Blocking Permeability," *J. Petroleum Technology,* pp 441-6, (April, 1964).

(7)　Irani, R.R. and Callis, F.C., "Calcium and Magnesium Sequestration by Sodium and Potassium Polyphosphates," *J. Am. Oil Chemists' Society,* 39, pp 156-9 (1961).

(8)　Bell, W.E. and Shaw, J.K., "Evaluation of Iron Sequestering Agents in Waterflooding," *Producers Monthly,* 22, No. 5 (March, 1958).

TERTIARY RECOVERY OF CRUDE OIL
MISCIBLE TECHNIQUES

The information in this chapter is based on:

> *Basic Research Needs for Tertiary Oil Recovery,* prepared by R.S. Schecter and W.H. Wade, of the University of Texas at Austin, for National Science Foundation Workshop, under Contract GP44165, June 26–27, 1974.

> *Evaluation of Thermal Methods for Recovery of Viscous Oils in Missouri and Kansas,* BuMines Report OFR 60–74, prepared by M.D. Arnold and A.H. Harvey, of Department of Mining, Petroleum and Geological Engineering, University of Missouri at Rolla, for U.S. Bureau of Mines, June 1974.

> *The Benefits/Costs of Tertiary Oil Recovery,* BuMines Report OFR 4–75, prepared by Arthur D. Little, Inc., for U.S. Bureau of Mines, December 1974.

> *The Estimated Recovery Potential of Conventional Source Domestic Crude Oil,* prepared by J.W. Devanney III, R. Ciliano and R.J. Stewart, of Mathematica, Inc., for U.S. Environmental Protection Agency, under Contract 68-01-2445, May 1975.

> *Review of Secondary and Tertiary Recovery of Crude Oil,* FEA Report G–75/482, prepared by Lewin and Associates, Inc., for Federal Energy Administration, June 1975.

> *The Potential and Economics of Enhanced Oil Recovery,* FEA Report B–76/221, prepared by Lewin and Associates, Inc., for Federal Energy Administration, April 1976.

> *Potential Environmental Consequences of Tertiary Oil Recovery,* prepared by C. Braxton, R. Stephens, C. Muller, J. White, J. Post, J. Norton, M. Goldberg, and P. Stevenson, of Energy Resources Co., Inc., for U.S. Environmental Protection Agency, under Contract 68-01-1912, July 1976.

> *Tertiary Oil Recovery: Potential Application and Constraints,* PNL Report RAP–25, prepared by C.A. Geffen, of Battelle's Pacific

Northwest Laboratory for U.S. Department of Energy, June
1978.

INTRODUCTION

Tertiary recovery of crude oil refers to chemical and/or thermal treatment of
oil reservoirs to increase the production of crude oil beyond those amounts recoverable by natural reservoir energy or by artificial maintenance of reservoir
energy. Such high recovery techniques seek to overcome the inefficiencies of
conventional primary and secondary methods.

In this chapter the nonthermal methods will be discussed. They encompass the
miscible methods, namely hydrocarbon miscible flooding and carbon dioxide
miscible flooding, and the pseudomiscible methods, namely polymer-augmented
waterflooding, micellar polymer flooding, and alkaline flooding, which are three
methods of chemically augmenting oil production.

In the search for increasing recovery efficiency, few concepts have intrigued
petroleum engineers as much as that of miscible displacement, which ultimately
promises 100% displacement of oil in place.

The basic concept of miscible flooding was proposed in 1927, but little field
development took place until about 1960. Miscible or near-miscible displacement may be realized by injecting a fluid that mixes with the oil, namely liquid hydrocarbons and carbon dioxide.

HYDROCARBON MISCIBLE FLOODING

Principles, Procedures and Projects

One of the key problems in oil recovery in oil-wet reservoirs is overcoming the
surface tension forces which tend to bind the oil to the rock. In water-wet
reservoirs surface tension forces act to create bubbles of oil which can block
pore passages as the bubbles resist the increased surface area associated with
squeezing through these passages. These surface tension forces are the primary
reason why reservoirs become increasingly impermeable to oil relative to water
as water saturation increases.

If the interfacial tension can be reduced between the oil and the driving fluid,
then these forces are reduced. This is the object of miscible floods. One such
form of miscible flood involves using a slug of light hydrocarbons under high
pressure followed by a waterflood.

Gas is usually not very efficient in displacing oil. The exception to this occurs
when the leading gas in the reservoir is made miscible with the trailing edge of
the oil bank. This kind of miscibility results from three major processes:

    (1)   High-Pressure Gas: Applicable to reservoirs containing low-
        viscosity oil and capable of supporting high reservoir pres-
        sures (in excess of 3,000 psi). The leading edge of the dry
        gas evaporates the more volatile hydrocarbons from the
        crude oil until a small bank of these hydrocarbons is formed

between the reservoir oil and the dry gas creating a self-sustaining miscible slug-type displacement.

(2) Enriched Gas: A similar miscible bank may be achieved by injecting gas containing a substantial portion of volatile hydrocarbons (predominantly propane and butanes). The reservoir oil is enriched by the extraction of the liquefied petroleum gas (LPG) components from the leading edge of the gas. This forms a mutually soluble bank of inter-mediate molecular weight material between the reservoir oil and the injected gas.

(3) Miscible Slug Process: Finally, intermediate molecular weight hydrocarbons (LPG) may be injected into the reservoir before the gas injection starts. The LPG slug process, while the most simple and direct of the hydro-carbon processes, may bypass much of the oil in hetero-geneous reservoirs.

Thus, hydrocarbon miscible flooding is the injection of light to intermediate weight hydrocarbons such as dry gas, propane, butane and LPG to mix with crude oil forming a bank of oil that is driven toward producing wells.

According to the *Oil and Gas Journal,* there have been thirteen field-wide hydro-carbon floods, four of which have been terminated (1). Three of these com-pleted floods were termed successful. While under ideal laboratory conditions miscible hydrocarbon flooding can recover all the remaining oil, the completed projects report residual oil saturations of one-half to two-thirds of the initial saturation.

Cost-Performance Characteristics

The biggest high-pressure hydrocarbon flood undertaken in the United States has been the Hunt project in 22,600 acres of the Fairway Field. This is a deep 10,000-foot limestone formation with moderately low permeability (22 milli-darcys), moderate porosity (11%), and an average thickness of 70 feet. Original oil in place is estimated at 400 MM barrels. The oil is very light (48° API), vol-atile (current GOR 3,317 ft^3/bbl) and fluid (viscosity 0.15 cp at reservoir con-ditions). Gas is being injected at 4,500 psi. In a 1966 prospectus, Hunt claimed:

Recovery Method	Incremental Recovery, MM bbl	Cumulative Recovery, % OOIP
Primary	96	24
Waterflood	+106	51
High-pressure gas	+178	70

In Hunt's 1966 prospectus, the cost of the project was placed at $4 MM includ-ing $1 MM for compression equipment and $2 MM for gas. This would have made this project extremely attractive at the time. As of January 1, 1975, 124 MM barrels had been produced from this field. The *OGJ* ultimate recover-able for Fairway is 210 MM barrels, which assumes a successful waterflood but does not count the projected increase over waterflood due to the high-pressure gas.

Limitations

Hydrocarbon miscible flooding suffers from several problems. Since high pres-sures are required, the method is limited to fairly deep fields both to avoid the cost of repressuring and the threat of blowing through the overburden. The method is energy-intensive both because the slug is hydrocarbon and more importantly because of the amount of compression required. Most important, however, the method suffers from a very poor mobility ratio. As discussed in the second chapter under Immiscible Gas Injection, the mobility ratio is a measure of the ease with which the driving fluid moves through the formation relative to that with which the reservoir fluid moves.

If this ratio is high, the driving fluid tends to finger the oil and channel through high permeability streaks. The net effect is that the driving fluid begins to recirculate rather than expulse oil. This is much the same problem that plagues waterflood. The resultant implication is that miscible hydrocarbon flooding is limited to very high API oils where the mobility ratio is not so unfavorable.

Future Prospects

Less than 12% of the discovered remaining oil in place is lighter than 40° API. Because of the problems with this method mentioned above and the lack of interest operators have expressed in it, the potential of hydrocarbon recovery seems poor. In addition, increases in gas prices have further discouraged applications of hydrocarbon miscible flooding. It would, therefore, be imprudent to count on hydrocarbon miscible for more than a billion or so barrels of additional recovery.

CO_2 MISCIBLE FLOODING

To meet the ultimate objective of enhanced recovery, the reach for the elusive 100% recovery, the invading fluid must be completely miscible with the reservoir oil, leaving no residual oil in the invaded region. There are only a few fluids which are completely miscible with both oil and water and reasonable enough in cost to be considered for tertiary recovery. Three fluids have been tried: alcohol, hydrocarbon gases, and carbon dioxide. While in the past mixtures of alcohols, usually propyl and butyl alcohols, were used, the higher molecular weight alcohols lose their water solubility while the lower alcohols, ethyl and methyl, are not miscible with many of the crude oils. As noted earlier, rising gas prices have made hydrocarbon miscible methods uneconomical. Thus, the miscible fluid now being considered most promising is carbon dioxide.

Carbon dioxide miscible flooding is the injection of CO_2 to dissolve in crude oil, swell it, reduce its viscosity, and vaporize it into the CO_2 phase, resulting in high displacement efficiency of the contacted oil. Banks of water alternated with injected CO_2 may be used to drive the oil toward producing wells. When CO_2 begins to appear at the producing well it is recovered, cleaned of impurities, pressurized and reinjected.

Principles and Procedures

Though carbon dioxide is not completely miscible with most crude oils, it is soluble with both crude oil and water at reservoir temperatures and pressures. When carbon dioxide is injected and mixing occurs, the viscosity of the crude oil is reduced. The carbon dioxide increases the bulk and relative permeability of the oil, swelling it so that reservoir pressure is increased and the oil flows more readily toward the production wells.

Three variations of CO_2 miscible flooding are presently employed by the petroleum industry for tertiary oil recovery: (1) injection of carbon dioxide in a slug, followed by water or carbonated water; (2) injection of carbonated water directly; and (3) injection of carbon dioxide at high pressure to achieve mixing directly with the reservoir oil and the formation of an oil-miscible slug in the formation.

The major factors contributing to oil recovery in carbon dioxide flooding are formation of a miscible slug in in situ modification of viscosity and changes in oil density and compressibility of fractions. By maintaining the slug in a single dense phase, its solubility with crude oil is increased considerably. Crude oil viscosities ranging from 5 to 90 cp can be reduced by a factor between 10 and 100 times under high injection pressures of carbon dioxide. After a carbon dioxide flood has been completed, the gas comes out of solution to some degree due to reduction in pressure, creating a further gas drive within the reservoir.

Projects—Performance and Costs

Carbonated waterfloods enjoyed some popularity several years ago. Most of these projects were disappointing, however, possibly because it was difficult to maintain an adequate carbon dioxide concentration at the displacement front. Renewed interest in carbon dioxide injection has been primarily in the use of a liquefied gas slug followed by water. Instances of apparently successful use of carbon dioxide injection have been reported for the Texas Sacroc, Wasson, and Crosset fields (2). For the location of technically feasible CO_2 miscible flooding projects in major oil fields of the U.S., California and Texas see Figures 3.1a, 3.1b, and 3.1c, respectively. The South Central and Southeast regions contain the largest concentration of potential projects using CO_2 miscible flooding.

These projects operate at relatively high pressures. Some investigators consider 1,000 psi to be a typical minimum pressure for miscible displacement by carbon dioxide, and this criterion would preclude use of the process in reservoirs which are only a few hundred feet deep.

It should also be noted that some beneficial effects can be achieved with carbon dioxide below the pressure required for miscibility. These effects on viscosity, formation volume factor, and density for an eastern Kansas crude oil are presented in Table 3.1 below.

In CO_2 projects, API gravities range from 32° to 42° with low viscosities. Depths are high, 4,300 to 9,000 feet, and permeabilities are low, 2 to 10 md.

Table 3.1: Effect of CO_2 on an Eastern Kansas Crude Oil

Carbon Dioxide Saturation Pressure (psia)	Density (g/cc)	Gas-Oil Ratio (scf/bbl at 60°F)	Viscosity (cp)	Formation Volume Factor (Bo)
15	0.9359	0	889	1.0000
100	0.9203	39	634	1.0180
200	0.9060	83	436	1.0345
300	0.8947	130	306	1.0470
400	0.8851	183	217	1.0571
500	0.8770	243	155	1.0656
600	0.8700	316	110	1.0725

Note: Pressure base, 14.65 psia at 60°F. Paraffin formation nil. No emulsion observed.

Source: BuMines OFR 60–74

Figure 3.1: Technically Feasible CO₂ Miscible Injection Projects

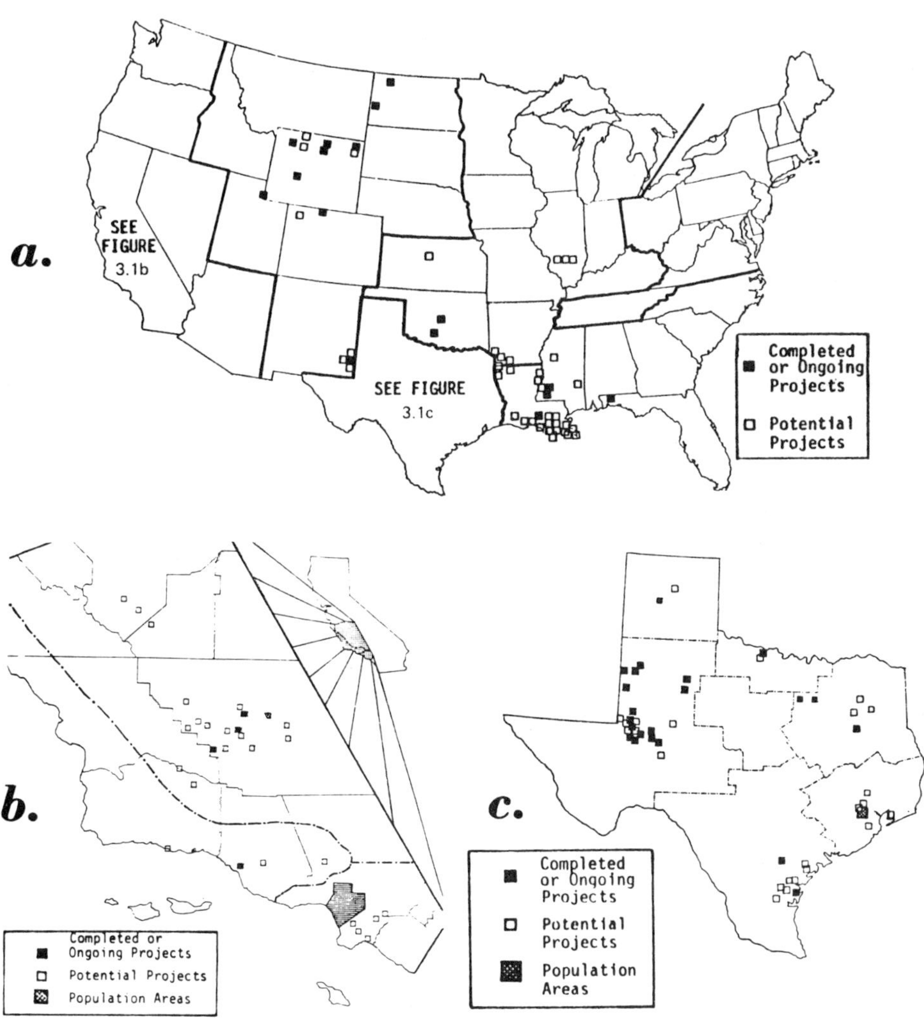

(a)   Major U.S. Oil Fields
(b)   Southern California
(c)   Texas

Source: EPA-68-01-1912

Sacroc Unit, Texas: The largest CO₂ project is the 50,000 acre Sacroc Unit encompassing the entire Kelly-Snyder Field in West Texas. This is a limestone reservoir at 6,700 feet in which the primary recovery mechanism was solution gas drive with no gas cap. Initial pressure was 3,122 psi, later adjusted to

2,400 psi; porosity, 7.6%; permeability (average), 9.4 md; net pay thickness, 213 feet; connate water saturation, 36%; gravity, 42°; viscosity, 0.375 cp at reservoir conditions.

Original oil in place was about 2.7 billion barrels, of which some 600 MM had been produced by the beginning of 1973. The field had been subject to a centerline drive water injection program. This waterflood has proved very effective with a residual water cut of 26% and a sweep efficiency of 72%. (Sweep efficiency is defined as the ratio of the volume of rock contacted by the displacing fluid to the total volume of rock subject to invasion by the displacing fluid.) After waterflood, however, about 1.5 billion barrels would be left.

The CO_2 flood is aimed at reducing residual oil saturation, not increasing sweep efficiency. (The CO_2 will not initially reach areas of the reservoir which are not reached by the waterflood due to the less favorable mobility ratio.) The Sacroc plan involved injecting 50 MM cfd of CO_2 over ten years. The CO_2 flood was to be followed by waterflood. CO_2 would be miscible with the light crude of 1,850 psi. Injection pressure would be 3,000 psi, volume 3 MM cfd/well.

The project's operator claimed lab tests indicated that residual oil saturation would be 0% in swept part of reservoir and that the CO_2 flood would net 156 MM barrels. In 1969, the projected cost of this CO_2 flood including the 200-mile, 18 to 20 inch pipeline and 5¢/hcf for the gas totaled $120 MM. In 1974 operators reported total CO_2 cost estimates had risen to $140 MM plus another $75 MM for the waterflood. Using these figures and computing a value at 10% return, the cost per unit of oil works out to be about $2.50 per barrel.

Chevron also had examined hydrocarbon miscible flood but claimed high-pressure gas would not work because reservoir pressure was too low for miscibility. Enriched gas would require a 15% slug to get 115 MM additional barrels. Chevron claimed it would cost $57 MM more than CO_2 and obtain 41 MM fewer barrels.

Gillock South Field, Texas: Another larger (5,900 acre) CO_2 project is that undertaken in 1971 by Amoco in the Gillock South Field. This also involved a light (38° API), low-viscosity (0.64 cp at 214°F) crude. The reservoir was at a depth of 9,000 feet with moderate pressures (3,300 psi). However, the field is a sandstone formation of high permeability (1,156 md) and porosity (30%), typical of the Gulf Coast. This field had been subject to a combined gas injection-water injection program, and primary plus secondary operations produced an aggregate recovery of 65% OOIP, 93 MM barrels in all.

The secondary operations accounted for 8.9 MM barrels over primary under good water drive, which Amoco claimed could be enhanced 65 to 78% by CO_2 flood for a net tertiary recovery of 11.2 MM barrels. Amoco, like Chevron, claimed all oil contacted by CO_2 would be recovered. The 78% then is essentially the sweep efficiency.

Amoco plans were to use six injection wells, injecting 10,000 reservoir barrels of CO_2 (500 tons or 8,500 Mcf surface) per day while phasing out the waterflood. The initial costs of the project were put at $1.5 MM and operating costs at $18 MM. Amoco planned to buy CO_2 from an ammonia plant in Texas City. About $15 MM of the operating costs were the CO_2, which, assuming a project life of

10 years, puts the cost of CO_2 at about \$6 per ton or 60¢ per reservoir barrel. If these figures are correct, the cost per additional barrel is then about \$1.50.

Crosset Field, Texas: A third large CO_2 project is that undertaken by Shell in the Crosset Field in West Texas. This reservoir is a fine-grained chalk. Permeability is low (3 md) but unusually uniform. This field is a rare exception in that it has never been waterflooded because water would cement this unusual formation. Porosity, at 21%, is moderate; API gravity is 42°, miscibility pressure is 1,650 psi. Shell planned to use a 93% CO_2, 7% CH_4 mixture from the Sacroc pipeline which passes within just a few miles of Crosset. Injection started in mid-1972. Shell planned to use 73 billion ft^3 over 10 years at a rate of 20 MM cfd and to reinject produced gas as is.

Shell also started a pilot project in the Little Creek Field in Mississippi, paying \$25/ton for CO_2 delivered to the field by truck.

Problems

Carbon dioxide miscible flooding is favored by high API gravity, requires a fair amount of pressure, and is relatively insensitive to permeability. A basic and continuing problem, however, has been mobility control, which indicates that CO_2 effectiveness is sharply dependent on reservoir homogeneity. The mobility ratio, while not as bad as that for high-pressure hydrocarbon, is still quite unfavorable for most reservoirs.

Unfavorable mobility ratios lead to an unstable slug/oil interface. This was dramatically illustrated by the premature breakthrough of CO_2 in the Sacroc project. The operators had calculated that initial breakthrough would occur two years after start of injection. Actual breakthrough occurred in two months. Once breakthrough occurs, there is very little one can do about it other than to begin recirculating and recompressing CO_2, and attempting a sandwich technique in which water and CO_2 injection are alternatively cycled at short intervals.

Poor mobility ratios lead to poor conformance and effectively confine the method to light crudes. Also, as gravity decreases, the pressure required to obtain miscibility increases. Certainly, any oil which was not reached by the preceding waterflood due to limited sweep efficiency will not be reached by CO_2 in the absence of additional drilling. No premature breakthrough has been reported by Shell in North Crosset, but as mentioned earlier, this reservoir exhibits exceptionally little variation in permeability.

One problem that has also been noticed in the use of carbon dioxide injection processes is the effect of precipitation of asphalt compound under carbon dioxide pressure. Variations in carbon dioxide flooding have been attempted which incorporate the use of foaming agents with the flood to achieve more mobility control with the gaseous medium. These techniques, however, have received few trials on a field-wide basis to date. Another basic problem discussed below concerns costs; the compression and transport of CO_2 is expensive and extremely energy-intensive.

Economic Aspects

CO_2 is a common by-product of industrial processes and would appear at first to be almost a free good. Unfortunately, however, the available CO_2 is rarely in the form or location desired. CO_2 concentrations in industrial stack gas run 10 to 13%. Practically pure CO_2 is available at certain chemical plants, such as ammonia facilities, but rarely in the volumes required.

Furthermore, by-product CO_2 is usually available at atmospheric pressure. The economics of pipelining CO_2 are such that it is best maintained above critical pressure. This generally requires pipeline injection pressure of 2,000 psi or more. For field injection into reservoirs, pressures of 3,000 psi are often indicated. It takes approximately 0.015 barrel of oil as fuel to compress 1,000 ft^3 of CO_2 from atmospheric pressure to 3,000 psi, or about 16 ¢/M cf in fuel costs.

CO_2 is available at pressure as a natural resource generally in association with gas. However, the geologic events required to form large quantities of CO_2 are relatively rare and always involve depths in excess of 15,000 ft which imply expensive wells. In the Val Verde Basin large quantities of CO_2 are produced in association with gas. Unfortunately, this is not a common phenomenon.

Transporting CO_2 is somewhat more expensive than transporting an equivalent amount of natural gas the same distance. Higher pressure and hence heavier walled pipe are indicated. The cost of transporting the Sacroc CO_2 200 miles by 18 to 20 inch pipeline has been placed at 25 ¢/M cf at 1972 prices. Total transport costs in terms of 1974 dollars were in the neighborhood of 20 ¢/M cf per 100 miles.

At 2,000 psi and normal temperatures, it takes about 820 surface cubic feet of CO_2 to equal one reservoir barrel; at 3,000 psi, about 1,125 ft^3. Typical production plans for CO_2 projects necessitate injecting 15 to 30% of pore volume, followed by water. Most of the operators report target recoveries of about 10% of the remaining oil in place. The costs per unit of oil recovered as a function of both distance to free source of CO_2 at atmospheric pressure and oil saturation are shown in Table 3.2. The amounts shown are exclusive of all costs (well maintenance, separation and waterflood, etc.) other than those of delivering and injecting the CO_2 and assume an equivalence of 1,000 ft^3 per reservoir barrel and a 20% pore volume slug. The obvious implication of these amounts, of course, is that the distance to the source of CO_2 is an extremely important variable in determining the prospects for tertiary yield from CO_2.

Table 3.2: Unit Cost of CO_2 Recovery as a Function of Oil Saturation and Distance to CO_2 Source

Distance to Source (miles)	Oil Saturation		
	30%	45%	60%
0	1.00	0.75	0.50
100	2.33	1.55	1.75
200	3.66	2.77	1.83
500	7.69	5.12	3.84
1,000	15.32	10.21	7.66

Source: EPA 68-01-2445

Production Schedule: It should also be noted that the total production schedule for CO_2 miscible flooding, as reported by Lewin and Associates, Inc., in their study prepared for the Federal Energy Administration, has an increasing rate beginning in the third year, a constant production rate for seven years and a gradual decline during the last ten years, to the economic limit. The production schedule for the incremental oil recoverable from CO_2 miscible is shown in Figure 3.2.

Figure 3.2: Production Schedule for CO_2 Miscible Flooding

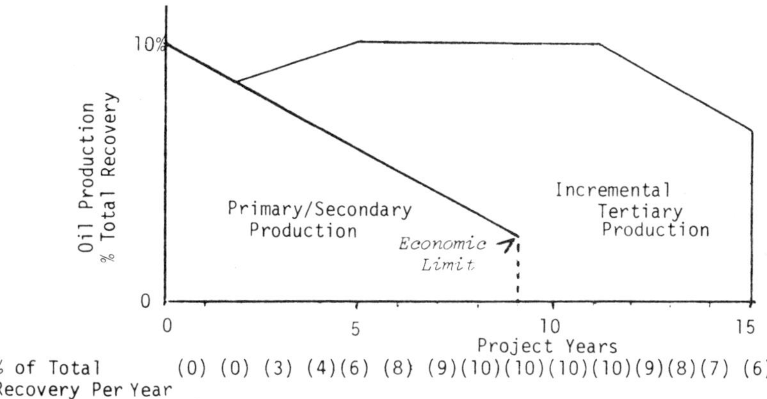

% of Total   (0) (0) (3) (4)(6) (8) (9)(10)(10)(10)(10)(9)(8)(7) (6)
Recovery Per Year

Source: FEA/B-76/221

Future Prospects

CO_2 appears to be the most promising of the methods which aim at miscibility. However, it will probably be limited to high-gravity fields in relatively close proximity to substantial sources of CO_2. Due to poor conformance it will be difficult even in these fields to obtain more than 10 to 15% of the oil remaining after waterflood. It remains to be seen how low in API gravity one can go with this technique. Mathematica, in their study for the Environmental Protection Agency (CO 68-01-2445), concluded that while there are a number of fields with high API gravity, gas drive, near CO_2 sources, for which a CO_2 project would be quite attractive, it would be imprudent to count on CO_2 for more than 2.5 to 5.0 billion barrels of additional oil over that represented in the API recoverables.

On the other hand, Lewin and Associates, Inc., in their study which was confined to the three states of California, Louisiana and Texas (which contain 64% of the oil remaining in the lower 48 states) estimated that CO_2 flooding in these three states could add 11.7 billion barrels of oil to our reserves by the year 2000. This represents a contribution by CO_2 methods of 38% of the 30.5 billion barrels of technically recoverable resources estimated by Lewin. [It should be noted that the reservoirs of these states are not necessarily representative of the other remaining states due to certain idiosyncrasies, e.g., the heaviness of California crudes, the fractures and salinity of Louisiana reservoirs, or the vast carbonate reservoirs of Texas, and such reservoirs may, therefore, be more receptive to certain enhanced oil recovery (EOR) techniques than others.]

To produce 10 billion barrels, approximately 45 trillion ft^3 (tcf) of CO_2 would be required over the next 30 years. While recycled CO_2 can substitute for about 20% of the requirement, natural sources would still need to supply 25 tcf and manufactured sources the remaining 10 tcf.

CO_2 supply appears to provide a major constraint to development of this method. While considerable quantities of naturally occurring CO_2 are produced with natural gas and CO_2 is a by-product of many chemical and refining processes, its limited value has led to CO_2 being considered a waste material. Recent changes in the economic value of CO_2 have led to a renewed interest for discovering and capturing CO_2 supplies (3). However, major improvements in transportation and recycling will need to be made before the full potential of this recovery technique can be realized.

POLYMER-AUGMENTED WATERFLOODING

Another type of enhanced recovery is polymer-augmented waterflooding, which is the addition of high-molecular-weight chemicals (polymers) to thicken water in waterflooding. This process decreases the mobility contrast between the drive water and reservoir fluids, thus reducing the tendency of water to bypass oil in less permeable portions of the reservoir.

Polymers

A polymer is a large chain molecule formed by thousands of repeating blocks called monomers. Polymers have long been used in the oil industry for drilling and fracturing fluids and water blocking agents. They are also used in the manufacture of paint and polishes. Polysaccharide biopolymers and partially hydrolyzed polyacrylamides are the polymers most commonly used in polymer flooding operations.

Polysaccharides used are the remains of a protective coating or shell grown by a microorganism. The polysaccharide is recovered by treating a culture of the microorganisms with isopropyl alcohol, followed by washing and drying.

Polyacrylamides are synthetically produced by the combination of carbon, hydrogen and oxygen into basic monomer units, which are polymerized to form a long chain molecule (4). Commercial polyacrylamides are provided in dry, powdered form or concentrated in a water-and-oil emulsion. The dry polymer must be dissolved at the well site. The emulsion polymers dissolve uniformly and easily in water and are easier to control; however, they are usually more expensive (5). The extent of hydrolyzation (a process which changes some of the amide monomer groups to carboxylated groups) that a polymer undergoes determines the chemical effect on fluid viscosity (4). Nonionic polyacrylamides usually have low degrees of hydrolysis and are compatible with a wide variety of brines. An anionic polyacrylamide, however, having a high degree of hydrolysis, will be compatible only with fresh-water or soft-water brines (5).

Polymers are nontoxic and noncorrosive (4). In most waterflood projects, the conversion to a polymer flood requires minimal capital expenditures, the principal additions being the polymer mixing and filtering equipment (6). The use of polymers reduces the producing water-oil ratios, so that more oil is produced while handling less water, reducing operating costs significantly (4).

Polyacrylamides should be mixed in fresh water since the magnitude of mobility ratio improvement achievable with these chemicals decreases with water salinity and divalent ion concentration. Polyacrylamides are also adversely affected by shearing, which may reduce the ability to decrease the permeability of the formation to water (7). Special care must be taken to reduce excessive shear in surface pumps, valves and wellhead equipment; high flow rates, particularly through perforations, should be avoided (4).

Polysaccharides are less sensitive to water salinity and less likely to shear during injection. However, these polymers may need to be filtered through micron-sized filters to prevent well plugging, and will normally require bactericides since they are also susceptible to bacterial attack. Polysaccharides, more expensive per pound than polyacrylamides, have been less extensively tested (7).

Principles, Procedures and Problems

This process employs an additive to the initial injection water to increase the viscosity of the displacing fluid. The influence of this thickened water on the efficiency of oil displacement is minimal at best since the thickened water pushes the connate water to form the actual displacing fluid. It does, however, cause the reservoir conformance (the fraction of the reservoir swept by the invading fluid) to be increased. It, therefore, follows that this process is most advantageous in reservoirs where the conformance will be poor either because of very adverse viscosity conditions or because of heterogeneities in the reservoir permeability. Thickened water can be used alone or it may be used as a following agent for any other miscible process.

Approximately 40% pore volume of a solution of polyacrylamide is usually injected for mobility control. The concentration of the polymer varies from 2,000 ppm initially to 100 ppm at the end of the injection (BuMines OFR 4–75).

A field test in Campbell County, Wyoming, in the Minnelusa geologic formation, has a projected recovery of between 3 and 5 barrels of additional oil per pound of injected anionic acrylamide. In one of the four patterns included in the testing area, 1,200,000 pounds of anionic polymer were required for injection with a projected oil recovery of 3 barrels per pound (8).

While the addition of water-soluble polymers to increase the effectiveness of waterflooding techniques is a fairly well-established industrial concept which has been shown to work both in the laboratory and in actual field tests, techniques and material systems need to be improved if polymer flooding is to become the general method for economically recovering oil on a commercial scale.

It was generally agreed by the participants of a National Science Foundation Workshop held in Austin, Texas, in 1974 (CO GP 44165) that a great deal of basic fundamental research is necessary. The following is a brief discussion of the research areas thought to be particularly important by the workshop participants.

Rheological Behavior of Polymer Solutions in Porous Media: Polymer flooding solutions may be almost completely viscous or somewhat viscoelastic in their rheological behavior, depending on the type and concentration of polymer employed. A basic question thus arises concerning the relationships between the efficiency of the polymer flood and the chemical composition and concentration of the polymer comprising it. Particularly, the effect of the viscoelastic and shear thinning characteristics of polymer solutions on the efficiency of oil recovery is unclear.

This uncertainty is directly related to the difficulty of properly characterizing viscoelastic flow in porous media. It has been suggested that a method of correlating flow through porous media with polymer solution properties measured in more simple flow fields would be particularly valuable.

Since the pore size and distribution of oil-bearing substrata vary from field to field and also within any particular field, it is expected that there may not be a single optimum polymer solution for all substrata. Rather there may be a relationship between the geometric properties of the porous medium and the polymer rheological properties required for optimum flood efficiency.

Regardless of the mechanism whereby the addition of polymer improves flooding efficiency, it is expected that deterioration of polymer molecular weight by shear forces particularly as it is injected into the porous medium and as it flows through the porous medium near the well bore is an important process parameter, neglect of which may lead to poor oil recovery. This expectation is based on the well-known observation that the rheological properties of polymer solutions can be irreversibly degraded by the application of mechanical shearing forces. What is not known is the extent of reduction of solution viscosity and elasticity as it flows through porous media and how this reduction is related to such system properties as polymer type, molecular weight, molecular weight distribution, polymer concentration, rock porosity and pore size distribution, and flood velocity.

Degradation of Polymer Solution Properties: In addition to the shear degradation of molecular weight caused by the injection into and the flow of the polymer solution through the porous medium, there are several other processes by which solution property degradation can occur.

Biological attack on some water-soluble polymers by aerobic microorganisms is well known, is usually controlled prior to injection by use of appropriate biotoxins, and presents no fundamental problems. Other processes, however, merit fundamental consideration both from the standpoint of characterizing the phenomena and of devising techniques for minimizing their effect.

Typical well temperatures can approach 175°F depending on their depth, so that the thermal stability of aqueous polymer solutions may be an important parameter in the reduction of solution properties. Thermal oxidative degradation of some polymer solutions has been observed in the laboratory and while little or no oxygen is expected in the well, it seems possible that the presence of heavy metal ions, such as are known to exist in some substrata, may catalyze the decomposition of polymer oxidation intermediates formed prior to injection of the polymer solution and effectively continue the degradation process even in the absence of oxygen.

The rheological properties of some polymer solutions are known to change markedly in the presence of electrolytes. Depending on the polymer system in question, electrolyte addition can cause the polymer to precipitate from solution or to change its molecular configuration in solution so as to cause a drastic reduction in solution viscosity and elasticity. There has been considerable concern over the salting-out effect in that polymer precipitation may actually lead to irreversible plugging of the well.

It has further been observed that in some critically stable polymer solutions, solutions in which the polymer is on the verge of precipitation, precipitation can be made to occur by pumping the solution through an orifice. Thus, well plugging may be caused by the combination of shear and electrolyte action. Work is definitely needed in this area to determine the envelope of appropriate polymer and electrolyte concentrations required to ensure nonplugging in a variety of substrata pore sizes, flooded at different rates.

Direct adsorption of the polymer on the porous media is a plausible explanation for the observed reduction in solution rheological properties. The extent to which this is the predominant mechanism in actual substrata is unclear. Fundamental studies of adsorption in the presence of complicating factors, such as pH and electrolyte concentration, are needed in order to establish whether the properties of the substrata are important to the process of polymer solution degradation.

Polymer floods are used to push a preceding surfactant flood (discussed later in this chapter) which is designed to wash out the oil droplets remaining in the rock after completion of primary recovery techniques. It has been reported that the two floods do not remain as separate entities during the tertiary recovery operation. Rather, the polymer solution travels at a faster linear rate through the porous medium, probably by a chromatography mechanism, and eventually merges with the surfactant flood so that a zone of intermixing of polymer and surfactant eventually occurs despite the injection process sequence.

It is known that surfactants can associate with water-soluble polymers. Such association presumably destroys the effectiveness of the surfactant and may also destroy the rheological properties of the polymer flood. The extent to which this can happen and its effect on the porous media and recovery process bears close study.

Analytical Problems: A large part of the difficulty in understanding and optimizing polymer flooding is due to the lack of well-defined measuring techniques. The following analytical problems warrant particular attention.

Methods of analysis for presence of polymers in injection and produced waters are especially needed. Such methods should be capable of detecting not only the concentration of the polymer (generally less than 0.1 wt %) but also its molecular weight and molecular weight distribution and ideally in the presence of contaminants such as crude oil and surfactant.

Analytical techniques capable of distinguishing between the combined effects of plugging due to physical entrapment, physical adsorption, excluded volume effects and salting-out phenomena need to be developed.

Methods for studying polymer degradation in dilute solutions need to be developed. Particularly needed are techniques capable of discriminating between the various possible degradation mechanisms cited above.

Fundamental Engineering Problems: The following problem areas have retarded wide-scale use of polymer flooding techniques. Successful implementation of polymer flooding will require considerable engineering effort in the following areas.

The ability to successfully simulate, or mathematically describe, the flow of polymer solutions in porous media, given all of the complex interactions likely to occur in a real well, is absolutely essential to the success of tertiary recovery. Scale-up techniques for applying laboratory data on plugging and other characteristics of polymer solutions to field conditions are definitely needed.

Polymer engineering, determination of relationships between molecular architecture and rheological properties in aqueous solutions flowing through porous media will be required to design the optimum polymer on a cost-performance basis. Improved techniques for optimizing dissolution of polymers while minimizing degradation of rheological properties need to be developed.

Potential and Production

Although the enhanced waterflood has a much greater potential for oil recovery than does a plain waterflood, two points should be considered. First, the additives must be purchased and injected at the beginning of the projects and second, the rate of oil recovery in the first few years of operation is not increased beyond that which would have been recovered from a standard waterflood. The extra oil recovery comes later in the life of the project.

Total Incremental Production: Based on a review of polymer-augmented waterflood projects by Lewin and Associates, Inc. (FEA/B–76/221) the use of polymer waterflooding improves waterflood recovery by 20 to 25%, when it is successful. It is concluded that a polymer used as a tertiary recovery technique will increase ultimate recovery by 10% over that which would have been recovered by primary and secondary alone. (For example, for a reservoir with a 30% recovery of oil in place, the revised recovery would be 33% for an incremental recovery of 3% of oil in place.) Thus, recovery projections are a function of the effectiveness of primary and secondary recovery.

Figure 3.3: Production Schedule for Polymer-Augmented Waterflood

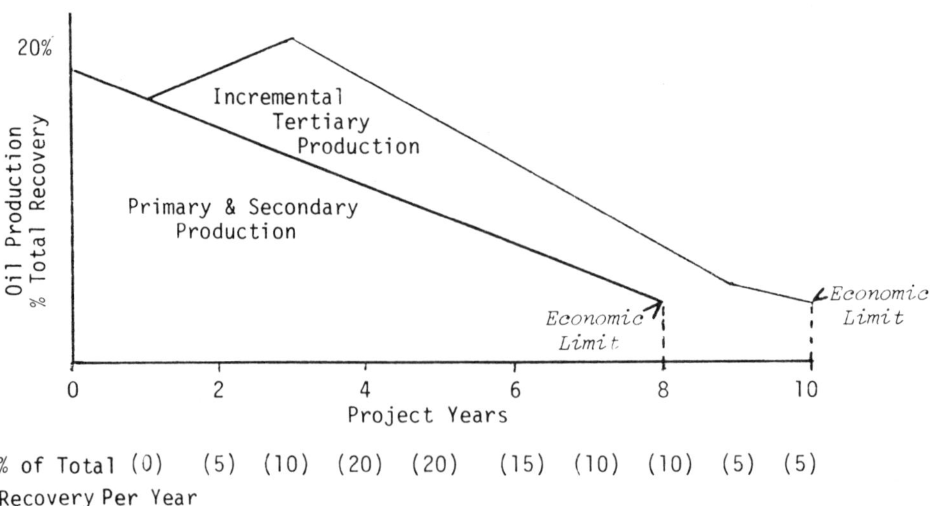

Source: FEA/B–76/221

Production Schedule: The production schedule used for polymer-augmented waterflood shows a response during the second year, an increasing production rate for two years and a gradual decline to the economic limit, as shown in Figure 3.3.

Lewin and Associates, Inc., estimate that less than 0.1 billion barrels of oil will be added to our potential reserves by 2000 in California, Louisiana and Texas through the application of polymer-augmented waterflooding, which is considerably less than 1% of the total contributions by enhanced recovery techniques.

Costs and Future Prospects

Unit Costs for Polymer Solution: The cost of polymer solution (at a concentration of 250 ppm) was $0.172 per barrel, based on the cost of the polymer at $2.00 per pound as of mid-1975. A higher cost polymer, capable of withstanding higher salinity, is used in the polymer-augmented waterflood. An important note in looking at polymer waterflooding is that it will generally be used with a new or an ongoing waterflooding project. Thus, the field development and operating costs will have been shared.

The supply of water-soluble polymers may fall short of demand by more than 100 MM lb by 1985; however, several factors should tend to mitigate this shortfall. The demand for polymers is expected to grow rapidly for other markets besides oil recovery, most notably in the use of polymers as flocculants in wastewater treatment systems. Hence, the marketplace for polymers is broad-based so that manufacturers can build additional capacity without having to rely on a single type of customer or application to support the plants. Also, other potentially less expensive processes for producing polymers are available.

The process of injecting a monomer into the oil-bearing formation under the proper conditions of pressure, temperature and catalysis to induce polymerization could be applied to mobility buffer solutions for tertiary recovery. Polymers have already been manufactured in this manner for other oil well treatment applications (9).

Lewin and Associates, Inc. (FEA/G-75/482) have concluded that augmented waterflooding requires considerable additional incentives for making the project economically feasible and competitive with waterflooding. The additional production from this enhanced recovery technique would need to be raised above $9.50 per barrel for the project to reach a 15% pretax and a 10% after-tax return. The entire production from the field would need to be released for miscible waterflooding to achieve an after-tax rate of return greater than that which could be realized from traditional waterflooding.

MICELLAR-POLYMER FLOODING

In setting up waterflood projects, the field is drilled usually in a regular pattern consisting of a series of injection wells surrounded by production wells or vice versa. When water is injected into the formation, oil is driven toward the production wells where it is produced. Initially, the movement of oil is caused by the water pushing it through the formation. Following water breakthrough, however, the oil must be produced by being mechanically entrained in the water drive. Since oil and water are immiscible (they do not mix but rather form two phases), the water will tend to bypass oil which is trapped and held by capillary forces within the interstices of the formation. Only the oil which is within or directly adjacent to the flow paths of the water will be drawn into the fluid migration.

Principles and Procedures

The technology of micellar-polymer flooding is similar in operation to a water-flood (although considerably more complicated), but it relies on chemical and physical forces to effect oil displacement. Certain chemicals are used which wash the oil out of the reservoir rock the same way that laundry detergent acts on greasy stains. Chemical techniques such as micellar-polymer flooding may produce about half of the recoverable tertiary oil. In a micellar flood, a slug containing a high concentration of surfactant is injected into the reservoir. The slug is called a micellar solution because the large concentration of surfactant causes the surfactant molecules to cling together in clusters called micelles. One of the unique properties of an oil-external microemulsion is that it is miscible with the oil and, therefore, displaces the oil in the formation by dissolving it in the slug.

A second displacement mechanism, common to low-tension floods (dilute sur-factant) or to the leading edge of the micellar flood is related to the reduction of the interfacial tension between oil and water as a result of the presence of the surfactant in the micellar slug. As the interfacial tension is reduced, the capillary forces acting to hold the oil within the reservoir are reduced. When the interfacial tension is reduced to a very low value, the capillary forces are no longer sufficient to retain the oil and, consequently, it flows freely out of the pores. A typical micellar flood project proceeds in four distinct phases as shown schematically in Figure 3.4, which are as follows.

(1) Preflush which is injected into the reservoir to adjust salinity of the formation to be compatible with the optimum sur-factant system. Final information on flow patterns may be obtained at this time. Chemical control of bacterial action may also be applied in this and subsequent steps in the pro-cess.

(2) Micellar solution (slug) including surfactant, cosurfactant, hydrocarbons, electrolytes, which in combination accom-plish the miscible displacement of oil from the formations.

(3) Mobility control solution (buffer), a low salinity water/poly-mer solution used to uniformly push the micellar slug through the formation in as close to a piston-like flow as possible.

(4) Water drive which pushes the micellar slug and mobility buffer through the reservoir to the production well.

Each of these steps is discussed below.

Preflush: In preparation for a micellar flood, the reservoir engineer must be con-cerned with how the slug and buffer solution will move through the reservoir. Selection of the surfactant system that will be most appropriate for the reservoir being flooded is also of concern. Initially, core samples are studied to determine the optimum micellar system which may include cosurfactant and electrolyte.

Once this is known, a brine solution having the optimum concentration of electrolyte is injected into the well to (a) provide a tracer analysis of the project flow characteristics, (b) to adjust salinity if necessary, and (c) to reduce the concentration of divalent ions in the reservoir which may adversely affect the slug. In the presence of a high concentration of these divalent ions, the surfactant in the slug quickly loses its effectiveness through complexing or through formation of precipitates within the reservoir. Adsorption of selective ions within the reservoir clays has also been a problem.

Figure 3.4: Micellar-Polymer Flooding Process

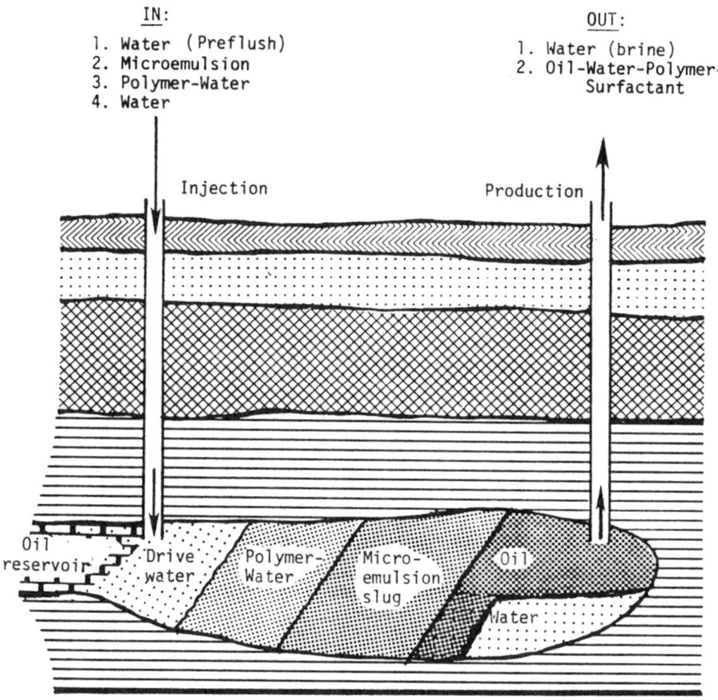

Source: EPA-68-01-1912

Micellar Solution (Slug): At this stage in the development of micellar-polymer flooding, the design of the surfactant system for a particular reservoir is still very much an empirical art. The slug is composed of hydrocarbon, brine and surfactant. A cosurfactant, such as alcohol, and salts are also added to control stability and viscosity of the fluid.

In Table 3.3, typical compositions of microemulsions are given. The surfactant systems are generally classed as either low- or high-water and oil-external or water-external. Such a classification system is used more for convenience than for providing a chemical description of the microemulsion system. For example, a given surfactant could be used to form both oil-external and water-external emulsions depending upon the concentrations of brine and hydrocarbon included in the initial mixture (10).

The choice of surfactant and cosurfactant depends upon cost and flood design. Economic considerations usually restrict the choice of surfactant to petroleum sulfonates and the choice of cosurfactants to light alcohols. Natural petroleum sulfonates used for most micellar floods have a distribution of equivalent weights and will not be single compounds. Petroleum sulfonates with equivalent weights greater than 400 g are predominantly oil-soluble and those with equivalent weights of less than 400 g are predominantly water-soluble microemulsions.

Table 3.3: Typical Microemulsion Compositions*, **

. . . . .Volume Percent. . . . .

Constituent	Low-Water*** Oil-External	High-Water† Oil-External	High-Water† Water-External	Low-Water*** Water-External
Surfactant	6–10	3–6	3–5	6–12
Cosurfactant	2–4	0.01–20	0.01–20	3–25
Hydrocarbon	30–80	4–40	2–50	20–60
Electrolyte	0.001–5	0.001–4	0.001–4	3–5
Water	5–55	55–80	30–95	25–40

*W.B. Bleakley, *Oil & Gas Journal* 69(48) (1971), p 50.
**Private communication, Union Oil of California, January 13, 1976.
***A low-water emulsion slug is comprised of less than approximately 50% water.
   Some operators believe that such slugs are not economically feasible.
†A high-water emulsion slug is comprised of more than approximately 50% water.

Source: EPA–68–01–1912

Mobility Control Solution and Water Drive: A mobility control solution is required to push the microemulsion slug through the reservoir. Polymer is added to water to increase the viscosity of the mobility control solution. The solution exhibits unique flow properties. At high linear flow rates, such as occur near the injection well, the polymer solution gives low resistance to flow while at low linear flow rates further from the injection it gives high resistance to flow. The high resistance to flow causes the polymer solution to invade a larger portion of the reservoir and move through the reservoir more uniformly than water alone would do.

A properly chosen polymer solution will force the microemulsion ahead evenly to contact a large part of the reservoir. The ratio of the part of the reservoir contacted to the total reservoir volume is known as the sweep efficiency of the micellar-polymer flood. In field operations a sweep efficiency of up to 75% has been achieved. The polymer used in a mobility buffer may also plug off parts of the oil reservoir by adsorbing onto the sand grains. Polyacrylamides exert mobility control by both methods. Polyethylene oxides and polysaccharides only change the viscosity of the mobility buffer.

The concentration of polymer in solution is gradually decreased as the injection of the mobility buffer progresses. This has helped to improve the economics of micellar-polymer flooding because less polymer can be used overall and the process can then be completed by water injection. Polymer consumption will average about one pound per barrel of oil produced by the project. Very low concentrations of the polymer will appear in the produced water because of either adsorption or dilution.

When a polymer solution is injected alone and not in combination with a micellar slug, polymer concentrations will be evident in every barrel of water produced (11). However, this has not been observed in field operations to date.

Performance

The efficiency of micellar-polymer flooding is dependent upon a complexity of factors. Among the most important variables are how much of the oil in the rock actually comes into contact with the surfactant in the slug and how evenly the mobility buffer pushes the oil and slug throughout the reservoir. Although there is a growing body of knowledge on the principles of micellar-polymer flooding, and the process may result in recovery of substantial additional quantities of oil from known fields, models to predict reliably the performance in various reservoirs are still under development.

Many of the early laboratory studies bear little resemblance to the field results because improper scaling or oversimplification of the reservoir has so changed the parameters of the experiment that no comparison is possible. Many combinations of chemicals for micellar solutions have yet to be studied in the field. In view of these considerations, and the fact that much of the required reservoir information has not been measured so that the reservoirs may be evaluated, it is nearly impossible to predict performance of micellar-polymer solutions with any quantitative accuracy.

The difficulty of predicting micellar-polymer flooding efficiencies has been one factor which has hindered wide-scale adoption of this method of oil recovery. Considering the large number of parameters which affect micellar-polymer flooding, qualitative estimates of the efficiency of the process can be made on the basis of important field characteristics. For a given oil saturation, the quality of injection water and the dissolved solids concentrations in the reservoir brine is an important characteristic.

Table 3.4: Typical Operating Range of Important Technical Parameters
for Micellar-Polymer Flooding

Oil viscosity	30 cp or less
Reservoir temperature	up to 200°F
Permeability*	more than 20-50 md
Total dissolved solids in brine**	less than 50,000 mg/l
Divalent ions in brine or reservoir clays***	less than 500 mg/l

*The ratio of variation in permeability should be less than 7 to 1.

**T.M. Geffen's original limit of 5,000 mg/l has been raised as a result of further research.

***Divalent ions which are typically found in oil field waters are calcium, magnesium, barium and strontium.

Source: EPA-68-01-1912

Table 3.4 lists a typical range of reservoir conditions which are suitable for micellar-polymer flooding. It is based on a paper by T.M. Geffen, Amoco Production Company, "Improved Oil Recovery Expectations When Applying Available Technology," API Division of Production meeting, Denver, Colorado, April 9-11, 1973. In the table factors such as oil saturation and porosity influence the economic feasibility of the process and laboratory studies are necessary to determine the effect of these parameters on technical feasibility.

Table 3.5 illustrates the efficiency of micellar-polymer flooding for different water quality conditions in reservoirs where the other important parameters are favorable. The highest efficiencies of recovery are achieved when the total dissolved solids and divalent ion concentrations in both the formation water and the injection water are low (total dissolved solids: <50,000 mg/l and divalent ions <500 mg/l). Poor quality water within the formation may be overcome by preflushing the formation with injection water of high quality, and good efficiencies of recovery are possible.

Fair to poor efficiencies of recovery are achieved when the quality of injection water is low due to high concentrations of total dissolved solids and divalent ions. Injection water problems in this case can sometimes be overcome by constituting the micellar slug with a higher proportion of hydrocarbon or by counteracting the adverse effects of the dissolved chemicals through chemical treatment. When the quality of injection water and the formation water are low, in terms of dissolved chemical concentrations, poor recovery efficiencies can be expected, and in such cases, a field-wide micellar-polymer flood would not be attempted. The recovery efficiency is equal to the percent of oil in place in the reservoir before tertiary recovery is started which is produced by the flood. Actual recovery efficiency in a given reservoir will also be influenced by the remaining oil saturation and other factors.

Table 3.5: Efficiency of Micellar-Polymer Flooding (Where Other Parameters are Favorable)

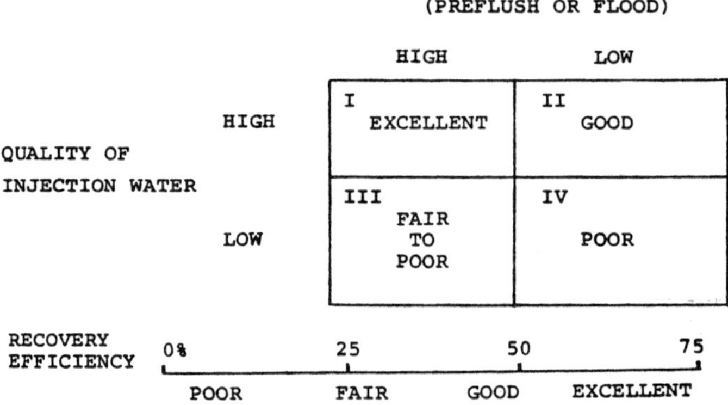

Source: EPA-68-01-1912

Research Problems

The National Science Foundation Workshop held in Austin, Texas, in June, 1974 (CO GP 44165) recommended that major research efforts of academic institutions should be directed toward elucidating the molecular mechanisms involved in the displacement of oil during tertiary oil recovery and in particular solving the problems listed below.

(a) Stability of micellar solution (microemulsion) under displacing conditions.

(b) Interaction of micellar solution with in-place oil and water in the reservoirs.

(c) Unusual rheological properties which micellar solutions may exhibit on mixing with resident reservoir fluids.

(d) Surfactant formulation—Considerable emphasis has been placed by various industrial laboratories on the use of petroleum sulfonates for the formulation of the micellar solution for tertiary oil recovery. However, other promising candidates including some of the nonionic detergents should be investigated for their efficiency in oil displacement. Even more surfactants could be economically attractive if it is possible to displace oil using lower concentrations.

(e) Thermal stability—Micellar solutions may be injected at a considerable depth in the reservoirs and, therefore, into regions of relatively high temperatures.

(f) Size of the surfactant slug—A basic question that remains to be answered is as follows. Is a low pore volume surfactant slug of high surfactant concentration or a high pore volume slug of low surfactant concentration better for oil recovery?

(g) Adsorption of the surfactant—The surfactant molecules are very likely to come out of the micellar solution and adsorb on the porous media at solid-liquid interface. This is a major problem in the utilization of micellar solution for oil displacement. The investigation should be directed toward the factors controlling the adsorption of surfactants on the solid-liquid interface. Various approaches should be explored to minimize such adsorption. Upon minimizing the adsorption it it is possible to extend the useful lifetime of the micellar solution or microemulsion slug.

(h) Ideal mobility buffer system—To push the surfactant slug or micellar solution, a low cost nonplugging polymer solution is often utilized. Unfortunately, there are only two current candidates for such a polymer solution: biopolymer, which is a polysaccharide, and polyacrylamide. Other polymers should be investigated for their viscosity behavior at high temperature and high salt concentration. For mobility control polymers, new approaches should be sought to minimize the permeability reduction due to mechanical entrapment of polymers as well as the salt effects on polymers. Adsorption of polymers on porous media is also a severe problem in using such mobility control fluids.

(i)   Ion exchange capacity of clays should be investigated in order to delineate their importance in the stability of surfactant solution in the porous media.

(j)   A detrimental interaction between surfactant and polymer has been suggested by oil displacement studies, as discussed under Polymer-Augmented Waterflooding. This should be explored in detail to find out the structural and rheological changes associated with such surfactant-polymer interaction.

(k)   Meaningful oil displacement test—It is felt that urgent attention should be given to devising meaningful oil displacement tests using small core plugs. The procedure should be standardized so that investigators in various laboratories can compare their results. It has been pointed out by various investigators that a standard procedure for adsorption, oil displacement, and oil recovery should be used by various groups working in this area.

(l)   Mechanism of oil displacement—In model systems the efforts should be made to elucidate in detail the mechanism by which trapped oil is displaced in the porous media. The role of low interfacial tension, hydrodynamic pressure difference, the role of pore geometry, and the role of interfacial viscosity of the oil-water interface should be established.

Projects

For the location of technically feasible micellar-polymer flooding projects in major oil fields of the United States, California and Texas see Figures 3.5a, 3.5b, and 3.5c. The figures show that such projects are most common in Texas. The North-Central and South-Central Regions also have a significant number of micellar-polymer flooding sites.

Maraflood Process: The surfactant/polymer technique, under extensive testing by Marathon (Maraflood) and Union Oil (Uniflood) is considered by many to have the greatest long-term potential for recovering significant amounts of tertiary oil. In the Maraflood process, a slug consists of 7 to 12% surfactant (petroleum sulfonates), 1 to 3% cosurfactant (generally an alcohol), 20 to 50% hydrocarbons, and 25 to 70% water.

The cosurfactant is a small molecule surfactant which tends to arrange itself between the molecules of petroleum sulfonate on the surface of the micelles stabilizing the structure and extending the range of concentrations and temperatures over which the microemulsions will form. Both lease crudes and light fractions have been used for the hydrocarbon component.

Production Schedule

The production schedule used for the surfactant/polymer has a response during the third year (with much of this due to the injection of polymer), followed by much greater production for the next two years and a decline during the last two years, to the economic limit. Thus, the production schedule for the oil recoverable from surfactant/polymer is as shown in Figure 3.6.

Figure 3.5: Technically Feasible Micellar-Polymer Flooding Projects

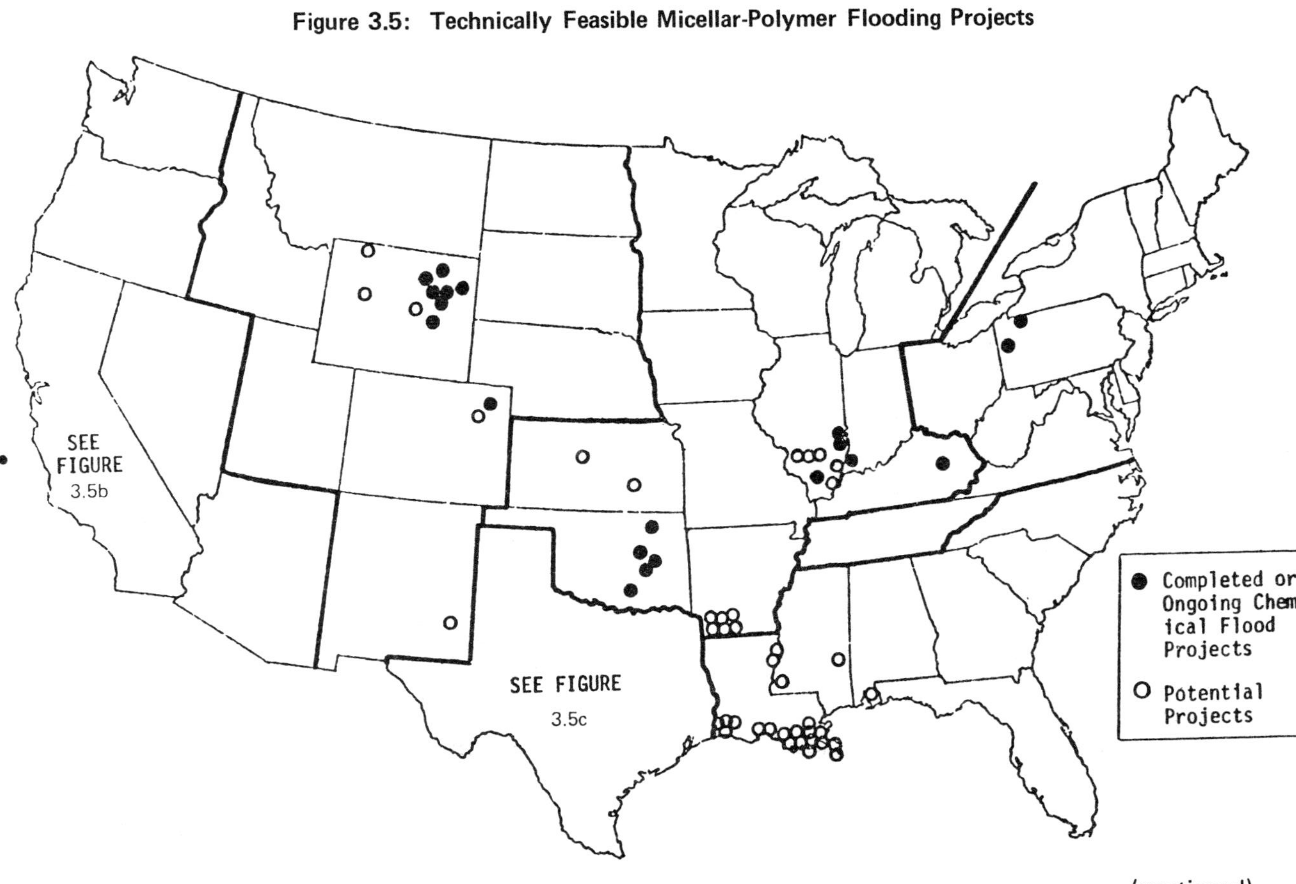

(continued)

Figure 3.5: (continued) 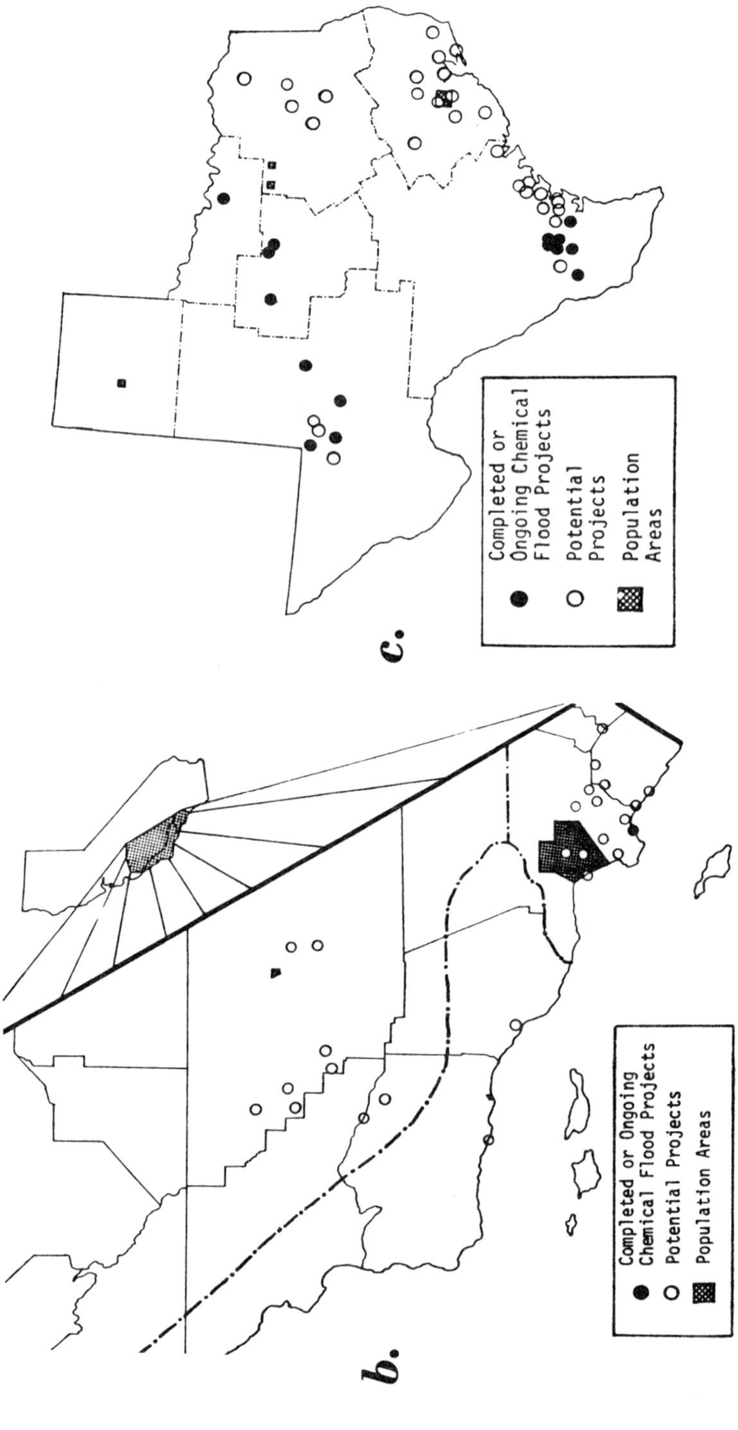
c.
Completed or
Ongoing Chemical
Flood Projects
Potential
Projects
Population
Areas
b.
Completed or Ongoing
Chemical Flood Projects
Potential Projects
Population Areas
(a) Major U.S. Oil Field
(b) Southern California
(c) Texas
Source: EPA-68-01-1912

Figure 3.6: Production Schedule for Surfactant/Polymer Flooding

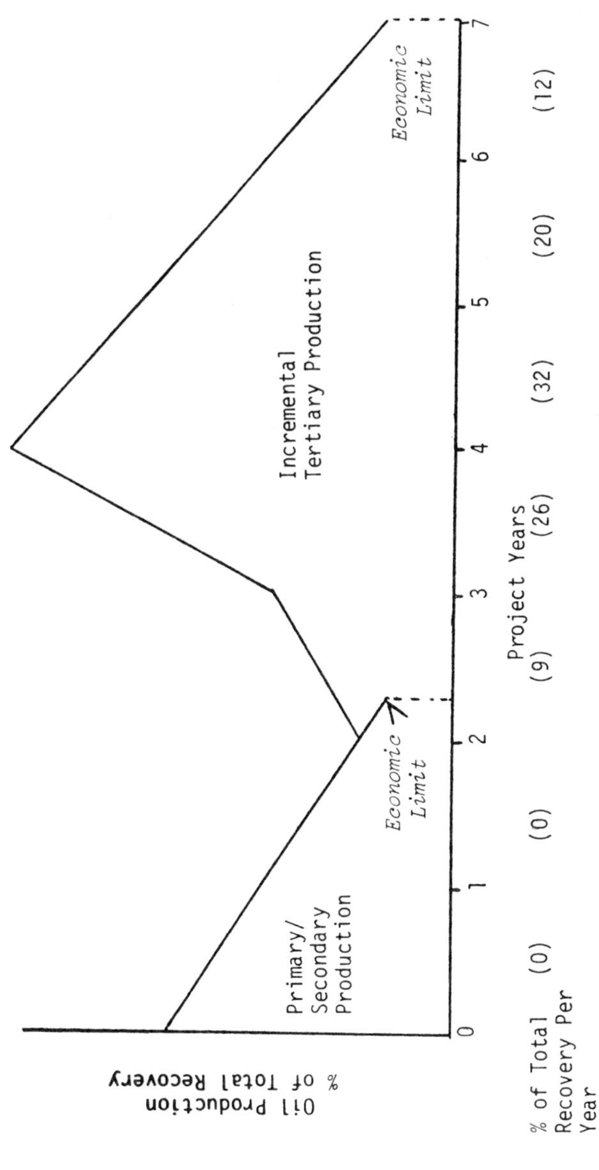

Source: FEA/B-76/221

Cost-Performance Characteristics

The ingredients of the slug are rather expensive. According to Mathematica (EPA–68–01–2445), petroleum sulfonates have been available at about 32¢/lb of active ingredients. Commercial sulfonates generally contain about 62% active ingredients. Large-scale micellar flooding would, in fact, put a severe strain on sulfonate production facilities. Marathon's facility aims at reducing the cost to 20¢/lb active. The cost of the cosurfactant has been about 30¢/lb.

Hence, assuming typical slug concentrations of 8% active surfactant, 2% cosurfactant and 25% lease crude, the slug cost would be about $12.00 per barrel at 30¢/lb surfactant cost, $9.50 per barrel at 20¢/lb surfactant cost.

The cost of the polymer-thickened waterflood (often called the mobility buffer) depends primarily on the viscosity required at the front of the flood. This viscosity in turn depends on the viscosity of the slug, which is determined by the mobilities of both the reservoir oil and reservoir water being displaced in front of the slug at the ambient oil and water saturations in this zone.

For average waterflooded, medium-gravity fields, this implies slug viscosities in the range of 20 to 30 cp. In order to obtain such viscosities in the mobility buffer, polymer concentrations of 1,000 ppm or more are required depending on reservoir salts, flow velocity, etc. Polymer costs have been in the neighborhood of $2.00 per pound although very large-scale usage could bring this down.

Table 3.6: Micellar Flooding Costs

Basis: 100 bbl of pore volume
Primary and waterflood recovery: 30 bbl
Micellar flood recovery: 30 bbl

	. . . . . Costs ($)	
	Low	High
Chemicals		
0.8 bbl petroleum sulfonate,		
$45 to $65/bbl	36	52.00
7.2 bbl water, 0 to 3¢/bbl	–	0.21
40 bbl polyacrylamide solution,		
$0.20 to $0.60/bbl (0.4 lb		
polymer/bbl, $0.50 to $1.50/lb)	8	24.00
40 bbl water, 0 to 3¢/bbl	–	1.20
150 bbl water, 0 to 3¢/bbl	–	4.50
Total	44	81.91
Equipment		
Pumps, tanks, filter, at $1.50/bbl		
recovered	45	45
Well liners, addition drilling	0	unknown
Pumping cost at 0.5¢ to 1¢/bbl	1	2
Total costs	90	129
Unit costs of successful flood		
(30 bbl recovered)	3	4.30
Overall unit cost (25 to 50%		
success rate)	6.00	17.20

Source: BuMines OFR 4–75

Polymer costs of $2.00 per pound and concentrations of 1,000 ppm equate to an expensive $7.00 per barrel injected. Thus, the polymer concentration is slowly reduced as the mobility buffer is injected in an attempt to balance unfavorable mobility ratios with economics. Some results are tending to mobility buffers of about 1% pore volume in which the average polymer concentration is about 750 ppm. In a study by Arthur D. Little, costs for micellar flooding were outlined as shown in Table 3.6.

In many reservoir situations, particularly those having a natural fluid drive, water is available at no cost. In situations where the water must be pumped in or alternatively must be treated in some way to control the presence of calcium and magnesium salts, the costs could be as high as 3¢ per barrel. Pumping costs are only a small proportion of the cost of micellar flooding. A cost estimate of 0.5¢ per barrel has been used.

The equipment required for micellar flooding will depend upon the specific reservoir but will typically include: metering pumps, storage tanks for polymer and petroleum sulfonate, and diatomaceous earth filters required to protect the micellar solution. The costs for equipment are based upon the estimated equipment costs for the Maraflood project at the Bradford field. These costs are shown in Table 3.7.

Table 3.7: Costs of Maraflood Process, Bradford Field, PA

Reservoir pore volume: 1.5×10^6 bbl
Remaining oil in place: 600,000 bbl

	Low		High
Chemical cost, $			
79,000 bbl micellar solution	359,550		519,350
1,257,000 bbl polymer	176,810		530,430
Equipment costs, $			
Metering pumps		15,000	
Storage tanks		200,000	
Filter		50,000	
Total		265,000	
Well costs, $			
Relining 41 wells at $1,000/well		41,000	
Pumping costs, $		10,000	
Total costs, $	492,810		846,430
Additional oil recovery, bbl		130,000	
Recovery, %		22	
Unit cost per barrel, $	3.79		6.51

Source: BuMines OFR 4-75

The ultimate recovery of the remaining oil in the reservoir using micellar flooding is dependent upon the sweep efficiency of the flood. An efficiency of approximately 50% has been assumed. Hence, in a 100 barrel reservoir approximately 30 barrels of oil will be removed by primary waterflooding techniques, approximately 30 barrels of oil by micellar flooding (assuming a 90% recovery efficiency and 50% sweep efficiency), leaving approximately 40 barrels of oil trapped in the reservoir.

In the case of the Maraflood demonstration at the Bradford field, the estimated oil recovery is 130,000 barrels compared to a pore volume of 872,000 barrels. This is equivalent to an ultimate recovery of approximately 35% of the remaining oil in place or 15% of the original oil in place.

Future Prospects

Mathematica's Forecasts: The effectiveness of this method is sharply dependent on oil saturation since the process is basically pore volume limited. Essentially the same volumes of slug and mobility buffer have to be injected whether the reservoir contains 20 or 80% oil. Since any field well suited to micellar flooding will also be well suited to the much less expensive waterflooding, residual oil saturation of 45% or less is the viable target range of this method. According to Mathematica, even assuming low end polymer and surfactant costs, at 45% residual oil saturation and 50% recovery, the process barely breaks even at 1975 prices.

Furthermore, the performance of pilot tests have proven disappointing, and the major manufacturer of the polymer has downgraded its polymer-thickened waterflood division despite the rise in price. This polymer has cost about $1.50 per pound, or very roughly $500 per barrel. This again implies that it must be effective at low concentrations to be economic. Evidence to date indicates that such effectiveness has not been observed by field operators.

The costs of this method are about equally split between the slug and the mobility buffer, with well and water injection costs being small. Since this method like most tertiary methods has heavy front-end investment costs, perhaps the most fruitful avenue for increasing net present values is infill drilling to gain more rapid recovery of the oil.

Mathematica would not count on micellar for more than a marginal amount of additional oil (perhaps less than 3 billion barrels) over that represented in the API reserves. Much of this oil could be almost as costly to the nation as imported OPEC crude.

Lewin and Associates' Forecasts: On the other hand, Lewin and Associates, Inc. predict a contribution from the surfactant/polymer technique of 8.8 billion barrels of oil by the year 2000 from the three states of California, Louisiana and Texas. This represents 29% of total potential reserves. They predict that surfactant/polymer flooding will climb slowly through the late 1980s at which time it will accelerate sharply, reaching 1.4 MM barrels per day in 1996.

To achieve this, tertiary oil recovery will need to spawn and nurture many new ventures in chemicals. Approximately 75 billion pounds of sulfonates, 7 billion pounds of polymers, and 10 billion pounds of alcohols would be required over the next 30 years to produce the almost 9 billion barrels of tertiary oil attributable to surfactant/polymer recovery.

However, sulfonate supply should not pose a near-term constraint. The domestic chemical industry already has the capacity to produce 500 MM pounds of sulfonate per year with additional major plants coming on stream. Moreover, new plant construction has a relatively short construction lead time (3 years) and the major demand for sulfonates will not come until after mid-1985.

Polymer requirements could pose some constraints given that the current production of polyacrylamides and polysaccharides is estimated to be only about 300 MM pounds per year (12).

GURC Forecasts: GURC considers Mathematica's forecasts overly pessimistic and predict that once the technology advances there will be a decrease in costs of materials so that oil production by micellar-polymer flooding could add 21 billion barrels by 1985 to our nation's supply.

ALKALINE FLOODING

Alkaline flooding uses caustic chemicals such as sodium hydroxide (NaOH) or sodium silicate (NaSi) to reduce the interfacial tension between the injected fluids and the reservoir oil. These chemicals, added to injection water, form surfactants (detergent-like substances which lower the tension at the interface of the oil and water) within the reservoir by the neutralization of petroleum acids (7). The first patent on the injection of caustic to improve waterflood recovery was issued in 1927; there have since been many laboratory experiments and a few field tests (13).

Principles and Procedures

The mechanisms by which alkaline, or caustic flooding improves oil recovery include entrainment, which reduces interfacial tension, entrapment, which improves sweep efficiency, and wettability reversal, which improves mobility ratios. Each of these mechanisms requires somewhat different initial conditions with respect to reservoir oil, rock and injection water properties, and may or may not be included in any one caustic flooding project (13).

Entrainment and entrapment mechanisms form surfactant emulsions in situ by reactions between the acids in reservoir oil and the injected alkaline chemicals. Since the lower API gravity crude oils seem to have a higher content of natural petroleum acids, alkaline flooding processes seem almost exclusively relegated to the recovery of moderately viscous, low API gravity crude oils (7). The number of naturally occurring emulsifiers (primarily oil-soluble carboxylic acids that are converted to sodium soap in contact with caustic solution) in a single crude oil may range into the hundreds or higher. The surfactants which are created at the crude oil-water interface reduce interfacial tension and form an emulsion of the reservoir oil. This emulsion lowers the mobility of the injected water and improves process sweep efficiency (14).

In the entrainment mechanism, oil droplets are suspended or entrained in the alkaline drive water by the formation of an emulsion. As the continuously flowing alkaline-water phase nears the producing well, the entrained oil is produced. However, it is difficult to maintain low interfacial tension while moving the alkaline mixture through reservoir rock, since much of the caustic is lost by reactions with the reservoir rock (15). Nearly all reservoir rocks will tend to react with and neutralize sodium hydroxide to some extent. Limestone and dolomite rocks are fairly unreactive, and the reaction of the caustic with the silica component in sandstone is too slow and incomplete to present much of a problem. The clay component of reservoir rocks, however, has a significantly rapid and complete reaction with the caustic (14).

Sufficient caustic must be injected to compensate for losses through chemical reactions (one of which is the ion exchange between the Ca^{++} and Mg^{++} of clay minerals and the Na^+ of the caustic solution)(15).

Entrapment mechanisms are designed to improve process sweep efficiency. Residual oil is emulsified in the reservoir and begins to move toward the producing well with the alkaline drive water. On the way, the emulsion is trapped in pore capillary restrictions which are too small for the oil emulsion droplets to penetrate, diverting the flow of alkaline water to other open flow channels. This process results in a reduced water mobility that improves both vertical and areal sweep efficiency, and is especially important in waterflooding viscous oils, where waterflood sweep efficiency is usually very poor.

Although the emulsified trapped oil is not recovered, and no significant reduction in the capillary-retained residual oil saturation is achieved, the process is useful because in heterogenous reservoirs of heavy oils where sweep efficiency is very poor, an improvement in vertical and areal sweep efficiency through an improved mobility ratio can be much more important economically than recovery of residual oil from the small volume of reservoir normally swept (13).

The third mechanism of improved recovery by caustic flooding is wettability reversal. Reservoir wettability depends on the extent to which naturally occurring polar compounds are present in crude oil. Those polar compounds adsorb on hydrophilic (water-attracting) mineral surfaces, making them less hydrophilic or even hydrophobic (water-repelling)(16). The changes in wettability are caused by adsorption of soap molecules (which were formed by interfacial reaction) onto the solid surface (17).

Sandstone wettability is affected by a variety of compounds associated with the higher molecular weight fractions of crude oil; limestone wettability is affected mostly by basic nitrogen compounds (16). On water-wet reservoir rock surfaces, the residual oil is present in immobile and discontinuous droplets. In an oil-wet porous medium, the residual oil is present as a film on the rock, with droplets of reservoir water held immobile. A reversal from oil-wet to water-wet in a region where oil is still flowing changes the relative oil and water permeability and provides a more favorable mobility ratio, improving oil recovery. In converting from water-wet to oil-wet rock, a discontinuous residual oil is converted to a continuous wetting phase, providing a flow path for what would otherwise be trapped oil (13).

The efficiency of the alkaline flooding process will depend on the chemistry of the reservoir oil and water, the physics and chemistry of the reservoir rock and specific flooding conditions such as rate of injection and temperature (18).

Performance

Alkaline flooding processes are still in a testing state. The amount and concentration of caustic injected appear to vary depending on the recovery mechanism to be used. Concentrations are generally lowest for the emulsification (entrapment and entrainment) mechanisms, from 0.001 to 0.500 wt %. Higher concentrations ranging from 0.500 to 3.0 wt % or even as high as 15.0 wt %, are usually required for wettability reversal. A slug of caustic is nearly as effective as continuous injection of the chemical, although the volume injected must be large enough to combat losses from rock reaction (17).

Laboratory experiments have shown that the success of caustic flooding is highly dependent on the chemical and physical properties of the reservoir materials. The composition of the crude oil itself is important, since it is the nature of the polar compounds in the oil that determine whether caustic can improve recovery (17). Reaction of alkali with different acidic compounds will generate surfactants of different degrees of effectiveness; some alter wettability, while others cause changes in surface films or promote low-tension displacement. The relationship between crude oil composition and the mechanisms of oil recovery has not yet been determined, although it is known that the occurrence of different recovery mechanisms is related to the presence of general classes of chemical compounds (asphaltenes, acids, etc.)(16).

Chemical composition of both reservoir and injection water is also important. Factors such as pH and salinity can change the soap distribution between the oil and water phase and its properties at the interface, affecting flooding results. The concentration of polyvalent ions, such as calcium and magnesium, must also be controlled, since these ions will react with organic acids to form less surface-active soaps. Rock properties are another factor in caustic flooding, since reaction with the rock consumes caustic (17).

Field experience for alkaline flooding has been limited, although there are some ongoing field tests. The potential for this process seems to be greatest for recovering high viscosity crude oils.

Although caustic or alkaline flooding has been treated separately here, it is often grouped with the polymer flood or micellar category, indicating the necessity for standardizing definitions and nomenclature.

REFERENCES

(1) Bleakley, W.B., "Journal Survey Shows Recovery Projects Up," *Oil and Gas Journal*, p 69 (March 25, 1974).

(2) Holm, L.W., "Residual Oil–Can We Recover It Economically," *Petroleum Engineer*, (December 1973)

(3) Holm, L.W., "Status of CO_2 and Hydrocarbon Miscible Oil Recovery Methods," *Journal of Petroleum Technology* (January 1976).

(4) Herbeck, E.F., Heintz, R.C. and Hastings, J.R., "Polymer Flooding, Fundamentals of Tertiary Oil Recovery, Part 7," *Petroleum Engineer*, pp 48–59 (July 1976).

(5) Sparlin, D.D., "An Evaluation of Polyacrylamides for Reducing Water Production," *Journal of Petroleum Technology*, pp 906–914 (August 1976).

(6) Noran, D., "Enhanced Recovery Requires Special Equipment," *Oil and Gas Journal*, pp 50–56 (July 12, 1976).

(7) *Enhanced Oil Recovery*, National Petroleum Council (December 1976).

(8) Sloat, B., "How to Get More for Your Chemical Dollar," *Petroleum Engineer*, pp 20–23 (November 1977).

(9) McLaughlin, M.C., U.S. Patent 3,490,533, assigned to Halliburton Company (January 20, 1970).

(10) Healy, R.N. and Reed, R.L., "Physicochemical Aspects of Microemulsion Flooding," *Society of Petroleum Engineers Journal* 10 (1974).

(11) Calgon Corporation, Bulletin 14-100, p 12.

(12) *Chemicals for Microemulsion Flooding in EOR*, Gulf Universities Research Consortium (March 1976).

(13) Johnson, Jr., C.E., "Status of Caustic and Emulsion Methods," *Journal of Petroleum Technology*, pp 85-91 (January 1976).

(14) Jennings, Jr., H.Y., Johnson, Jr., C.E., and McAuliffe, C.D., "A Caustic Waterflooding Process for Heavy Oils," *Journal of Petroleum Technology*, pp 1344–1352 (December 1974).

(15) *ERDA Symposium on Enhanced Oil and Gas Recovery and Improved Drilling Methods*, Proceedings, Volume 1–Oil, Oil and Gas Journal, Camelot Inn, Tulsa, Oklahoma (August 30–31 and September 1, 1977).

(16) Erlich, R., Hasiba, H.H. and Raimondi, P., "Alkaline Waterflooding for Wettability Alteration–Evaluating a Potential Field Application," *Journal of Petroleum Technology*, pp 1335–1343 (December 1974).

(17) Cooke, Jr., C.E., Williams, R.E. and Kolodzie, P.A., "Oil Recovery by Alkaline Waterflooding," *Journal of Petroleum Technology*, pp 1365–1374 (December 1974).

(18) Linville, B. (ed.), *Contracts and Grants for Cooperative Research on Enhancement of Recovery of Oil and Gas*, Progress Review #10, April 1977, ERDA, BERC–77/2 (July 1977).

TERTIARY RECOVERY OF CRUDE OIL THERMAL TECHNIQUES

The information in this chapter is based on:

Basic Research Needs for Tertiary Oil Recovery, prepared by R.S. Schecter and W.H. Wade, of the University of Texas at Austin, for National Science Foundation Workshop, under Contract GP 44165, June 26-27, 1974.

Evaluation of Thermal Methods for Recovery of Viscous Oils in Missouri and Kansas, BuMines Report OFR 60-74, prepared by M.D. Arnold and A.H. Harvey, of Department of Mining, Petroleum and Geological Engineering, University of Missouri at Rolla, for U.S. Bureau of Mines, June 1974.

An Investigation of Primary Factors Affecting Federal Participation in R&D Pertaining to the Accelerated Production of Crude Oil, prepared by J.M. Sharp, of Gulf Universities Research Consortium, for National Science Foundation under Contract NSF-C-942 (GURC Report 140), September 15, 1974.

The Estimated Recovery Potential of Conventional Source Domestic Crude Oil, prepared by J.W. Devanney III, R. Ciliano and R.J. Stewart, of Mathematica, Inc., for U.S. Environmental Protection Agency, under Contract 68-01-2445, May 1975.

Review of Secondary and Tertiary Recovery of Crude Oil, FEA Report G-75/482, prepared by Lewin and Associates Inc., for Federal Energy Administration, June, 1975.

The Potential and Economics of Enhanced Oil Recovery, FEA Report B-76/221, prepared by Lewin and Associates, Inc., for Federal Energy Administration, April 1976.

Potential Environmental Consequences of Tertiary Oil Recovery, prepared by C. Braxton, R. Stephens, C. Muller, J. White, J. Post, J. Norton, M. Goldberg, and P. Stevenson, of Energy Resources Co., Inc., for U.S. Environmental Protection Agency, under Contract 68-01-1912, July 1976.

Tertiary Oil Recovery: Potential Application and Constraints,
PNL Report RAP-25, prepared by C.A. Geffen, of Battelle's
Pacific Northwest Laboratory for U.S. Department of Energy,
June 1978.

INTRODUCTION

Thermal recovery pertains to oil recovery processes in which heat plays a principal role. The most widely used thermal techniques are in situ combustion
(fireflooding), continuous injection of hot fluids such as steam, water, or gases,
and cyclic operations such as steam soaking. Because of the strong temperature
dependence of oil viscosity, these thermal methods find greatest application in
the recovery of extremely viscous, low API gravity crudes, for which the usual
displacement methods such as waterflooding are unfruitful.

Heat is applied to the crude to:

> (a)   reduce the viscosity of the crude
> (b)   activate a solution gas drive in some instances
> (c)   result in thermal expansion of the oil and hence increased
>   relative permeability
> (d)   create distillation and in some cases, thermal cracking of
>   the oil.

Types of Thermal Techniques

Thermal methods are generally of three types:

> (1)   Cyclic steam injection (sometimes called steam stimulation,
>   steam soak, or "huff and puff"). In this process, steam is
>   injected down a producing well to heat up the area
>   around the well bore and increase recovery of the oil
>   immediately adjacent to the well. After injection for a
>   short period, the well is placed back on production. This
>   is essentially a well bore stimulation technique, each well
>   responding independently.
> (2)   Steam drive (sometimes referred to as steam displacement
>   or steamflooding). Here steam is injected via injection
>   wells and the oil is displaced to surrounding producing
>   wells as in conventional fluid injection operations.
> (3)   In situ combustion (sometimes called fireflood). This proc-
>   ess involves in situ combustion of portions of the oil. Air
>   is pumped into the reservoir which either self-ignites or is
>   ignited, depending on reservoir temperature and composi-
>   tion. Heat and gases from the combustion pressurize the
>   reservoir, and decrease viscosity both by heating and crack-
>   ing. Often water is injected behind the fire front.

Certain thermal methods have been in use to such an extent that they are sometimes not classified as enhanced recovery methods (e.g., steam huff and puff and
hot water injection methods which have been used for many years and have accounted for about 200,000 barrels per day of domestic production).

Hot Water Flooding: Hot water flooding, which will not be discussed in depth in this chapter, is used occasionally to effect the recovery of viscous oil. The process seems best adapted to reservoirs that require only mild heating, and the sweep efficiency of the hot water flood may not be as high as for other thermal recovery methods, especially for high viscosity oils. As compared to steam injection, the hot water flood will operate at a lower surface pressure and will presumably be capable of utilizing a poorer quality water source. In high pressure reservoirs the hot water flood might be used in preference to steam injection in order to limit injection temperatures (since steam drive temperature is fixed by reservoir pressure). In some areas, hot water might be used in preference to steam for reservoirs that do not have adequate steam injectivity.

Another possibility would be to start the recovery process with hot water injection, then convert to steam injection after adequate injectivity has been established. This technique has been employed experimentally in western Missouri, and steam injectivity was found to increase after the injection of a substantial volume of hot water. In general, however, hot water flooding has not been promising.

This poor performance is attributed to a number of factors. The same capital investment in heat generating equipment is needed as in steam injection processes. Hence, the same environmental problems exist as for other methods which can achieve high recovery in a given field. Furthermore, well bore heat losses are high—up to 30%—while overall heat loss has attained 60%. Oil recovery efficiency has been low—generally, approximately 10%—though in a few fields up to 20% of the remaining oil-in-place has been recovered.

EXTENT OF THERMAL RECOVERY PROJECTS

Figure 4.1a illustrates the location of technically feasible thermal tertiary recovery projects in major oil fields of the United States, while Figures 4.lb and 4.1c depict the location of such projects in southern California and Texas, respectively. The concentration of potential thermal recovery projects is obviously greatest in the oil provinces of California in the Pacific West region.

CYCLIC STEAM INJECTION

Principles and Procedures

Cyclic steam injection (also called huff and puff and steam stimulation) is the most commonly applied thermal recovery process. High-quality steam (about 80% quality), produced in generators at the surface, is injected directly into the reservoir through the production wells to heat the surrounding area. The condensation and cooling of the steam heats the reservoir rock and oil, reducing the oil viscosity and thus increasing production rates. After two or three weeks, the steam injection is stopped and the heated oil is produced from these same wells. (See Figure 4.2). After the hot oil production has ended a new cycle may be initiated. The time period of the cycles is on the order of six to twelve weeks, but may continue for as long as a year.

Figure 4.1: Technically Feasible Thermal Recovery Projects

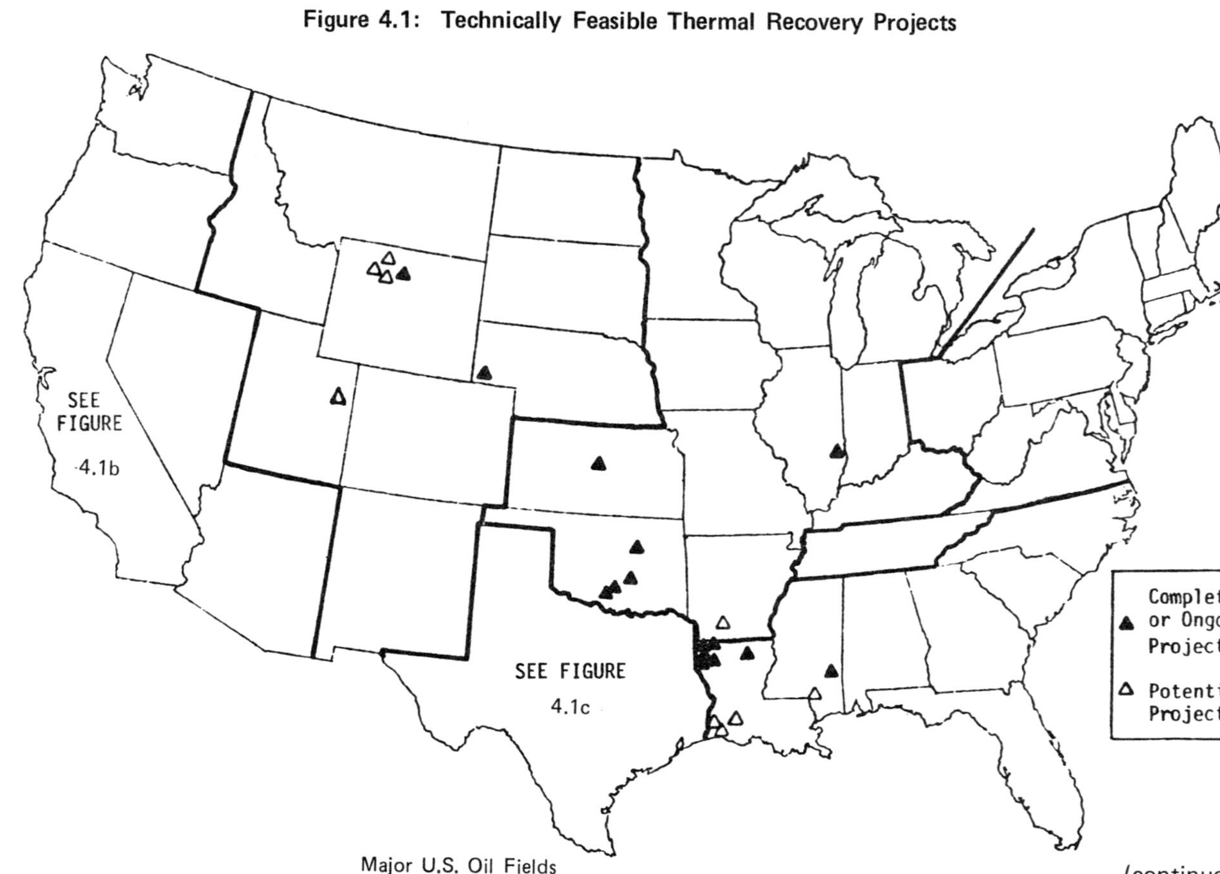

Major U.S. Oil Fields

(continued)

Figure 4.1:　(continued)

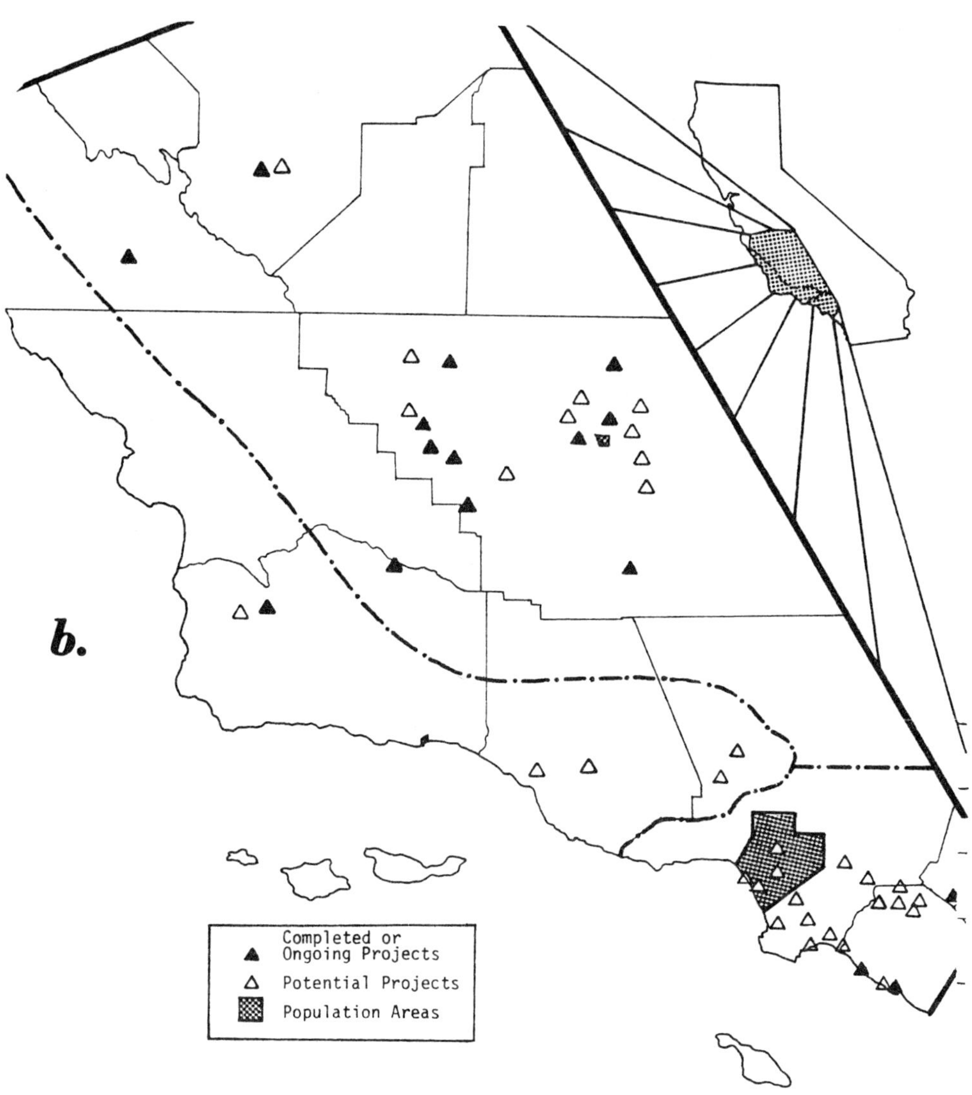

Southern California

(continued)

Figure 4.1: (continued)

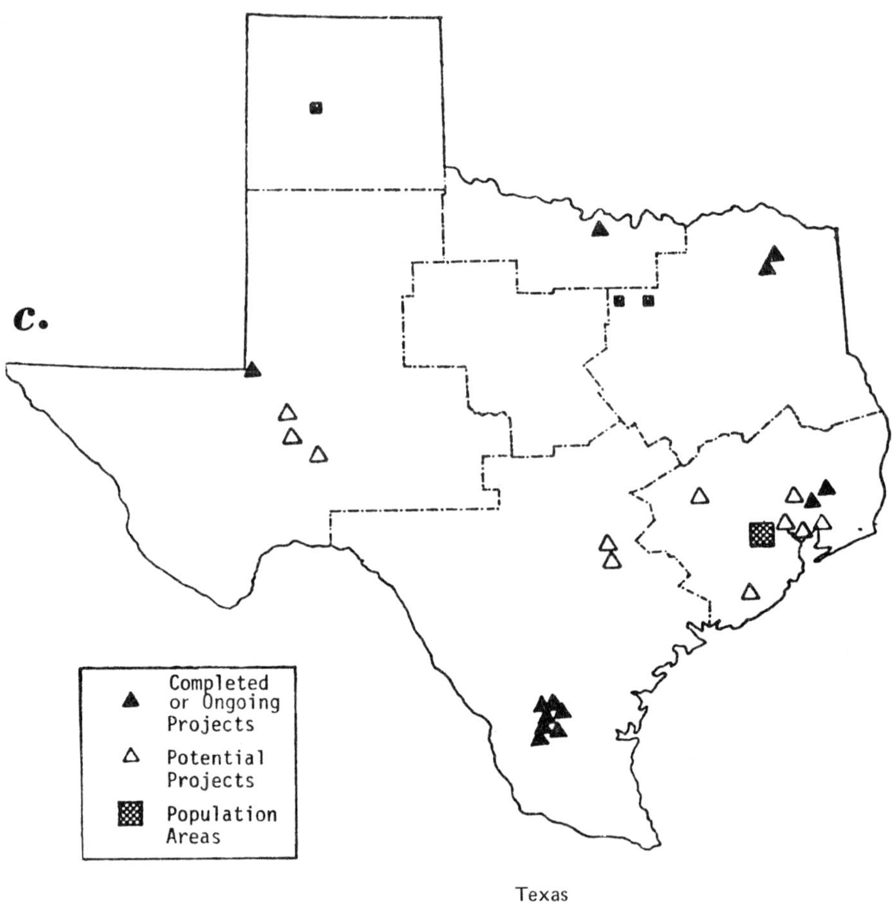

Source: EPA-68-01-1912

The optimum number of steaming cycles is somewhat dependent upon reservoir type. The reservoirs are usually shallow and the producing wells are drilled on very close spacing because the heat does not penetrate from the wells. In fact, because the injected heat only affects the reservoir area near the injection well, the overall recovery of these steam soak projects tends to be low (1).

In California reservoirs, which are generally steep, dipping structures, many cycles are possible as the heated, less viscous oil flows downward by gravity drainage (discussed in a later chaper) to producing wells. Cases have been reported of more than 20 steaming cycles being used in these dipping reservoirs. In flat reservoirs, where gravity does not aid flow, the number of profitable cycles is much more limited.

Figure 4.2: Schematic of the Cyclic Steam Stimulation Process

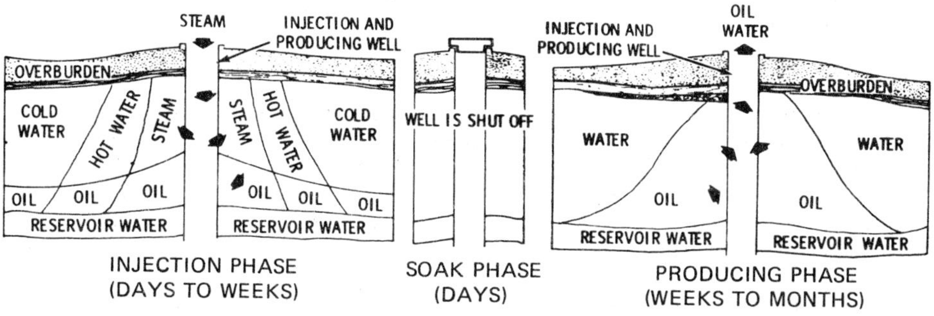

Source: PNL RAP-25

Cyclic steam injection generally becomes less efficient as the number of steaming cycles increases. A good indicator of project performance is the produced oil to injected water ratio. Best results indicate early cycles recovering as much as 30 barrels of oil per barrel of water injected as steam (2).

The volume of steam injected ranges from 5,000 to 15,000 barrels (feedwater volume) (3). Generally, only 20 to 40% of this injected water is recovered during the first production cycle. However, this proportion increases rapidly, and by the last cycle, most of the injected water has been produced (4). After several steaming cycles, the project will usually be converted to a steam drive (3).

Advantages and Disadvantages

The value of the process lies not so much in improving the ultimate recovery as in increasing the production rate. The average increase in oil production rate is from 10 to 30 times, although increases of as much as 100 times the pretreatment rate have been reported in some reservoirs (1). Furthermore, in comparison with most methods for viscous oil production, cyclic steam injection requires only a small investment; additionally, the cyclic process yields a response that is almost immediate upon cessation of steam injection. The primary benefits of the process are the reduction of oil viscosity near the well and the cleaning of the well bore.

Since the process does not normally produce a significant increase in reservoir pressure, it is necessary that the reservoir have sufficient energy to produce the oil after it has been reduced in viscosity. Therefore, most very low-pressure reservoirs are not particularly good candidates for cyclic steam injection. Shallow reservoirs and thick sand sections are favorable from the standpoint of heat loss. However, some apparently poor candidates for steam stimulation may yield a satisfactory response to the process because of the cleaning action of steam on the well bore.

Steam stimulation, by its very nature, is a limited process. After several cycles (perhaps over 2 to 3 years), the area in the immediate vicinity of a well has been flushed out and the wells fail to respond to further stimulation. It is rarely possible to recover more than 10% of the original oil-in-place via steam stimulation. Therefore, the real payoff via thermal methods will have to come from steam drive or in situ combustion. However, steam stimulation is almost universally the first step in applying thermal measures to a field.

History and Extent of Use

The effectiveness of steam in recovering heavy oils was foreseen early; the first field tests were conducted in the 1920s and the 1930s. Initial use of cyclic steam, in 1930 to 1950, was to remove paraffin from the well bore. Use of cyclic steam gained prominence in the early 1960s in California with Shell's success, and it has now progressed technically to an economic, reliable recovery technique.

Steam stimulation projects (many of them dating back to the mid sixties) are in place on 19 of the largest California fields. The average starting date of these projects was 1966. Even allowing for a period of field testing, most of them are well along toward completion. Of the 44 domestic steam stimulation projects listed as active by the *Oil and Gas Journal* in 1973, in 32 cases the operator reported the project was at least half-finished. In fact, much of this oil may be already in the API reserves estimates, if, in fact, it has not already been produced.

Limitations and Economic Aspects

Oil recovery efficiency depends on the thermal properties and compositions of the reservoir fluid of the formation. Since the kinematic viscosity of the oil in the formation decreases linearly as temperature is increased, oil production is directly related to the temperatures achieved in the reservoir. The major technical limitation of steam processes is the depth of the oil-bearing strata. At greater depths, higher temperatures and pressures are needed for maximum effective usage of the steam.

These operating conditions can cause damage to the well's casing and cement. The major factor contributing to inefficiency of steam injection processes is the loss of heat. Oil-bearing sands thinner than 20 feet are not favorable because too much heat escapes to rocks above and below. There is less loss of heat in the well when pumping into shallow formations. Surface losses are on the order of 5%; well bore heat losses may be in the range of 10 to 15% of the total number of British thermal units (Btu) injected. Unfortunately, insulation is expensive and can only be justified for long term projects.

The important economic consideration for the thermal methods of tertiary oil recovery is the cost of fuel. Steam processes are sensitive to the cost of fuel required to generate the steam. Similarly, most limitations to the thermal methods are based upon economic inefficiencies of heat loss rather than on technical factors such as reservoir chemistry. In particular, thin formations with low porosity are unfavorable for the thermal methods because too much heat is lost through the surfaces above and below the formations and it is too difficult to inject the steam or air into a tight formation without prior fracturing.

Kern River Project: The Kern River field in California produces 12° to 16° gravity oil from 500 to 1,300 foot depths. Original oil in place is about four billion barrels. This field has been the subject of more steam stimulation and steam drive investment than any other field in the country. Over 2,500 wells have been subject to steam stimulation. In 1970 after some 5 years of pilot tests the principal operator in the field, Getty, estimated the ultimate recovery including complete steam drive at 1.4 billion barrels.

The 1975 *OGJ* estimate of the ultimate recovery from this field was 1.5 billion barrels. Clearly, the *OGJ* estimate includes not only steam stimulation but also projected steam drive oil. Thus, it is difficult to separate the recovery contribution of steam stimulation alone.

The *OGJ* estimate for this field was increased by about 450 million barrels between January 1974 and January 1975. In the year before, it was increased by 50 million barrels. Under the assumption that the *Oil and Gas Journal* and API estimates are similar, it appears that the lag between the decision to implement an enhanced recovery project and the appearance of the projected incremental recovery in the API reserves estimates is at most 1 or 2 years.

Future Prospects

Cyclic steam injection suffers from the same problem as waterflood with respect to the likelihood of the crude price rise adding significant new volume to reserves. That it was economically attractive at pre-1973 crude prices is clearly indicated by the average age of the steam stimulation projects. The Interstate Oil Compact Commission cited cost figures of $0.30 to $1.35 per barrel produced (5). T.M. Geffen claimed $0.75 to $1.25 (6). These are not economically unreasonable numbers at pre-1973 crude prices.

Finally, steam stimulation is basically limited to shallow fields. Close well spacings are required and well costs rise rapidly with depth as do injection pressures. Much more importantly, however, the natural temperature gradient generally makes steam stimulation superfluous at depths of more than 3 or 4 thousand feet. To put it another way, highly viscous heavy oil is only found at shallow depths. (One important exception is portions of the Venezuela heavy oil belt.) Therefore, little oil will be found in the future in nonfrontier areas to which steam stimulation can be applied.

STEAM DRIVE

Principles and Procedures

Steam drive (also referred to as steamflooding or steam displacement) involves the injection of steam into a group of outlying wells to push oil toward the production wells. In this process the heat is pushed into the perimeters of the reservoir to displace oil and reduce viscosity. To ensure high rates of production at the wellhead, steam flooding projects are typically conducted jointly with cyclic steam injection in the production wells.

The steam saturated zone, in the reservoir whose temperature is approximately that of the injected steam, moves oil to the production well by steam distillation of the oil, solvent extraction, and a gas drive.

As the steam is injected, a series of zones forms around each injection well (see Figure 4.3). The zone nearest to the well is steam, with a temperature approximately equal to that of the injected steam; this zone expands as more steam is injected into the well. The next zone contains condensed steam, or water. The temperature across this zone decreases from that of the steam to that of the reservoir; pressure also decreases across this zone until the reservoir pressure is reached (7). The condensed hot water at the steam front will tend to settle below the steam vapor, since it is much denser than the vapor. The steam thus flows in the upper part of the oil-bearing sand as the steam front moves out from the injection well.

The amount of upper reservoir layer which is actually contacted (the vertical invasion efficiency) seems to be dependent upon steam injection rates, well spacing, formation thickness and permeability distribution. The zone farthest away from the injection well is one of cooled water. An ordinary waterflood thus precedes warm and hot waterflooded regions. The relative importance of these zone mechanisms in the production of oil is shown in Table 4.1.

The flooding sequence in a steam drive results in a gradual reduction of residual oil saturation throughout the flooded area (3). In the steam zone, steam distillation contributes to the reduction of residual oil saturation. Oil saturation is reduced most, however, in the condensed water zone because of lower oil viscosities and high temperatures and crude swelling effects (2). Before the steam zone arrives at, or breaks through to, producing wells, additional heat may be supplied by intermittently steaming and producing the production wells, as in a cyclic steam project (7).

The sweep efficiency of the steam drive, or volumetric coverage, is determined to a certain extent by the arrangement of production wells around injection wells. The areal sweep of a steam flooding operation is fairly high in a relatively uniform sand. However, the vertical sweep is affected by any heterogeneities in reservoir structure, by the channeling, or fingering, of steam through depleted reservoir zones (which have already had oil removed) and by the gravity segregation of vapor and condensed steam. Since the condensed steam is not as effective a displacement agent as vapor (it does not reduce residual oil saturation to as low a value), the overall reservoir recovery is limited (7).

When the vertical distribution of steam in the reservoir presents problems, the injection of lower quality steam (less than 80% quality) may be desirable. The added volume of hot water in the injected fluid could aid recovery of oil from the lower strata of very thick permeable zones.

During steam drives, much of the heat injected as steam is tranferred to the reservoir rock. Added process efficiency can result if, in the final phases of the recovery operation, unheated water is injected to recover heat from the reservoir rock. The heat will then be carried forward as a hot waterflood. This method uses the heat more economically, since a complete process using only steam would require the use of fuel costing more than the value of the incremental oil that could be recovered with steam at that particular stage of production (3). The injection of unheated water can also prevent land subsidence, by taking the place of the recovered oil underground. If the flooding pattern is to be expanded from a pilot to a commercial scale, filling the old pattern with water will equalize underground pressures and prevent the movement of reservoir oil to already swept patterns.

Figure 4.3: Schematic of the Steamflooding Process

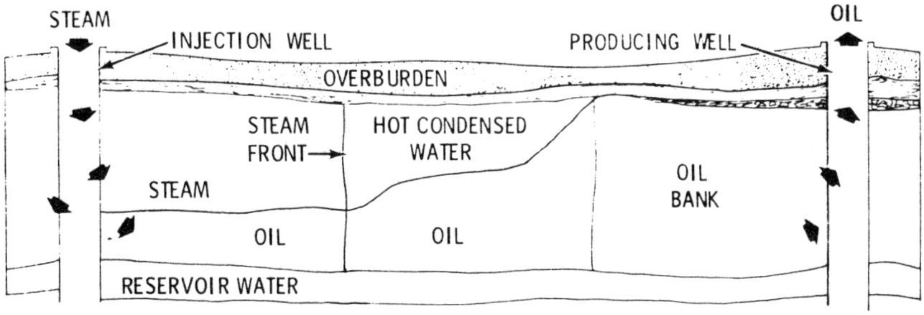

Source: PNL RAP-25

Table 4.1: Percentage of Oil Produced per Mechanism in a Steamflood*

	- - - - - - - - - - - - -Type of Crude - - - - - - - - - - - - - - -		
	Nondistillable	**25% Distillable**	**50% Distillable**
Hot water	87	75	68
Gas drive	5	4	3.5
Distillation	0	14	22.5
Solvent extraction	8	7	6
Total	100	100	100

*Based on an analysis of laboratory data presented by B.T. Willman et al, *Transactions of the AIME,* 222, (1961), p. 681.

Source: EPA-68-01-1912

Advantages and Disadvantages

Unanticipated side benefits of steam drive such as in-place distillation of the oil and improved gas drive make this technique the most likely near-term candidate for recovering substantial tertiary oil. Some experts see steam drive as the most universally applicable technique. Its oil recovery efficiency ranges from 35 to 50% of the reservoir oil in place, depending on oil and reservoir characteristics. Its major drawback is the amount of energy required relative to the amount produced.

Once such a drive is in full operation and has raised the temperature of the reservoir fluid, there is the same nonlinear physics of a waterflood with the exception that the water is more expensive and the mobility ratio even more unfavorable. That is, the relative permeability of the reservoir to oil drops as the steam/water saturation increases. Water/oil ratios at producing wells increase sharply and beyond a point the additional oil produced per additional unit of injected fluids drops off sharply. Most of the oil produced by steamfloods will be produced at low steam/oil ratios. Moreover, the additional oil which can be produced at higher steam/oil ratios is quite marginal and falls off sharply as this ratio increases.

In the application of cyclic steam injection, horizontal sweep efficiency is usually high for oils with viscosities up to about 1,000 cp at normal reservoir temperatures. Viscous bypassing may be significant for oils with a normal reservoir viscosity greater than 1,000 cp. Difficulties in starting the recovery process have been reported when the oil viscosity is very high.

Techniques for alleviating this problem include the injection of steam into an underlying aquifer, the hydraulic fracturing of steam injection wells, and the use of steam to stimulate producing wells. Chemical water treatment is usually required for steam generation, and the quality of available water is an important cost factor.

The success of a steaming project depends on the rapid, continued growth of a steam zone with resulting high oil displacement rates. To achieve this end, heat losses must be minimized. Heat losses during steam injection are a function of injection temperature, the equipment used and reservoir characteristics.

Heat losses increase as temperature increases. It is thus preferable to apply steam injection at lower pressures to minimize heat losses, since high pressures require higher temperatures to convert water to steam (2). This limits the process to relatively shallow reservoirs, usually less than 2,000 feet deep (7). The injection temperature is also affected by the permeability of the oil-bearing zone, since permeability affects the injection rates and pressures. Low-permeability zones require higher pressure and temperature injection, increasing process heat losses (2). Heat losses through surface lines and in injection well bores may be quite severe.

The equipment used in steaming projects basically consists of steam generators; water softeners and related equipment and materials (since fairly fresh water is required to make steam); downhole packers that are specially constructed for high temperature use; and insulation for lines and wellheads, to minimize heat losses (8). Heat losses may be further minimized by placing generators close to injection wells. Burying injection lines is normally sufficient to keep surface heat losses below 10%. In addition to depth, well bore heat losses depend on the type of well completion used (2).

The largest source of heat loss in a steaming operation is by conduction through the surrounding reservoir rock. The rate of heat loss depends on the surface area available for heat flow and increases with steam zone growth. Because of high heat losses of costly surface-generated heat, steaming cannot be used in deep, thin or low permeability reservoirs (2). In deep reservoirs, radial heat losses from the injection string and difficulties in maintaining the mechanical integrity of the string itself place a depth limitation on steam drive. In heterogenerous formations, the steam channels through the highly permeable zones and bypasses less conductive zones which may contain a large fraction of the oil.

Moderately thick formations are preferred to reduce heat losses to adjacent strata. High reservoir water saturations are undesirable because water production becomes very high when formation water and steam condensate are combined (7).

Isolating the geologic zone to be steamed, pumping of hot wells and excessive water production are a few of the major operating problems encountered. There are also difficulties in obtaining accurate production measurements in hot wells (9). Sand control, however, seems to be the major problem (both produced sand and sand-plugged lines), along with deleterious high temperature effects on casing and other tubular equipment (7). Failures of thermal packers are also a major operating problem. The current packers apparently cannot meet the extreme operating conditions encountered in a steam injection project (pressures up to 1,375 psi and temperatures of 590°F) (10).

Weather conditions can also adversely affect a project. Besides water freeze-ups, the cold weather severely affects the startup of air compressors. The lubricating oil becomes so viscous that the crankshaft cannot be spun fast enough to start (11). Other surface problems which have been mentioned in connection with steam injection projects are a scarcity of source water for steam generators, limited generator capacity, and limited availability of low sulfur fuel to power the generators (12).

Low sulfur fuel is necessary for steaming projects to meet air pollution regulations regarding emissions. Cement failure in conventionally completed wells is frequent in thermal recovery operations. New wells must be completed and equipped to operate at high temperature (2).

Production and Projects

The production schedule for steam drive has an increasing rate during the first year (with much of this due to the injection of cyclic steam), a constant production rate for the next 2 years, and a sharp decline during the last 3 years to the economic limit, as shown in Figure 4.4.

Figure 4.4: Production Schedule for Steam Drive

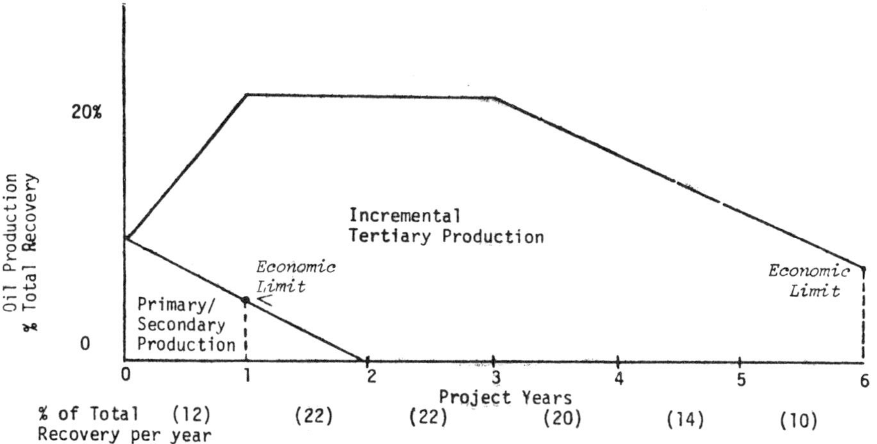

Source: FEA/B-76/221

Steam drives have been in place in the fields indicated in Table 4.2.

Table 4.2: Summary of Steam Drive Projects

Field*	Year Initiated**	OOIP (MM bbl)	Cum. Prod. as of 1/1/75 (MM bbl)
Kern River	1964	4,000	636
Midway Summit	1967	6,000***	1,197
S. Belridge	1970	1,000***	195
San Ardo	1963	1,500	274
Huntington Beach	1965	6,000	923
Yorba Linda	1964	100	30
Smackover†	1971	1,500	508
Winklemann Dome††	1964	100***	

   *All fields shown are in California unless otherwise noted.
  **Generally refers to start of pilot, not start of field-wide project.
***Estimated volumes.
   †Located in Arkansas.
  ††Located in Wyoming.

Source: EPA-68-01-2445

Economic Aspects

The principal aim of thermal methods is a decrease in viscosity. Increasing the temperature of a 15° API crude from roughly 90° to 300° will result in a reservoir fluid which has about the same viscosity as a 30° API oil in a shallow reservoir. In order to obtain this increase, operators have reported steam/oil ratios ranging between 3 to 10. It takes about 1 barrel of fuel to produce 15 barrels of 70% quality saturated steam at 200 psi surface.

The fuel used in the projects given in Table 4.2 above is generally lease crude or topped lease crude, although in some cases residual oil is used. In any case, the cost to the nation of this fuel is about $10 per barrel or 67¢ per barrel of steam. To this must be added the cost of the steam generator. A 50 MM Btu/hr steam generator capable of producing 700,000 barrels of steam per year has cost about $250,000 delivered.

Assuming a ten year life and a 10% cost of capital this works out to 6¢ per barrel of steam. Socal has reported maintenance costs of 5¢ to 8¢ per barrel of steam. The water reclamation plant at Kern River constructed at a cost of 1.7 million dollars can reclaim 75,000 barrels of water per day or about 2¢ per barrel. Smaller treatment plants might have a cost of 5¢ per barrel. In total, then, steam might cost about 85¢ per barrel.

Adding normal waterflood lifting costs (well workover, pumping, well maintenance, fluid distribution, etc) of $0.50 to $1.50 per barrel of crude produced leads to the aggregate costs shown in Table 4.3.

Table 4.3 illustrates the obvious point that steam drive is energy intensive and as such has not been helped all that much by the rise in crude prices. According to these numbers, a tripling in crude prices raises the breakeven steam/oil ratio from about 5:1 to about 10:1.

Table 4.3: Steam Drive Cost per Barrel of Recovered Crude as Function of Steam/Oil Ratio (all values $/bbl)

	Nonsteam Related Costs	3	5	7	10
		- - - - -Steam/Oil Ratio- - - - - -			
$3.00/bbl fuel cost	0.50	1.70	2.50	3.20	4.50
	1.00	2.20	3.00	3.70	5.00
	1.50	2.70	3.50	4.20	5.50
$10.00/bbl fuel cost	0.50	3.05	4.75	6.45	9.00
	1.00	3.55	5.25	6.95	9.50
	1.50	4.05	5.75	7.45	10.00

Source: EPA-68-01-2445

The project operators expected average steam/oil ratios of less than 5:1, which is consistent with the fact that the projects were either under way or at least being seriously considered before the price rise. Furthermore, they expected these projects to allow the fields to reach ultimate recoveries of 45 to 50% of original oil in place.

Study by Lewin and Associates, Inc.: To demonstrate the economics of steam-flooding, Lewin and Associates, Inc., selected a moderately large (50,000,000 barrels of original oil in place), heavy crude (14° API) oil field in California for study. The reservoir is shallow, with an average depth of 1,250 feet and is located in a sandstone formation with an average permeability of 500 md and porosity of 40%.

The field has been under primary production with a solution gas/gravity drive. The project would begin with cyclic steam injection for increasing the immediate rate of production and would follow this with a continuous steamflood in a pattern of injection wells.

> 5 million barrels of oil had been produced to date (1975) with the ultimate primary recovery projected at 6 million barrels (12% of the oil in place).

> An additional 13 million (26% of oil in place) was estimated as recoverable through the combination of steamflooding and cyclic steam injection, for a total recovery of 19 million barrels, 38% of the oil in place.

> Investment costs would be nearly $5 million, consisting of $176,000 for the cyclic steam injection and $4,750,000 spaced over 2 years for the steamflood.

> Operating costs were averaging slightly over $4.00 per barrel.

> Production taxes, consisting of ad valorem and severance taxes, were 8%, with the field under one-eighth royalty.

The project was realizing additional production immediately
from the cyclic steam injection and was reaching its maximum
rate of production 1 year after completing the investment in
the steamflood project.

The ROR calculation showed a negative rate of return for the project at $5.25
per barrel. Overall this project would require a crude price of $7.70 per barrel
to realize an after-tax return of 10%. Thus, steam drive, in combination with
cyclic steam injection appeared unattractive at the controlled price of $5.25 per
barrel.

Future Prospects

Mathematica's Forecasts: Mathematica (CO 68-01-2445) has estimated that there
are perhaps 40 billion barrels of oil remaining in already discovered reservoirs
which meet the requirements of steam drive (i.e., low API gravity, high permeability, shallow depth, etc). On average, these reservoirs have already produced
about 20% of the original oil in place. Assuming all these fields can be made
to produce 50% of their oil, this leaves some 12 billion barrels still to be recovered. Mathematica concluded, therefore, that steam drive may add as much
as 10 billion barrels to reserves.

Lewin and Associates' Forecasts: Similarly, Lewin and Associates, Inc., predict
a contribution to our potential reserves by the year 2000 through the application
of steam drive technology in the large oil-producing states of California, Louisiana
and Texas of 9.1 billion barrels of oil.

This represents, according to Lewin, 30% of total potential reserves. Furthermore, steam drive will provide the bulk of the tertiary production in the early
years, reaching 1 million barrels per day by 1985.

IN SITU COMBUSTION

Another thermal recovery technique is in situ combustion, also called fireflooding. It introduces heat into a reservoir through the injection of air into the reservoir to burn some of the oil in place.

History of Use

In situ combustion had its start in the early 1900s when underground combustion was accidentally started from air injection used to drive oil toward producing wells. Production increased and heat was noted but neither was attributed
to subsurface fire. The produced oil showed 10 to 15% carbon dioxide composition. Only much later was spontaneous combustion acknowledged as the cause
of these effects.

Russians did the first cognizant work in subsurface combustion. They ignited
shallow oil with glowing charcoal, fed into the well by injected air. Increased
production was reported in 1935, with the first English translation of the report
appearing in 1938.

First U.S. mention of in situ combustion was a 1923 patent granted to F.A.
Howard (Standard Development Company). The first test, by Magnolia Petroleum,

was reported in 1953. It produced only 80 barrels of oil but demonstrated the feasibility of this technique.

The oldest ongoing in situ combustion project in the nation was initiated in March, 1958, West Newport Field, Orange County, California. It is 30% finished as a technically successful and economically profitable project.

While early in situ combustion pilots in the United States were generally discouraging, more recent efforts show improved results; in situ combustion, however, continues to be a high-risk technique.

Principles and Procedures

By controlling the air injection, the amount of oil burned (and therefore heat generated) is kept within desired limits. The hot combustion gases thus generated flow toward production wells pushing or carrying the oil as they pass. In some cases, cyclic steam injection may be used in conjunction with projects of this type to increase oil recovery rates.

Typically, air is injected into the reservoir to create a gas saturation high enough to allow the large air flows necessary to sustain combustion. When the air flow is large enough, spontaneous combustion may take place or a downhole heater is used in the injection wall to initiate the combustion. The burning oil front should proceed slowly through the reservoir. Higher rates of air injection tend to burn excessive amounts of the oil in the reservoirs and may cause the burning front to override the oil.

A number of zones exist in the reservoir undergoing a fireflood. In each of these zones, dynamic processes occur which determine the rate and amount of additional oil recovered. The burned region is composed primarily of clean, finely grained sand. The action takes place at the burning zone. Here, heat breaks down the oil into coke, which catches fire, and lighter oils are vaporized or move ahead of the burning region. Temperatures in this zone range from 600° to 1200°F depending upon the reservoir's physical characteristics and the operating conditions of the project.

Just ahead of the burning front is where coke is produced by cracking and distillation of crude oil, which dissociates the lighter from the heavier hydrocarbons. The coke residual fractions are composed of high-boiling point hydrocarbons containing oxygen, sulfur, nitrogen and trace metals. Molecular weights of these residual fractions range from 300 to 900 grams per mol. These fractions can represent up to 20% of the crude oil. Their exact composition is not known.

Another zone created in a fireflood consists of light hydrocarbons, which are vaporized by the burning front or other hot regions of the process. As the gases are driven through parts of the reservoir they displace some oil and improve the oil recovery.

Types–Forward, Reverse and Wet Combustion

There are three basic types of in situ processes: conventional, or forward combustion; wet combustion, (sometimes called COFCAW) a combination of forward combustion and waterflood; and reverse combustion.

Forward Combustion: The forward combustion process requires the drilling of air injection wells and oil production wells. Although ignition in the reservoir may occur spontaneously at the injection wells, a heat source is usually employed to start combustion. The forward combustion process has been operated successfully in several localities.

In the forward combustion process there are a variety of zones formed within the reservoir (see Figure 4.5). There is a heat zone, which causes decomposition of the oil in place, and a combustion zone. In the heat zone, the lighter, volatile hydrocarbons are extracted from the reservoir oil and a tar or coke layer (the fuel to be burned) is left in the reservoir. In the combustion zone, the oxygen from the injected air contacts the coke and combustion occurs. This process moves forward through the reservoir rock with the combustion zone, or burning fire, providing heat to form a layer of coke ahead of it (13).

A burned zone through which the fire has progressed is left near the injection well. In this zone, all of the liquid has been removed from the rock, leaving only air in the pore spaces of the rock. The temperatures here range from near reservoir temperature at the injection well to a peak at the flame front (14). At the flame front, where the heavy fuel is burned, the temperatures normally range between 600° and 1200°F, although they may in some cases reach 1800°F.

Ahead of the flame is a vaporizing zone, which contains combustion products, vaporized light hydrocarbons, and steam, formed by the vaporization of reservoir water near the flame front. Water formed in the combustion reaction also contributes to this steam zone. The steam mobilizes and displaces much of the heavy oil, leaving a relatively low oil saturation of heavier fractions to be burned by the flame front. The coke that is burned is the least valuable, asphaltic portion of the crude (3). Temperatures across the vaporizing zone range from the high temperatures of combustion to that just necessary to boil water at the prevailing reservoir pressure, usually near 400°F.

The next zone is the condensing zone, where condensed light hydrocarbons mix with the reservoir oil to help displace that oil, aided by the hot waterflood caused by the condensed steam and the gas drive provided by combustion gases. Temperatures in this zone usually range from 50 to 200°F above the initial reservoir temperature (14). The condensed water tends to settle below the steam vapor and combustion gases. The combustion or burning front, however, because of the concentration of gas flow, will tend to move in the upper part of the reservoir, sweeping only the upper portion of the oil zone (3).

The oil bank, the final zone, is where the displaced oil accumulates. Pore space in this region is occupied by reservoir water, displaced oil and some of the combustion gases. Since the temperatures in this region are very near the initial reservoir temperatures, there is little improvement in oil viscosity. Ahead of this zone lies the undisturbed reservoir (14).

Once the steam front arrives at the producing well, the oil production rate increases because of the lowered oil viscosity. However, as the flame front nears the well, measures must be taken (such as circulating water through the well annulus) to protect the well from damage as production continues (3).

Figure 4.5:　Schematic of the Forward In Situ Combustion Process

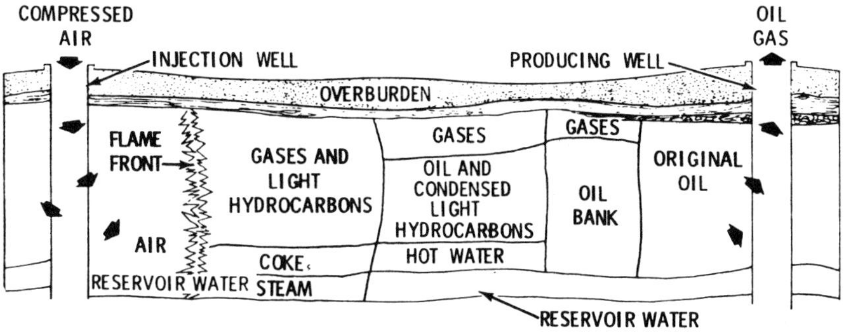

COMPRESSED AIR
OIL GAS
INJECTION WELL
PRODUCING WELL
OVERBURDEN
FLAME FRONT
GASES AND LIGHT HYDROCARBONS
GASES
GASES
ORIGINAL OIL
OIL AND CONDENSED LIGHT HYDROCARBONS
OIL BANK
AIR
COKE
HOT WATER
RESERVOIR WATER STEAM
RESERVOIR WATER

Source:　PNL RAP-25

The amount of coke which a particular reservoir crude oil will deposit upon being heated is one of the most important parameters in a fireflood.　Coke deposited as fuel is measured in lb/ft^3 of reservoir rock.　If this value is too low, combustion cannot be supported; if too high, flame movement is too slow since all the fuel must be burned before the flame advances.　The amount of fuel deposited will also determine the amount of air required to advance the flame through a specific volume of reservoir rock.　The more fuel that is deposited, the greater are the air requirements (14).

The amount of air injected per unit area of flame front, or air flux, is another important factor in combustion projects.　Air flux is measured in $scf/hr/ft^2$ of flame front and usually ranges from 10 to 30.　Below a minimum value (which is dependent on reservoir conditions and crude oil type) oxidation of the oil is too slow to generate the heat required for sustained combustion.　Increasing the air flux will increase the speed of the flame front but will also reduce the contact time of the injected air with the flame, and result in poor utilization of the oxygen in the air.

The forward combustion process does not utilize heat very efficiently.　Because of the poor heat-carrying capacity of air, only about 20% of the heat generated is moved ahead of the flame front where it works to improve oil recovery.　The remaining heat is absorbed by reservoir rock above and below the oil-bearing strata (14).　The wet combustion process was developed to improve the efficiency of the in situ processes.

As the flame moves from the air injection well to the producing well, the oil must pass through a cold region of the reservoir.　If the oil is very viscous, liquid blocking could occur, terminating the combustion process (15).　Reverse combustion is designed to alleviate this problem.

Reverse Combustion:　The reverse combustion process resembles forward combustion, except that the reservoir oil is ignited at the production wells rather than at the injection wells. A possible use of the process might appear to be in the recovery of tars, since the high temperature of reverse combustion can produce thermal cracking which increases oil API gravity and drastically reduces oil viscosity. Furthermore, the forward combustion process is not readily adaptable to recovery of tars, since condensation of vaporized hydrocarbons may impede or completely stop the flow of combustion gases from the oxidation zone to the production well.

In reverse combustion, the flame front travels in a direction opposite to that of air flow. Air is initially injected into a well, which will later be the producing well, to start combustion. After ignition has occurred, further air injection takes place through a different well. Combustion gases, steam and oil in the burning region flow through the previously burned area to the producing well (3). Thus, oil is constantly moving through heated areas of the reservoir. This method allows the production of oil which is too viscous to flow under reservoir conditions.

In actual application, the process has very low efficiency. A valuable fraction of the crude oil is burned as fuel, leaving the undesirable tar fractions in the region behind the combustion front. Also, if spontaneous combustion occurs near the air injection well, it could use up all the air required for reverse combustion (15). Thus, the process is limited to crudes of relatively low reactivity to oxygen, since it can continue only as long as spontaneous combustion does not occur at the injection well (7). The reservoir must also have adequate air permeability for the process to work. Reverse combustion is expected to have limited use, if any, because of its low efficiency (14).

Wet Combustion: An important variation of in situ combustion method is a process known as wet combustion. A fire is started in the formation. Then, water is injected alternately with air to transfer excess heat from the burned region through and ahead of the combustion front in the form of water vapor. The effects of added heat and steam drive further reduce the viscosity of the oil ahead of the combustion front. With this process it is possible to move thicker oils than with a dry fireflood and to operate at lower pressures with possibly less fuel.

Because less fuel may be burned, less air is required than with dry combustion. Though the process has not had wide application in the field, it is expected that recovery efficiency of 40 to 60% of the original oil in place will be possible for mobile crudes and highly porous rock formations. Somewhat lower oil recovery efficiency (on the order of 30 to 50%) is expected where porosities and initial oil saturations are not as high.

COFCAW Project — One process of wet combustion known as COFCAW or combination of forward combustion and waterflooding has been patented by Amoco Production Company (formerly Pan American Oil Company). While this process is still economically unproven, it has achieved success in a series of small pilots and limited success in the 980 acre field test at the Sloss Field in Nebraska, where results were economically discouraging.

In this process, the amount of coke left to be burned as a fuel by the flame front is substantially decreased. More oil is thus displaced from the reservoir and less air is required to burn a unit volume of the reservoir. Water injection in a ratio of 0.3 to 0.5 barrel of water per thousand cubic feet of air can reduce fuel and air requirements by as much as 30 to 50% (3). A water-air ratio is usually chosen to move the heat forward at about the same rate as the combustion front, while maintaining high combustion temperatures. Water can be injected with air at ratios of up to 1000 to 2000 scf of air per barrel of water without completely quenching the fire front. The injected water develops a steam zone in front of the combustion zone. The major benefit of the COFCAW process is that there can be a threefold reduction in the air required to produce 1 barrel of crude oil (1).

The wet combustion, or COFCAW, process uses heat more efficiently than dry combustion, taking advantage of the high heat content of water, as well as its latent heat of vaporization, to recover energy that would otherwise remain stored in the burned-out region. However, although the process does have a higher thermal efficiency than dry combustion processes, it does not avoid the problem of minimum oil mobility in the cold regions of the reservoir (15).

Some problems have been encountered with respect to combustion sweep efficiencies. This large pilot was conducted by Amoco over the greater part of the 39° API Sloss Field (16). This operation reported a volumetric sweep efficiency of only 14%. The project (now concluded) recovered about 0.8 million of the 5.9 million barrels remaining after waterflood. The residual oil saturation at the beginning of this project was only 30%.

Limitations

There have not yet been as many in situ combustion projects initiated in comparison to steam flooding. The combustion projects attempted, however, have covered fields exhibiting a wider range of API gravities, depths, and permeabilities than with steam flooding. Almost all of these projects, though, have been begun on a pilot test basis.

In situ combustion is undoubtedly the most difficult of the tertiary processes to model and predict. Published reports are conflicting and noncommittal. In very general terms, however, the economics of combustion are not all that different from that of steam drive. In situ combustion involves the maintenance of an ignited flame front in the reservoir fed by the reservoir crude itself. The fuel for this method then is essentially a free good since not only is no out-of-pocket cost incurred with its use, but also there exists no alternative opportunity cost for the consumed crude.

Some of this advantage is lost, however, in the fact that the injection air (about 11,000 ft^3/bbl burned) must be compressed. Also, the high temperatures at the fire front lead to greater losses to the formation. This is generally countered by injecting water either simultaneously with the air or after, thus transferring heat from the formation back to the driving fluid.

Most of the operating problems encountered with in situ combustion projects relate to the high temperatures involved. The thermal stresses on cement, tubular goods and pumping equipment increase the frequency of equipment failure. Sand production creates pumping problems and severe wear. The composition of the combustion gases, together with high temperatures, accelerate corrosion problems (13). Other problems which have been encountered in in situ combustion operations include poor air distribution (which results in poor vertical and areal sweep) and inabilities to keep the burning front going when the spacing between injection and producing wells becomes too large (perhaps a 10-acre spacing of wells) (9). A possible environmental problem may be emissions of combustion gases from the well, although there are as yet no data on this problem. It is uncertain whether these gases are regulated by air pollution laws.

Firefloods in Venezuela have been applied to hardly depleted, deep, thick, permeable, heavy oil reservoirs, with quoted target recoveries of about 50%. In these instances, the firefood is operating in much the same fashion as a steam drive. It appears, therefore, that firefood has an advantage over steam drive as heavy oil reservoirs get deeper. Certainly, firefood cannot be used in very shallow reservoirs for the pressures generated would blow through the overburden.

Though fireflooding is applicable to a wide range of reservoir conditions, some oils do not form coke, which is the fuel for the burn, in adequate quantities. Asphaltic crudes, as a general rule, seem to burn better than paraffinic oils for a given gravity due to the former's high levels of bitumen.

Peculiar characteristics within the oil-bearing strata also affect in situ combustion. Some oils which leave high deposits of coking fuel could make the process uneconomical due to the large volume of air needed to maintain combustion. An increase in the rate of input of air produces higher combustion temperatures and therefore a higher efficiency. At a low air input, the percentage of the oxygen utilized is low and therefore the process is uneconomical because unnecessary costs are incurred in pumping air.

Production

The 6 year production schedule used by Lewin and Associates, Inc., for in situ combustion shows an increasing response until the third year, and a gradual decline from year three to the economic limit in year six. The production schedule for the oil recoverable from in situ combustion is demonstrated in Figure 4.6.

Figure 4.6: Production Schedule for In Situ Combustion

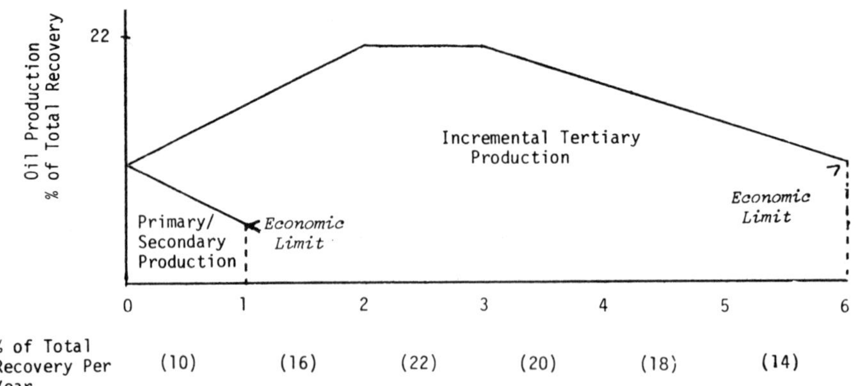

Source: FEA/B-76/221

Economic Aspects

In situ combustion projects require very high capital investment. The cost of operation and maintenance of air compressors, however, seems to be the biggest expense (7). The size of the compressors is determined by the desired flame front movement within the reservoir and the necessary air injection rates (14). Large volumes of air at high pressures must be handled in in situ combustion recovery. Air requirements ranging from 4,000 to 20,000 scf/bbl of oil recovered have been reported.

Compressor operating costs can range from 40 cents to more than $8 per barrel of oil produced; the energy of compression can represent as much as 25% of the heating value of the recovered oil (17). The air to oil ratio is thus a good indicator of economic project operation. Economically successful projects have air to oil ratios of less than 20 thousand scf/bbl (14).

The amount of coke available for combustion (fuel content), deposited as a result of distillation and thermal cracking, can also affect air requirements. If the fuel content is too low, combustion cannot be self-sustained; if too high, there will be a high air requirement and excessive power costs, along with a low oil production. Field projects have shown the fuel content to be a complex function of reservoir and crude oil properties, including reservoir thickness, permeability, oil saturation, viscosity and depth (15).

A study by Lewin and Associates, Inc., examined 15 field wide fireflood projects. Only 5 of 15 of these projects were reported as being successful and profitable under existing technology and prices; thus, the study acknowledges the higher risks associated with this technique. Since Lewin's rates of return have not been probabilistically adjusted for risk or failure rates, the economic analysis in this instance seems to call for a higher rate of return than for other exotic enhanced recovery projects.

Lewin concluded that in situ combustion would appear marginally attractive at a $5.25 per barrel level, but given the higher risk associated with this technique, the fireflood project would require a price of $7.60 per barrel to reach a 20% after-tax rate of return.

A hypothetical moderately large field having 52,000,000 barrels of oil in place was selected for the study by Lewin. The reservoir was shallow with a depth ranging between 1,000 to 1,800 feet and had permeability of 1,100 millidarcys and porosity of 37%. The crude was heavy, of 15% API gravity and low sulfur content.

> 5,200,000 barrels (10% of the oil in place) was projected as recoverable through primary production.
>
> An additional 20,800,000 barrels (40% of the oil in place) was expected to be recoverable by using in situ combustion.
>
> Investment costs would be $15.5 million, consisting of $6.4 million for drilling injection wells, $3.0 million for production equipment, $2.5 million for injection facilities, steam and air compression (for cyclic steam) and $200,000 per year (for 18 years) for replacing well casings and equipment damaged by heat.

Operating costs would average $2.70 per barrel.

Additional production from the project would become available after one year, reaching peak production in the third year.

Future Prospects

From the point of view of projecting domestic supply from heavy oil reservoirs, with the exception of extremely shallow deposits, both steam drive and fireflood are going after the same type of low API gravity oil. Thus despite their relative economics, it would be double-counting to credit the recovery potential for these heavy oil deposits to each of these thermal methods separately.

The use of firefloods on lighter crudes in previously waterflooded reservoirs is a bit different situation. Most of these projects have been discontinued, but it is a potential means of going after truly tertiary oil. Considering the lack of combustion activity in this area relative to that for CO_2 and micellar, it appears that its economics are less favorable than those of these miscible methods. Therefore, combustion would not make a significant contribution to supply from medium- to light-oil reservoirs.

Through the application of in situ combustion technology in California, Texas and Louisiana, only 0.9 billion barrels of oil are forecasted by Lewin and Associates, Inc. to be added to our potential reserves by the year 2000. This is equivalent to 3% of total potential reserves.

NECESSITY FOR MATERIALS RESEARCH

Because of the high temperatures encountered within the wells, problems of corrosion and matrix unconsolidation can become more severe in thermal recovery processes. Thus, there is a need to develop materials (possibly nonmetallic) which can withstand the rigorous conditions encountered. The solutions have involved materials that are in short supply and expensive.

Typical of the effects encountered is the oxidation and corrosion of tubular material in producing wells, caused by the flue gases produced by the in situ combustion process. Materials should be developed to resist these detrimental reactions. Other materials are needed which can be injected into the porous matrix to prevent unconsolidation from occurring while at the same time retaining matrix permeability.

Similar injection materials might find application in selectively blocking off high permeability streaks within a reservoir so that more efficient fluid displacement might be achieved. While some materials have been developed for injection into reservoirs for the above mentioned purposes, their use in thermal recovery may be precluded by the high temperature level peculiar to the process. Thus development of new materials particularly suited to high temperature environments is desirable.

CONCLUSIONS

Thermal recovery methods are designed to increase oil recovery by reducing the viscosity of oil and enabling it to flow more readily toward producing wells.

The technology for these processes is fairly well developed, and both steamflooding and in situ combustion have been employed on commercial scales. Thus, thermal enhanced recovery methods are said to be among the lowest risk investments of the tertiary recovery techniques.

The success of any tertiary recovery project depends quite heavily on matching the process desired with an appropriate reservoir. Not all methods work equally well in all reservoirs. Thermal methods seem to work best when applied to heavier crude oils, usually those of less than 25° API gravity, and in sandstone reservoirs (18). Screening parameters have been developed which help to decide whether thermal methods will work well in any particular reservoir. The most critical parameters for steam drive seem to be reservoir depth, zone thickness and crude oil viscosity.

Virtually all of the oil produced by steam injection is from high saturation, high porosity, shallow reservoirs, containing low API gravity crude oils (19). A reservoir depth of less than 3,000 ft minimizes heat losses, and the latent heat of steam is highest at low pressures. Thus, more heat per pound of steam injected can be transported in low-pressure, shallow reservoirs than in the deeper zones which are at higher pressures (2). Other favorable factors for steam injection projects are existing wells which are adaptable to steam injection; an available fuel supply for steam generation; and an available supply of water which is cheap, slightly alkaline, free of H_2S, oil, dissolved iron and turbidity. Some of the factors which increase the risk of steam drive operations include low residual oil saturation and extensive reservoir fractures (6).

The majority of steamflood projects are located in California. As of 1975, there were 33 active steamflood projects in the United States, 26 of which are in California. Most of the remaining projects are in the Texas-Louisiana area. The largest project is in the Kern River field in California and has 781 injectors. Nineteen of the projects have been deemed to be profitable; the others are either indeterminate at this time or are economic failures (7).

In situ combustion processes must also be carefully designed for formation characteristics (3). The process has been applied mainly to reservoirs with high viscosity crude oils, although it has been used for tertiary recovery in reservoirs with oils, in the gravity range of 20 to 40° API (14).

In situ combustion was once throught to be a very promising recovery technique. However, operational and economic problems have resulted in a declining application of the process, from 38 active projects in 1970 to 21 active projects in 1975. Of the projects ongoing in 1977, eight have been termed profitable and seven unprofitable; others were yet to be evaluated (7).

The largest potential for thermal recovery on an extensive scale is in California's Central Valley; about 90% of the entire enhanced oil recovery potential is in California (18).

REFERENCES

(1) Howell, W.D., Johnson, F.S. and Johansen, R.T., "Enhanced Oil Recovery Methods," CONF-750306-3, presented at the 15th ASME Resource Recovery Symposium, Albuquerque, New Mexico (March 6-7, 1975).

(2) Herbeck, E.F., Heintz R.C. and Hastings, J.R., "Thermal Recovery by Hot Fluid Injection," Fundamentals of Tertiary Oil Recovery, Part 8, Petroleum Engineer, pp. 24-34, (August, 1976).

(3) *Enhanced Oil Recovery,* National Petroleum Council, (December, 1976).

(4) Rivero, R.T. and Heintz, R.C., "Resteaming Time Determination - Case History of a Steam-Soak Well in Midway-Sunset," *Journal of Petroleum Technology,* pp. 665-671, (June, 1975).

(5) *Secondary and Tertiary Oil Recovery Processes,* Interstate Oil Compact Commission, Oklahoma City, (1974).

(6) Geffen, T.M. "Oil Production to Expect from Known Technology," *Oil and Gas Journal,* pp 66-76, (May 7, 1973).

(7) *Management Plan for Enhanced Oil Recovery,* Volumes I and II. Program Strategy, Petroleum and Natural Gas Program, ERDA, (February, 1977).

(8) "Enhanced Recovery Requires Special Equipment," *The Oil and Gas Journal,* pp. 50-56, (July 12, 1976).

(9) "How the Industry is Recovering More Oil," *Oil and Gas Journal,* (1975).

(10) "Diversified Equipment Assists Enhanced Recovery," *The Oil and Gas Journal,* pp. 75-80, (October 17, 1977).

(11) Linville, B., ed, *Contract and Grants for Cooperative Research on Enhancement of Recovery of Oil and Gas.* Progress Review #11, July 1977, published August 1977, BERC-77/3.

(12) Noran, D. "Enhanced Recovery Action is Worldwide," *Oil and Gas Journal,* (April 5, 1976).

(13) *ERDA Sumposium on Enhanced Oil and Gas Recovery,* Proceedings, Volume I- Oil. Petroleum Publishing Co., Camelot Inn, Tulsa, Oklahoma, (September 9-10,1976),

(14) Herbeck, E.F., Heintz, R.C. and Hastings, J.R., "Thermal Recovery by In Situ Combustion," Fundamentals of Tertiary Oil Recovery. Part 9. *Petroleum Engineer.* pp. 46-56, (February, 1977).

(15) Chu, C. "A Study of Fireflood Field Projects," *Journal of Petroleum Technology,* pp. 111-120, (February, 1977).

(16) Buxton, T. and Pollock, C., "The Sloss COFCAW Project—Further Evaluation of Performance During and After Air Injection," *Preprints—1974 Improved Oil Recovery Symposium,* Society of Petroleum Engineers, (April, 1974).

(17) Stinson, D.L., Carpenter, H.C. and Cegielski, J.M., "Power Recovery from In Situ Combustion Exhaust Gases," *Journal of Petroleum Technology,* pp. 645-650, (June, 1976).

(18) Elkins, L.E., *Estimated Contribution of Enhanced Oil Recovery to USA Domestic Crude Oil Supply,* (1977).

(19) Doscher T.M. and Wise, F.A., "Enhanced Crude Oil Recovery Potential- An Estimate," *Journal of Petroleum Technology,* pp. 575-585, (May, 1976).

MINING OF PETROLEUM

The information in this chapter is based on:

> *Mining for Petroleum: Feasibility Study,* BuMines Report OFR
> 56-79, prepared by J.S. Hutchins, E. Bond and D.M. Bass of
> Energy Development Consultants, Inc., for U.S. Bureau of Mines,
> July 1978.

> *Oil Mining: A Technical and Economic Feasibility Study of Oil
> Production by Mining Methods,* BuMines Report OFR 55-79, pre-
> pared by A. Edey, B.A. Kennedy and L.A. Readdy of Golder
> Associates, Inc. for U.S. Bureau of Mines, October 1978.

The mining of rock outcrops for their oil content is recorded from antiquity, and was a feature of numerous local economies prior to the Industrial Revolution. The discovery of oil in circumstances where it was producible from wells in large quantity and at low cost, however, tended to overshadow such ventures and often destroy their economic reason for existence. Since that era, relatively few oil mining ventures have been undertaken and many, if not all, of those have had a special reason for their existence.

Very little attention is normally drawn to the fact that the majority of oil that has been or will be discovered will remain in the ground due to inherent limitations of production through surface wells. Best sources in the petroleum industry estimate that in the producing or producible fields alone, a total of over 300 billion barrels of oil will remain in the ground on completion of production, some two-thirds of the original quantity. Even such a large figure is, in fact, an underestimate of the oil in the ground which is the possible target for oil mining efforts. Numerous other hydrocarbon-impregnated rocks are known to exist which are not included in these figures since they have never been considered feasible for conventional production. Such deposits would typically include oils whose viscosity, or the nature of the host rock, inhibits flow to wells. Extreme examples of this are the diatomaceous deposits whose extent has never been demonstrated, or probably even seriously indicated, and the tar sands and asphaltic deposits lying near surface.

HISTORY OF PETROLEUM MINING

A number of foreign countries have investigated and conducted successful mining operations in depleted oil zones. Germany, for example, attempted to increase its supplies of petroleum by underground mining methods during World War I. In 1935, German oil mining was carried out on a more commercial basis. For many years, France successfully and economically recovered oil by mining zones that would not produce by conventional wells. More recently, the USSR has undertaken the mining of oil in the Yarega Field, which is presently producing 1.5 million barrels annually.

Although foreign countries are experiencing some success in recovering oil by underground mining methods, little research has been performed in the U.S. regarding oil mining. The study of the methods used in foreign countries alone is not sufficient to evaluate the practicality of oil mining in the U.S. because reservoir and geologic characteristics vary so greatly that no one method can be applied to all reservoirs.

In 1932, the U.S. Bureau of Mines published Bulletin 351, *Mining Petroleum by Underground Methods,* by G.S. Rice, which concluded that under certain conditions, oil mining might be preferred over conventional recovery methods. This conclusion was based on studies begun in 1923 by Rice, Chief Mining Engineer, U.S. Bureau of Mines, of oil mining methods employed in Europe at oil sand mines at Pechelbronn, Alsace, France, and Wietze, Hannover, Germany. Later studies, in varying detail, were conducted by Uren, Sack, Bruderer and Louis, and Robertson Research International Limited.

In the past five decades since the publication of Bulletin 351, tremendous technical advances have been made in mining and petroleum technology and designs. In light of these technical advances, it is essential to reassess the 1932 USBM study and to detail present-day mining technology and equipment that could be used economically to mine petroleum in the U.S. Mining technology alone, however, cannot be unilaterally considered without understanding the mechanics of the oil reservoir itself. Therefore, the mining engineer must first determine from petroleum engineering technology what is mechanically required for increasing ultimate recovery and must be reactive to those reservoir technology constraints.

PETROLEUM ENGINEERING TECHNOLOGY

Nature and Occurrence of Petroleum

A key to unlocking the origin of petroleum is to note the environmental conditions of petroleum accumulations as they exist today. Crude oils are complex mixtures of many different hydrocarbons and impurities, and no two reservoirs contain crude oil of exactly the same composition and in the same physical state. In geological terms, petroleum is not a mineral because it varies so widely in composition. It can, however, be called an organic mineraloid, although the terms mineral fuels and fossil fuels have become widely accepted.

Generally, petroleum occurs in sediments chiefly marine in origin. In terms of fossil fuels, the fossil organic matter had to be buried quickly and in some manner protected from decay. Continental, or nonmarine, sediments, therefore, would not be a likely source bed for petroleum because continental sediments are subjected to oxidation and other weathering agents. Organic material is constantly settling to the bottom of seas along with silt, clay and other water-transported materials where there is not enough oxygen to support decay and disintegration of organic matter.

Insoluble organic matter has been found to be nearly universal in the sediments. It took modern analytical technology to discover that organic matter soluble in organic solvents also occurs almost universally. Analytical investigators have reported concentrations of the soluble petroleum hydrocarbons in the sediments ranging from 10 to 50 barrels per acre-foot. More commonly, the sediments contain less than 10 barrels per acre-foot.

The presence of porphyrins indicates that the origin of petroleum is probably a low-temperature phenomenon, since porphyrins are destroyed at temperatures above 392°F (200°C). Reservoir temperatures rarely exceed 225°F, although temperatures up to 300°F occur in some deep reservoirs. In most areas, reservoir temperatures are a function of depth and the local temperature gradient.

Petroleum hydrocarbons occur in nature as solids, liquids and gases. The solid, semisolid and liquid hydrocarbons are of primary interest here. Solid hydrocarbons exist as asphaltic bitumens (tar sands) and kerogen (oil shale). Semisolid hydrocarbons are usually referred to as heavy or viscous oils and usually are found at low temperatures and pressures.

Figure 5.1: Map Showing Major U.S. Petroleum Regions

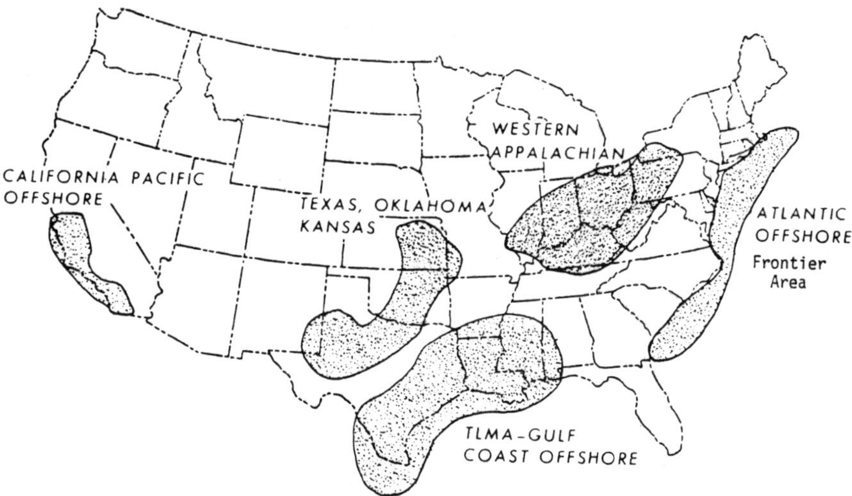

Source: BuMines OFR 56-79

In general, an accumulation of petroleum requires a source bed, a transporting formation and a reservoir trap. It is generally accepted that marine, organic shales and clays are the most common source beds for petroleum. Prerequisites for a reservoir are that the host material must be porous and permeable and that it must be contained or sealed in some manner to prevent escape of the hydrocarbons. A permeable carrier bed, then, is required for the hydrocarbons to migrate from the source bed to the reservoir. A special carrier bed is not required where the reservoir is in contact with the source bed and acts as the carrier formation.

The major liquid petroleum producing regions in the U.S. are shown in Figure 5.1. Areas containing shallow oil fields are shown in Figure 5.2.(1).

Figure 5.2: Map Showing the Locations of Shallow Oil Fields in the U.S. (1)

Source: BuMines OFR 56-79

Solid Petroleum: A reservoir that contains heavy, nonflowing oil, or asphaltic hydrocarbons (tar) could at one time have been a conventional, pressured liquid oil with gas in solution. Because of natural tectonic forces or by becoming overpressured, the reservoir cap becomes cracked and broken, allowing the gas and the lighter oil fractions to escape. As a result of the gas fractions escaping the reservoir, the viscosity of the remaining liquid increases, which retards the ability of the fluid to move. Regions in the U.S. that contain petroliferous rock are shown in Figure 5.3 (1).

Another natural phenomenon affecting conventional reservoirs is erosion. As the vertical depth from the surface to the reservoir is diminished by erosion, the pressure is reduced and gas comes out of solution. If the gas is allowed to escape the trap, the remaining oil will be more viscous. After the reservoir is exposed by erosion, the reservoir becomes an inspissated (drying up) deposit. Exposure at the surface permits the hydrocarbon near the surface to oxidize, leaving only the very heaviest or asphaltic portion of the original fluid in what is commonly called a tar sand.

Figure 5.3: Regions in U.S. Containing Petroliferous Rock (1)

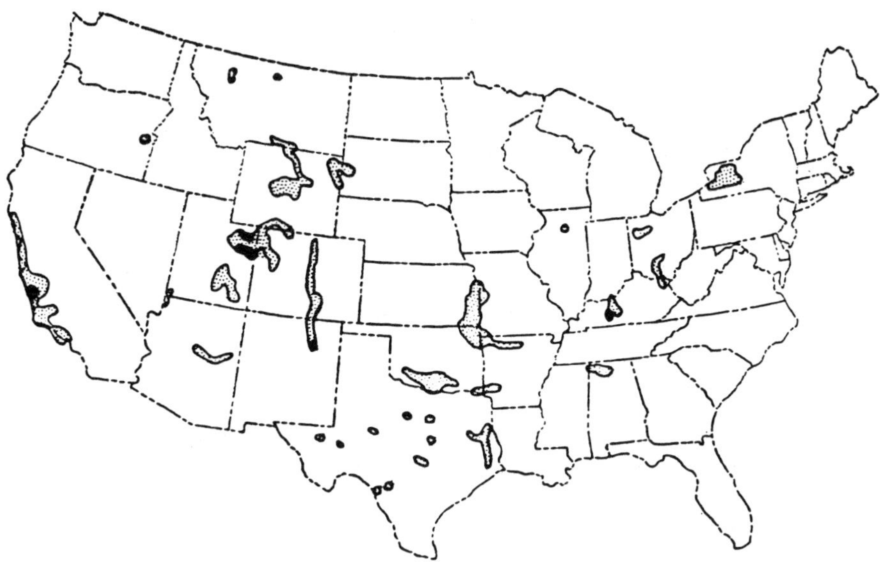

Source: BuMines OFR 56-79

Solid hydrocarbons disseminated in nonreservoir rocks were probably deposited and formed with the sediment. The hydrocarbons may have been a part of the sedimentary source material or they may have been formed by diagenetic forces after deposition on organic material deposited with the sediment.

MINING ENGINEERING TECHNOLOGY

Types of Reservoirs Suitable to Mining

Reservoirs which may be candidates for a mining recovery process can be classified in two broad general areas: depleted and virgin.

Depleted Reservoirs: The depleted category probably is the most prevalent and contains the greatest volume of potentially recoverably petroleum. If considered worldwide, however, this assumption would probably be false because tar sands, oil shale and viscous crude reservoirs are still in the virgin state and probably contain the greatest amount of petroleum potentially recoverable by the mining process.

The depleted reservoirs can be classified into two general categories: those which have undergone some type of secondary recovery process and those which have only been primarily depleted. The former category probably contains the greatest amount of remaining petroleum because they represent the greatest number of petroleum reservoirs. Numerous primary depleted reservoirs exist that contain viscous crude oil. Some of these reservoirs have undergone steam stimulation, but not a further recovery process. The U.S. volume of oil remaining in these reservoirs is probably in excess of 200 billion barrels (primary-depleted) and 300 billion barrels (secondary processes).

The depleted systems which have undergone secondary recovery processes were flooded in the late 1930s and early 1940s and are located at relatively shallow depths at low pressures. Some of these reservoirs were placed on a vacuum during World War II and, hence, would contain only small amounts of gas in solution in the oil. As a result, only small amounts of any gas remain in the reservoir to be utilized as a displacing fluid. The other type of reservoirs which have undergone secondary recovery will have been flooded at pressures in excess of 500 psi and will have considerable gas in solution and possibly free gas to assist in the removal of fluids from the formation. These depleted reservoirs will probably contain oil having the following properties:

    Viscosity—1 to 10 cp
    Gravity—greater than 25°API
    Pressures—less than 300 psi
    Gas in solution—between 10 and 800 scf per reservoir barrel
    Oil saturation—between 10 and 40% of the pore space
    Porosity—between 15 and 30%

Whether these reservoirs can be economically produced by a mining process will be a function of the product of the formation thickness, formation porosity and residual oil saturation, as shown below:

$$\text{Oil in place} \ = \ \text{area} \ \times \ \text{thickness} \ \times \ \text{porosity} \ \times \ \text{oil saturation}$$

Factors of depth and mineability of formations above or below the oil zone itself will be major factors in the economics and technical feasibility of any such process.

The reservoirs which have undergone primary depletion with or without steam stimulation will normally contain a much higher percentage of in place oil at the time any mining process may be initiated. These reservoirs will normally contain a more viscous oil, be at relatively low pressures, and at relatively shallow depths. These reservoirs are represented by the higher viscosity oil reservoirs in California, Venezuela and Canada. They are presently being produced, but with great difficulty. There are probably over 500 billion barrels of oil in this class of reservoirs in the world.

Virgin Reservoirs: Virgin reservoirs which might be susceptible to underground mining are represented by the known tar sands and the very viscous oil deposits throughout the world. It is known that very extensive reserves of petroleum exist in these deposits in the U.S., Canada and South America—some 1,000 billion barrels. These virgin deposits are susceptible to both strip mining and to underground recovery.

The tar sands which are being produced in Canada by strip mining have been exposed to conventional petroleum recovery mechanisms with very little success. The petroleum of these tar sands changes into a very viscous oil with depth; the sands have not been treated with combinations of known recovery technology to make them productive. They have been overlooked as a potential source of petroleum production primarily because of the quality of the petroleum and their location.

Figure 5.4: Characteristics of Depleted and Virgin Reservoirs

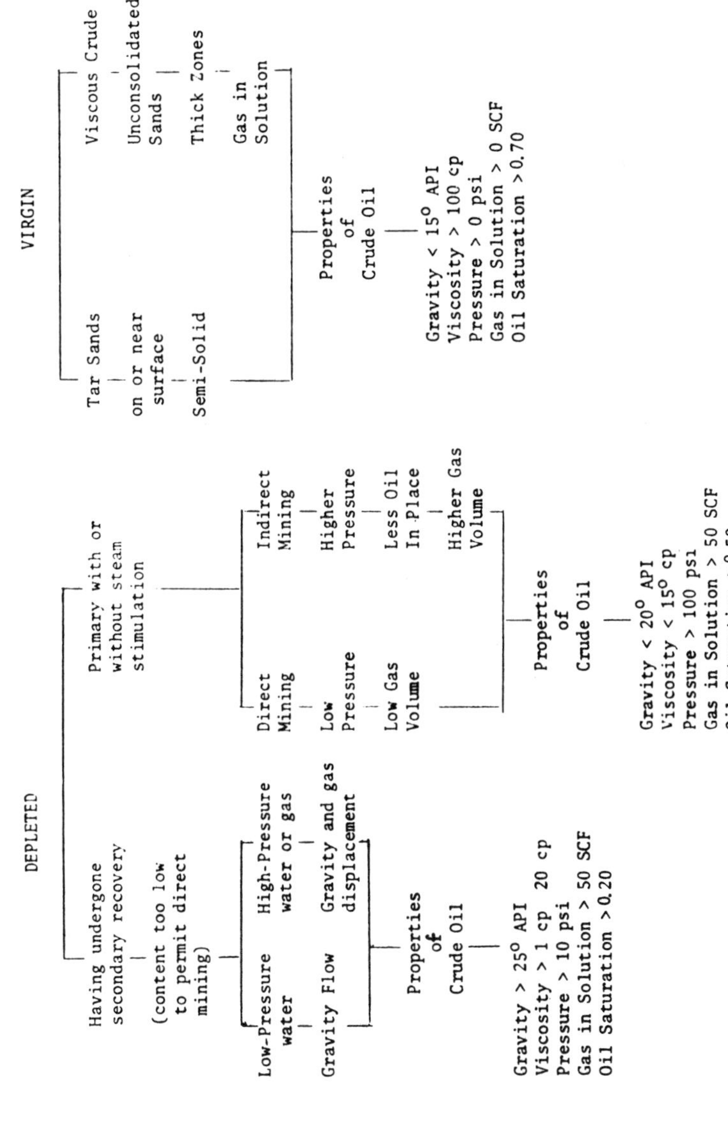

Source: BuMines OFR 56-79

In many cases these reservoirs represent a petroleum deposit which is directly mineable. As the depth of these deposits increases, the contained petroleum is less viscous and has gas in solution so that removal of the petroleum-containing formation is not feasible.

Conditions for Mineability: Certain conditions must be present to mine petroleum. The formation must contain sufficient petroleum to make it economically feasible and the formations above and below the hydrocarbon reservoir must be competent enough to permit the driving of tunnels within that formation. These two conditions are not met by all of the petroleum-producing formations within this country. In the multisand fields in the Gulf of Mexico and in the salt dome areas of Louisiana and Texas, the confining zones above and below producing formations are not thick enough or strong enough to allow the construction of mine workings.

Some of the most productive zones susceptible to mining are the thick limestone reservoirs which themselves are competent and which have over- and underlying competent formations. (Incompetent formations are those which tend to slump or cave into any voids created when oil becomes mobile.) These limestone reservoirs are fairly prevalent in the Appalachian states, West Texas, and New Mexico and contain large quantities of hydrocarbon material. The recovery from these limestone systems, whether by primary or secondary methods, is usually small. These formations have not behaved as existing engineering theory predicted, so a tremendous amount of oil is still present in most of these limestone formations.

The characteristics of depleted and virgin reservoirs are shown in Figure 5.4.

Mining for petroleum will be performed using one of two basic techniques. One technique is to mine the hydrocarbon as an ore, that is, to remove the hydrocarbon and the reservoir material together and separate the hydrocarbon from the rock in a surface facility. The other method is to use mining technology to get close to a hydrocarbon reservoir, then use petroleum technology to extract the hydrocarbon leaving the reservoir rock in place. This second method can be termed in situ mining.

Surface Mining

Physical extraction of hydrocarbon ore by surface mining methods is conducted on a large scale in Canada. The extensive Alberta oil sands are water-wet, which aids in separating the hydrocarbon from the host rock. In the U.S., much of the reservoir material containing similar hydrocarbons is oil-wet, which makes the separation process much more difficult.

Open pit and strip mining are the two fundamental methods of surface mining. The choice of mining method and the mine configuration is dependent upon the size, shape and resource concentration of the ore body and on the topographic features of the ground surface. Open pit mining generally is used for ore bodies of limited lateral extent which are relatively thick vertically. Open pit methods are common for near-surface ore bodies or copper and iron. Problems basic to open pit mining include waste disposal, slope stability and dust control.

Strip mining commonly is used for thin near-surface ore bodies such as coal that may extend laterally a great distance. Overburden and waste disposal is handled by dumping the overburden and waste rock back into the mined-out strips. Dust control commonly is accomplished by wetting the area with water. The extensive Canadian projects of mining the Alberta oil sands and the Athabasca tar sands have shown that ore extraction surface mining of hydrocarbons is only a marginally economic venture.

Equipment common to surface mining includes various types of excavators and haulage equipment. Excavating equipment includes power shovels, drag lines, scrapers and bucket-wheel excavators. Haulage equipment includes such choices as bulldozers, scrapers, trucks, trains and conveyors. The cost of stripping in open pit mining has remained fairly constant because of increased productivity due to improved technology. The cost of underground mining, however, has risen steadily. More than 90% of metal produced in the U.S. today originates in open pits.

Underground Mining

There are many complex mining systems and combinations of mining methods in use for physically extracting the various kinds of ore under different geological settings. One system involves shaft sinking by conventional methods and the other is shaft sinking using drilling and boring equipment. Both vertical shafts and sloped openings are possible.

In conventional shaft sinking, the sinking cycle consists of three operations: drilling and blasting, mucking, and installing concrete lining and support. The shaft sinking technology best suited for mining for petroleum probably is drilling. The optimum locations for a shaft may be through the center of a petroleum reservoir if reservoir pressures can be controlled using blind hole drilling equipment. Once the shaft has been lined and cemented, it would be safe for miners to enter and to create levels below or above the reservoir as the situation permits.

For underground in situ mining of petroleum, tunneling will provide access to a reservoir and not actually mine any of the hydrocarbon reservoir. Tunneling equipment used for petroleum mining probably will be the newest mechanized continuous miners.

Because of the potential for gas problems or other flammable material in mining for petroleum, it would be advantageous to use equipment or a mining system other than the drill-blast-mucking system. In many cases, the geologic strata either underlying or overlying the hydrocarbon reservoir will probably be shale or other suitable medium for continuous mining equipment.

Ground support techniques and materials technology have advanced steadily in the past few years. Mining greater depths, demands for greater extraction rates and new mining methods and systems have created additional problems. Both government and industry have responded to these support difficulties by developing more efficient roof hazard monitoring/detection systems and increasing the use of resin in rock anchors, sponsoring research in prereinforcing and backfilling technology, improving shotcrete systems and by increasing the use of computers in the design simulation control of ground support.

Some type of roof support generally is required to prevent roof collapse and walls falling that would endanger life and equipment thereby interrupting the operations. The rock is sometimes strong enough to support itself, but man-made structures and devices usually are installed to insure stable roof and wall conditions.

The techniques and equipment used in ground control and in tunneling must be designed for the strata that is normally adjacent to a petroleum reservoir. If a petroleum mine, for example, is to have a life of 20 years, the ground support must be well-designed and planned so that the mine will remain open and safe for that duration.

Ventilation will be extremely important in mining for petroleum. The development openings will be close to the hydrocarbon reservoir so extra care is necessary to prevent the accumulation of toxic, suffocating or explosive gases. In situ mining for petroleum discussed later will only expose limited rock surfaces in the development openings. If the hydrocarbon reservoir were mined as ore as in conventional mining, it would be much more dangerous due to toxic, suffocating or flammable gases. Prior to mining, the following parameters should be measured in order to estimate the inflow of the gases into the development opening: gas pressure; permeability of the rock; and the rate of water inflow and the content of hydrocarbons and toxic gases in the water. The geological conditions such as fault zones have to be mapped accurately and caution must be exercised prior to mining these particular areas to avoid the sudden inflow of large quantity of fluids.

Due to the natural geothermal gradient, a temperature of about 140°F will be reached within depths of 6,000 to 7,000 ft. For temperatures above 140°F, cooling of the mine through the ventilation air will become much more expensive. For shallow oil mining, temperature control probably can be achieved by using ordinary ventilation. At great depths, however, additional thermal control procedures will be necessary, using a refrigerator plant or cold service water.

The view that mining is a highly dangerous occupation is gradually on the decline. Government and industry research devoted to improving health and safety condition in mines has produced many ideas, devices and systems to reduce hazardous conditions which have produced immediate and long-term improvements on the overall health and safety of both underground and surface mining operations. Explosive gases and roof control are likely to be the most serious problems in oil mining.

PETROLEUM MINING TECHNOLOGY

Introduction

The petroleum mining methods described by Hutchins, Bond and Bass in their *Mining for Petroleum: Feasibility Study* resulted from combining the most current technologies of petroleum and mining engineering. Methods and techniques are given for mining conventional liquid hydrocarbon reservoirs and for heavy, viscous oil and tar sands. In general, petroleum technology is used for the extraction of the hydrocarbon from its reservoir and mining technology is used for close access to the reservoir. One of the techniques, however, is a mining engineering technique, but the crude production can be calculated using petroleum engineering theory.

Basically, two processes are described, one for conventional liquid hydrocarbon reservoirs and one for heavy oil and tar sands. Each process has several mining applications, depending upon the depth, configuration and other physical and chemical properties of the reservoir. Drip Drainage is the name given to the basic process for mining for petroleum from a conventional petroleum reservoir, and Open Surface Flotation (Flip-Flop) is the term for the process described for heavy oil and tar sands. Because conventional petroleum reservoirs contain the hydrocarbons compatible with existing refining, processing and consumer facilities, the gravity drainage process and its various applications will be discussed first.

No attempt was made in their study to design a mine or mine workings for any particular hydrocarbon reservoir or for any selected geographical area. The objectives of the study were to review U.S. Bureau of Mines Bulletin 351, *Mining Petroleum by Underground Methods,* George S. Rice (1932) and to develop technically and economically feasible concepts for mining petroleum.

The project team, after careful study of available technology and of the problems involved, decided not to concentrate on processes that involve removing reservoir material as ore and also not to emphasize surface mining methods.

Surface ore extraction mining methods have not been emphasized because surface mining feasibility and economics are straightforward and relatively simple after a location has been chosen. One of the applications discussed later begins as a surface mining technique, but it is an in situ process.

Petroleum Mining Targets

Nearly any liquid, viscous or solid hydrocarbon reserve is a potential petroleum mining target for the two processes developed by the project team. For liquid hydrocarbons, the logical initial targets are the shallower depleted or nearly depleted oil fields. For initial demonstration and development, the choice of a shallow reserve will cut shaft costs and would significantly cut the initial development time. For heavy oil and tar sands, any surface exposure or near surface occurrence of adequate size will be amenable to the surface process described in this report. For heavy oil and tar sands reserves at depth, the overburden ratio must conform to certain mining constraints. For deeper heavy oil deposits, cavern jetting or other technology must be further developed.

Conventional Liquid Petroleum: The major constraints in mining for liquid petroleum by Drip Drainage will be the physical competency and thickness of the rock strata immediately above and below the reservoir and the temperature of the mine workings.

Temperatures of mine workings are a function not only of depth, but of geothermal gradient for the site specific geographical area of the U.S. Geothermal gradients (the rise in temperature with depth) vary markedly over the U.S. In areas of substantial oil reserves in the lower 48, the "hot" areas are the Texas-Louisiana Coast and the southwestern California Los Angeles-San Joaquin Valley deposits. Mining (at depth, not surface mining) in these areas will produce high mine operating temperatures in the range of 140°F at depths of around 4,000 ft. There are "cool" areas containing large petroleum deposits, such as those located in the Permian Basin of West Texas-New Mexico. A mine in some of these locations can be operated at around 7,000 ft before a 140°F temperature limitation is reached. Because mine operating temperatures and the human working environment are a limitation, the geothermal gradient in "cool" areas will make available to conventional mining sizable reserves at greater depths. Figure 5.5 is an isotherm map of the U.S. which generally indicates the "cool" and "hot" areas and represents the temperature limitations to mining before excessive ventilation costs (above 140°F) are encountered.

Figure 5.5: Isotherm Map of the U.S.

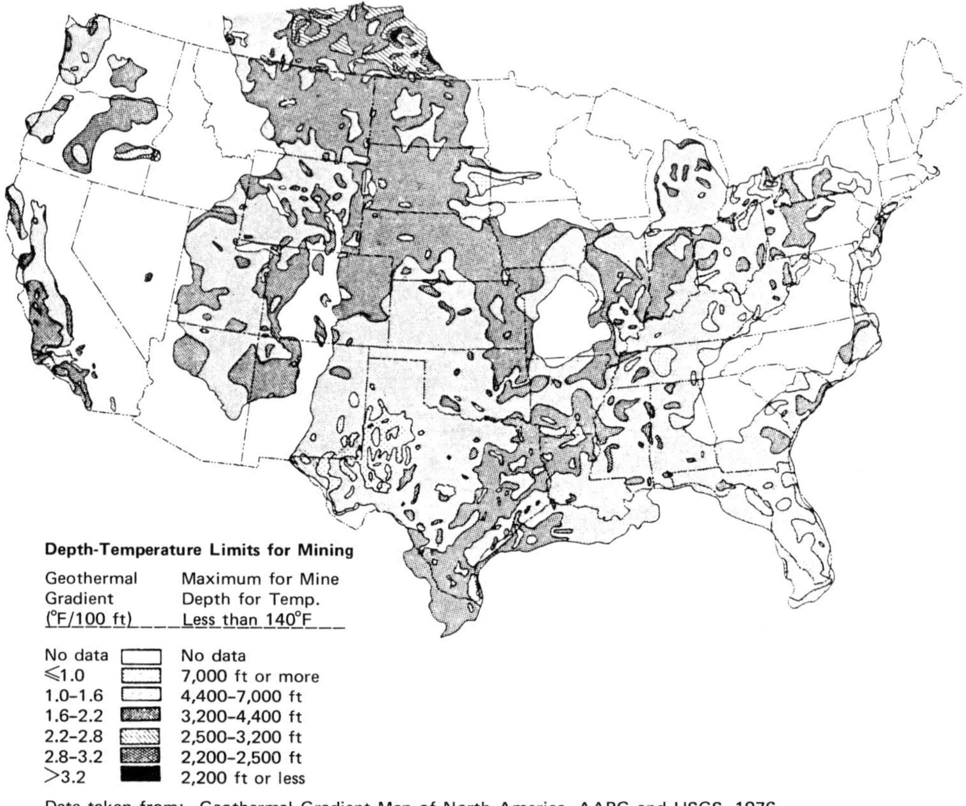

Source: BuMines OFR 56–79

Reservoir targets below the 140°F isotherm are available to mining for petroleum if the value of the petroleum reserve is sufficient to overcome the costs associated with added ventilation and larger shafts and tunnels required to accommodate higher air flows. The geothermal gradient, therefore, is not an absolute or technical barrier to mining at lower depths, but is a limitation because of economics. As greater mining depths are projected, earth pressure increases require smaller tunnel diameters to accommodate the overburden stresses. Since the two factors, added ventilation requiring larger tunnels and overburden pressure requiring smaller tunnels work against each other, there is a practical depth limit to mining for petroleum.

Table 5.1: Largest U.S. Oilfields

. . . . .Thousand Barrels. . . . .

Field	Location	Estimated Cumulative Production	Remaining Proved Reserves	Original Oil-in-Place	Unrecoverable Reserves by Present Technology	Unrecoverable Oil-in-Place (%)
Prudhoe Bay	AK	3,170	9,396,830	23,800,000	14,400,000	60.5
Spraberry	WTNM	431,091	103,537	8,598,550	8,063,922	93.8
Wilmington	CA (LA)	1,807,458	677,542	9,693,100	7,208,100	74.4
Huntington	CA (LA)	–	–	6,000,000	4,940,000	82.3
Panhandle*	WTNM	–	–	6,060,000	4,651,000	76.7
Midway Sunset**	CA (SJ)	–	–	6,185,000	4,554,500	73.6
Coalinga**	CA (SJ)	–	–	4,505,000	3,801,000	84.4
Wasson*	WTNM	–	–	4,747,792	3,230,327	68.0
Kern River**	CA (SJ)	–	–	4,062,000	3,082,000	75.9
Ventura	CA (Coast)	–	–	3,500,000	2,600,000	74.3
McElroy*	WTNM	–	–	2,544,015	2,033,900	79.9
Yates	WTNM	–	–	4,000,000	2,000,000	50.0
Elk Hills	CA (SJ)	–	–	2,815,321	1,793,141	63.7
Sho-Vel-Tum	OK	–	–	3,100,000	1,722,558	55.6
Eunice*	WTNM	–	–	2,000,260	1,621,445	81.6
Goldsmith*	WTNM	–	–	1,823,130	1,208,923	66.3
Belridge South**	CA (SJ)	–	–	1,400,000	1,905,000	78.2
Vacuum*	WTNM	–	–	1,425,426	1,069,808	75.1
San Ardo	CA (Coast)	–	–	1,450,000	1,050,000	72.4
H. Glasscock*	WTNM	–	–	1,421,100	1,035,839	72.9
Rangley	CO	–	–	1,681,575	955,577	56.8
Bay Marchand	LA (S)	–	–	1,569,073	952,851	61.0
Fullerton*	WTNM	–	–	1,322,570	948,359	71.7
Sooner Trend	OK	–	–	1,202,778	922,448	76.7
Slaughter*	WTNM	–	–	1,810,000	864,000	47.7
Salt Creek	WY	–	–	1,478,980	846,885	57.3
Kelly-Snyder*	WTNM	–	–	2,160,886	827,814	38.3
McArthur River	AK	–	–	1,336,000	825,001	61.8
Cowden North*	WTNM	–	–	1,208,750	762,650	63.1
McKittrick**	CA (SJ)	–	–	992,900	731,275	73.7
Golden Trend	OK	–	–	1,170,000	715,345	61.1
Levelland*	WTNM	–	–	1,012,000	659,200	65.1
Seminole*	WTNM	–	–	1,008,800	597,200	59.2
Healdton	OK	–	–	940,000	594,491	63.2
Oregon Basin	WY	–	–	905,600	572,429	63.2
Cailou Island	LA (S)	–	–	1,218,993	539,973	44.3
Hobbs*	WTNM	–	–	815,024	537,001	65.9
Elk Basin*	WY–MT	–	–	1,058,705	535,515	50.6
Dos Cuadras*	CA (Coast)	–	–	750,000	535,000	71.3
Foster*	WTNM	–	–	755,000	521,750	69.1

*West Texas-New Mexico "cool" area oil fields available to gravity drainage mining for petroleum.

**Shallow California heavy oil deposits available for Flip-Flop mining for petroleum.

Source: BuMines OFR 56-79

Each case will be calculated separately by the petroleum and mining industries as the depth limit can be modified by refrigeration, tunnel bracing, or other costly mechanical means, if the economic petroleum target is sufficiently attractive. As the free market price of oil rises, additional reserves will be economically brought into the viable range of mining for petroleum.

Mining for petroleum may be thought of as a tertiary petroleum recovery method, but could be used as a primary or secondary recovery process. As a tertiary technique, the mining target is the approximate two-thirds of oil in place which is unrecoverable by conventional primary and secondary methods. Underground mining for petroleum can be used upon primary depletion, replacing conventional secondary methods, thereby adding its 10 to 15% of oil in place to the mining target. In certain situations, mining for petroleum by underground methods could be economically effective, bypassing all conventional production methods including primary production development by surface wells. There are a great number of large depleted fields in the U.S.

The magnitude of the potential of mining for petroleum is partially indicated by Table 5.1, which is a listing of the largest U.S. oil fields in descending order of the amount of unrecoverable oil remaining in place (2). Half of these 40 largest fields can be categorized into two very specific types and locations of reservoirs: West Texas-New Mexico Permian Basin San Andres (about 4,000 ft) geothermally "cool" conventional oil and California San Joaquin Valley shallow viscous oil. Therefore, this chapter addresses two methods of recovery by mining for petroleum: one method for each of these large reserve categories. For the remaining 20 fields listed in Table 5.1, at least one of the two methods described herein is a potential recovery method. Therefore, it is felt that mining for petroleum has a very broad potential to recover a large percentage of the 300 billion barrels unrecoverable oil in the U.S.

The estimates within this report are exclusive of presently undiscovered conventional oil which, according to U.S. Geological Survey Circular 725, approximates 80 to 100 billion barrels of recoverable original oil in place. Since current technology recovers only 32%, a remaining undiscovered 160 to 200 billion barrels in the future could be available to recovery by methods described herein. Most of this undiscovered oil, however, is at sufficient depth that reservoir access is not possible with current mining technology or it is offshore where mining would be very costly.

Heavy Oil and Tar Sands: A significant portion of U.S. hydrocarbon reserves are locked in heavy oil and tar sands. While not as great as liquid petroleum, heavy oil and tar sands reserves are still several times the 30 billion barrels remaining producible reserves attributed to the U.S. There are few profitable commercial ventures in the U.S. for extracting hydrocarbons, although much research is being conducted.

Mining for Conventional Liquid Petroleum

There are many oil reservoirs within the continental U.S. which have been pressure-depleted, depleted by gas-driven secondary recovery or waterflooded. The residual oil in place varies with the different reservoirs and the only ones to which mining presently would apply probably would be those with residual oil contents in excess of 50%. The oil remaining in most of these reservoirs will have a viscosity of probably less than 10 cp and a solution gas content of less than 500 ft^3/bbl. The remaining reservoir pressure would probably be less than 2,000 psi and the depth would be less than 4,000 ft.

The mining/production processes will be one of two types: (a) to recover all the fluid within the reservoir, that is, gas, oil and water; and (b) to recover only the oil and whatever gas is in solution in that oil. The selection of the mining/production process will be a function of the reservoir, the condition in which it has been abandoned and the auxiliary types of energies which might be involved.

For purposes of this discussion, natural reservoir energies may be separated into (1) forces acting on the reservoir, and (2) forces acting within the reservoir. The forces acting on a conventional reservoir can be measured in terms of pressure (psi). These forces include (A) the hydraulic pressure due to the weight of a column of water (water-drive); (B) the weight of the overburden (overburden pressure); (C) entrapped gas cap pressures and entrained gas (gas drive); and (D) other less important pressure sources such as those due to temperature change.

The forces acting within the reservoir include those of capillarity and bouyancy forces due to gravity. The effects of these forces cannot be measured in terms of hydraulic reservoir pressure. The effect of capillarity is a primary force in the migration and accumulation of crude oil and bouyancy causes the separation of gas, oil and water.

In primary crude oil production, wells are completed so that only the forces acting upon the reservoir that are measurable in terms of pressure are used. In many cases, the forces acting on the reservoir are great enough that if a well is drilled into the reservoir oil will flow at the surface. In other cases, the oil must be pumped from the well by mechanical means. As oil is produced from a reservoir, the reservoir pressure declines, resulting in a decline of oil production. At some point, the well becomes marginally economic and secondary production techniques are initiated.

The Gravity Drainage Process: In the process of gravity drainage extraction of liquid crude oil, the wells will be so completed that only the forces acting within the reservoir are used. The forces acting on the reservoir are left intact, perhaps maintained or increased. A large number of closely spaced wells can be drilled into a reservoir from an underlying tunnel more economically than the same number of wells from the surface. In addition, only one pumping system is required in underground drainage, whereas at the surface each well must have a pumping system. The objective of using a large number of wells is to produce each well slowly so that the gas-oil and water-oil interfaces move toward each other efficiently. By maintaining the reservoir pressures because of forces acting on the reservoir, it is then assured that the oil production is being provided by the internal forces due to gravity (the bouyancy effect) and capillarity.

Although the production rate of each well will be low in gravity drainage, and the oil interfaces will move very slowly, production will be over a large area, thereby providing an economical volume of oil extraction. After conventional production has ceased, up to 90% of the oil remaining in place may be recoverable by gravity drainage extraction.

Gravity drainage of petroleum from an underlying mine tunnel is illustrated as

Figure 5.6. In the illustration, each well penetrating the reservoir has its own pressure gauge and valve and is hooked to a common header pipe drainage to a single sump. Only one pumping system for all the wells is necessary to transport the crude oil to the surface. A computer system probably would be used to control individual well flows and pressures.

Figure 5.6: Gravity Drainage of Oil from Tunnel Underlying a Reservoir

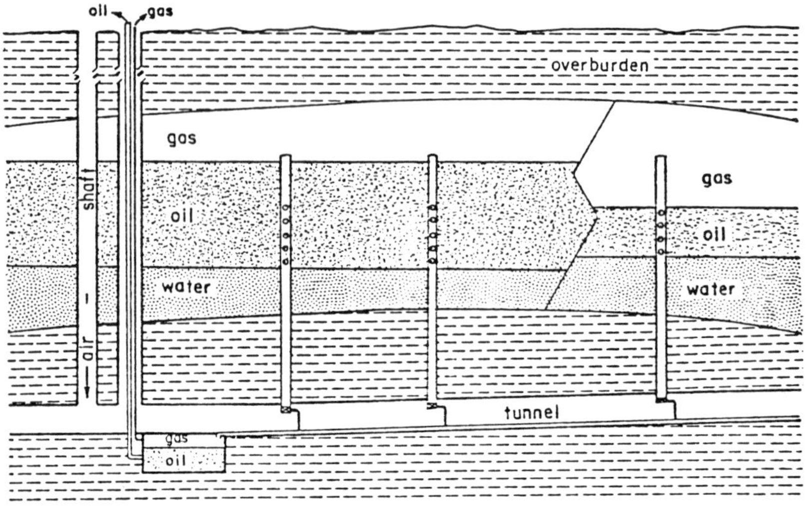

Source: BuMines OFR 56-79

Fluid Distribution of Waterflooded Reservoirs — There are many fluid distribution systems possible in an oil reservoir that has undergone waterflooding, perhaps even in a single reservoir. Some of the possible fluid distributions are:

(1)   Water and low residual oil saturation occupying the lower parts of the formation and original (high) oil saturation in the upper parts;

(2)   100% water in some part of the lower section of the formation with water and oil in capillary equilibrium in the upper parts of the formation;

(3)   Some of the areal region having a dispersed oil and water saturation throughout the vertical height and other parts of the areal region with original oil saturation throughout the vertical height; and

(4)   A dispersed oil and water saturation throughout the vertical height and areal region.

Fluid distribution (1) would be obtained using Dietz's method of calculating oil displacement by water. Fluid distribution (2) is the result of fluid redistribution by gravitational separation after the cessation of water-injection and is time-dependent. The resulting fluid locations are essentially the same as in distribution (1). Fluid distribution (3) would be calculated using the Frontal Advance theory in conjunction with areal coverage prediction procedures. Fluid distribution (4) results from the application of the Frontal Advance method only.

Fluid Distribution of Gas-Driven or Solution Gas Driven Reservoirs — There are two possible fluid distributions in a gas-driven or solution gas driven reservoir that are predictable under petroleum theory. These distributions are as follows:

(1) A high oil saturation in the structurally lower parts of the reservoir and a very high gas saturation in the upper part of the structure; and

(2) A uniform saturation of oil and gas throughout the entire reservoir.

In the first distribution, the oil and gas are distributed in accordance with gravity and capillary forces. Generally, this is the same as finding a gas reservoir with a thin oil column. Under the conditions of the second fluid distribution, the reservoir permeability with respect to the oil is so low that the reservoir essentially cannot be produced. This type reservoir could be produced by mining the rock as ore.

Fluid Distribution Flow Characteristics of Waterflooded Reservoirs — In the case of fluid distributions (1) and (2), where the upper section of the formation has a very high oil saturation, well completions would be designed to produce only from the region of high oil saturation. In this type completion, if one were to take whatever fluids could be produced utilizing all the energies, the equations for flow would be given below:

$$(1) \qquad \text{Radial flow:} \quad Q_r = \frac{Kh_o}{\mu_o} \times \frac{(P_r - P_w)}{\ln(0.472 r_e / r_w)}$$

$$(2) \qquad \text{Linear or slot flow:} \quad Q_l = \frac{\overline{Kh_o}W}{\mu_o} \times \frac{(P_l - P_w)}{L}$$

where:

Q_r and Q_l are volumetric flow rates for radial and linear or slot flow, respectively (ft^3/hr);

K and $\overline{K}$ are proportionality constants including permeability to oil;

h_o is height of oil saturated formation (ft);

P_r and P_l are formation pressures midway between producing surfaces (psi);

P_w is back pressure at producing surface (psi);

r_e and L are half the distance between producing wells or surfaces (ft);

r_w is radius of a producing well (ft);

W is length of a linear producing surface (ft); and

μ_o is oil viscosity (cp).

By utilizing all the pressure within a system for production purposes one would expect to produce oil, water and gas in the process. If the reservoir were under an active water-drive it would be anticipated that relatively large volumes of water would be handled. In addition, the reservoir would eventually be pressure-depleted with the possibility of limited subsidence. Regardless of whether or not the total energy within the system is used for production purposes, the production process would eventually reduce itself to the gravity drainage flow system which will be discussed later.

In the case of fluid distribution (3), it would be required to complete all producing surfaces within the region of the highest oil saturation. In this case, the flow equations would be the same as for distributions (1) and (2), except that the height of the oil zone (h_o) would be the total thickness of the formation instead of a partial thickness wherein the oil saturation was at a maximum.

For fluid distribution (4), little choice would be had of selectively completing wells to prevent the production of the water within the formation. In fact, if fluid distribution (4) does exist, such a reservoir would represent a poor prospect for in situ petroleum mining processes.

Fluid Distribution Flow Characteristics of Gas Driven or Solution Gas Driven Reservoirs — Based on the two gas-oil distributions previously discussed, the first fluid distribution is the most significant. With this distribution, the mining development would be in regions containing very high oil saturation. If well completions in this region are properly located, the reservoir can be depleted by utilizing both the gas energy at first and gravity drainage production in later stages. If not completed properly, the gas energy will be rapidly (and inefficiently) depleted and the majority of the oil will then have to be recovered using only gravity drainage.

The equations describing the fluid flow for gas-oil systems are identical to those for a water-oil system wherein fluid distributions (1) and (2) for the water-oil system are utilized. Both the total energy equation and the gravity drainage equation are the same.

If gas-oil distribution (2) exists the equations for flow would be as follows:

$$(3) \qquad \text{Total energy, radial: } Q_r = \frac{K k_{ro} h_t}{\mu_o} \times \frac{(P_r - P_w)}{\ln(0.472 r_e r_w)}$$

$$(4) \qquad \text{Total energy, linear: } Q_l = \frac{K k_{ro} h_t}{\mu_o} \times \frac{(P_l - P_w)}{L}$$

where:

Q_r and Q_l are volumetric flow rates for radial and linear or slot flow, respectively (ft^3/hr);

k_{ro} is oil fraction of the total permeability; and

h_t is total formation thickness (ft).

$$(5) \qquad \text{Gravity only, linear: } Q_o = \frac{1.1407 \times 10^{-4}}{2} \times \frac{k_o W}{\mu_o} \times \frac{\gamma_o h^2 (y^2 - x^2)}{L}$$

where:

Q_o is volumetric flow rate gravity drainage (ft³/hr);

k_o is oil permeability (md);

W is length of slot (ft);

γ_o is specific gravity oil (fraction);

h is saturated thickness (ft);

y is portion of total height that contributes to gravity flow (fraction);

x is portion of height (h) through which fluid flows into slot (fraction); and

L is total length of flow (ft).

(6)   Gravity only, radial: $Q = 7.1659 \times 10^{-4} \times \dfrac{k_o}{\mu_o} \times \dfrac{h^2_o\, \gamma_o (y^2 - x^2)}{\ln(0.472 r_e r_w)}$

where:

r_e is radius of outer drainage boundary (ft); and

r_w is radius of drain hole (ft).

Gravity Redistribution of Reservoir Fluids — The fluids in an untapped petroleum reservoir under equilibrium conditions are separated by density according to natural laws. If the reservoir is disturbed, such as by primary and secondary production, the fluids, by the same natural laws, will redistribute themselves in time in an attempt to reach to and then maintain equilibrium. The equations below represent this redistribution process.

(7) $S_w = \dfrac{1}{(ah + 1)^{1/2}}$

(8) $ah_w S_{wm} + 2(ah_w + 1)^{1/2} = aS_{wi}h_t + 2(ah_t + 1)^{1/2}$

(9) $S_w = \dfrac{1}{(ah + 1)}$

(10) $ah_w S_{wm} - \ln(ah_w + 1) = aS_{wi}h_t - \ln(ah_t + 1)$

(11) $S_w = \dfrac{1}{(ah + 1)^2}$

(12) $ah_w S_{wm} + \dfrac{1}{(ah_w + 1)} = aS_{wi}h_t = \dfrac{1}{(ah_t + 1)}$

(13) $S_w = \dfrac{1}{(ah)^2 + 1}$

(14) $ah_w S_{wm} - \tan^{-1}(ah_w) = aS_{wi}h_t - \tan^{-1}(ah_t)$

where:

S_w is water-saturation (fraction);

S_{wi} is initial average water saturation over the total liquid saturated thickness (fraction);

S_{wm} is maximum water-saturation obtainable by capillary redistribution, usually 100% as a first try (fraction);

a is the difference in liquid specific gravities, $(\gamma_w - \gamma_o) \times 0.433$, (fraction);

h_w is the height to which S_{wi} applies (ft); and

h_t is the total height of liquid saturated zone (ft).

Gravity Flow into Drainage Hole (Well) or Slot — It is hoped that fluid distribution (2) (waterflooded reservoir) would be found most often. For this distribution, approximate flow equations can be derived for gravity flow of oil into a drainage hole or slot. Based on a fluid distribution as shown in Figure 5.7, the flow into a slot can be represented by equation (15).

Figure 5.7: Fluid Drainage Configuration

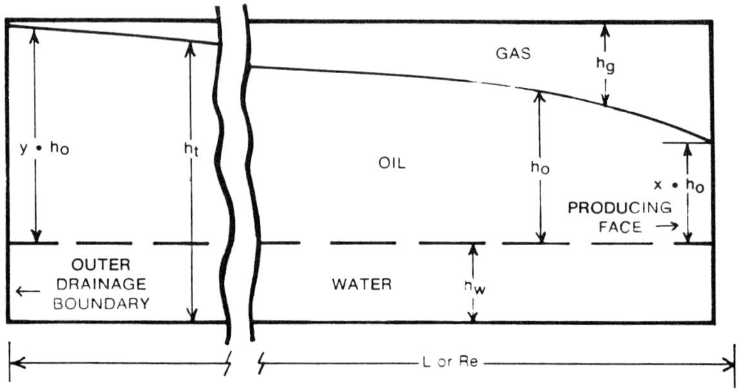

Source: BuMines OFR 56-79

$$(15) \qquad \frac{Q_o}{W} = \frac{1.1407 \times 10^{-4}}{2} \times \frac{k_o}{\mu_o} \times \frac{\gamma_o h^2 (y^2 - x^2)}{L}$$

where:

W is the length of the slot (ft);

Q_o is the oil flow rate (ft³/hr);

k_o is the oil permeability (md);

h is the height of the oil column (ft);

x is the height of (h) through which the fluid flows into the slot (fraction);

y is the total height which contributes to gravity flow (fraction);

y_o is the specific gravity of the oil (fraction);

μ_o is the oil viscosity (cp); and

L is the total flow length (ft).

The variables k_o, μ_o, γ_o and h are controlled by the reservoir selected for development. In a waterflooded reservoir very little can be done to alter their in situ value. The variable L is controlled by the frequency or spacing of the slots or wells placed within the reservoir and, hence, is a function of the mine design. The variables y and x are controlled by the manner of completion in the slot such that the reservoir can be controlled so as to produce at the maximum drainage rate.

Based on a controllable factor, equation (15) can be rearranged such that

$$(16) \qquad \frac{Q_o \mu_o \times 8{,}776.55}{W k_o \gamma_o} = \frac{(y^2 - x^2)h^2}{L}$$

From equation (16) it is seen that the rate of production for a given reservoir is inversely proportional to the distance between the slots. This inverse proportionality creates an economic relationship between the time cost of money and capital cost of construction. The greater the capital cost, the shorter the time required to recover the petroleum.

Again, using fluid distribution (2), the flow into a radial borehole would be given by equation (17).

$$(17) \qquad Q = 7.1659 \times 10^{-4} \times \frac{k_o h^2 \gamma_o (y^2 - x^2)}{\mu_o \ln(0.472 r_e r_w)}$$

where:

    r_e is the outer drainage boundary (ft);

    r_w is the radius of the drain hole (ft);

    and all other terms are as previously defined.

Equation (17) can be rearranged to illustrate the effect of in place factors and design criteria on a particular reservoir as shown in equation (18).

$$(18) \qquad \frac{1{,}395.5 Q \mu_o}{k_o \gamma_o} = \frac{h^2(y^2 - x^2)}{\ln(0.472 r_e r_w)}$$

Petroleum Mine Design Considerations — Production of a reservoir by gravity drainage may be done by means of wells or slots and also by exposing and fracturing selected portions of the underside of the reservoir. In order to choose the most economical mining method and to properly design the mine workings, a predevelopment phase will be necessary to collect and evaluate all relevant information. Here, mining and petroleum are interrelated and information collection and evaluation techniques must be taken from both disciplines.

If a reservoir is chosen that has been depleted by conventional primary and secondary recovery methods, much of the petroleum engineering will already have been done. For example, the reservoir pressure, temperature, oil and rock characteristics and geologic configuration will have been determined previously as a result of conventional oil exploration and production. Much more information, however, will be necessary to design a mine. Additional current information on the reservoir will be required as well as more specific information relative to the geological strata both overlying and underlying the reservoir.

The geology of the overburden, the reservoir and the underlying strata will have to be mapped in detail and must include stratigraphy, structure and lithology. This information should be obtained from drill hole data and correlated seismic profiles. This information will be necessary to choose the proper shaft location. Detailed data on shaft construction, tunneling, pressure control, casing and other phases of petroleum mining are given in the original report.

Well Production from Tunnels Beneath the Reservoir: Completing wells from a level of drifts beneath the reservoir is the most economical application of gravity drainage. In this case, only one pumping system is required rather than the need for installing a pump in each well as is necessary in wells drilled from the surface. Figure 5.6 illustrates crude oil production by wells from an underlying tunnel.

Well Production from Tunnels Driven Above the Reservoir: If the rock underlying the reservoir is not competent enough to support mine workings, tunnels can be driven above the reservoir. In this application of gravity drainage, a pump must be installed in each well, but much smaller pumps than would be needed from the ground surface, since the lift height is minimal, equaling approximately 100 ft as opposed to several thousand feet.

Well Production from Tunnels Driven Alongside the Reservoir: Horizontal wells drilled into a petroleum reservoir from tunnels driven alongside is another application for gravity drainage. If the horizontal wells are drilled on a slight incline, the oil produced would drain out by gravity, eliminating the need for a pump in each well. This application could be more economical than drilling from overlying tunnels if the areal extent of the reservoir can be reached by the horizontal drilling.

Well Production from Tunneling Combinations: Applying gravity drainage for petroleum extraction using tunnels at various levels above, below and alongside the reservoir may have its advantages. For example, Figure 5.8 illustrates the use of tunnels both overlying and underlying the reservoir. The wells drilled from the lower tunnel are used for crude oil production, while the wells drilled from the upper tunnel can be used for reservoir pressure maintenance, sweeping fluid injection or steam injection.

Figure 5.8: Reservoir Maintenance and Well Production from Tunnels Above and Below the Reservoir

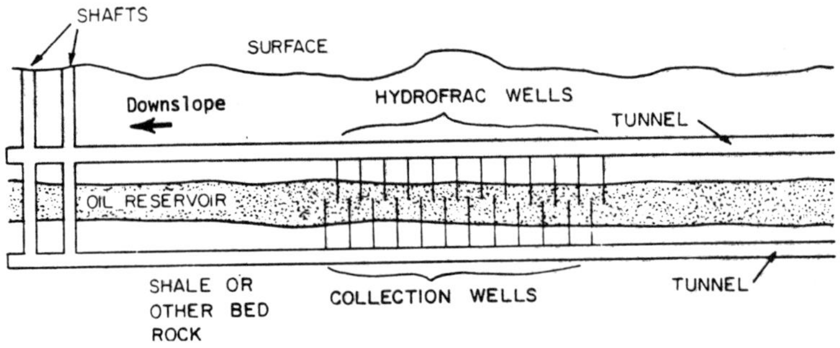

Source: BuMines OFR 56-79

Oil Production by Means Other Than Wells — To increase the fractured permeability of the reservoir, three methods for substantially fracturing the reservoir and allowing fluids to drain through the tunnels to a common collecting point are fracture caving, modified fracture caving and collapsed slot (tunnel). Applicability is limited to very unique situations. Generally, massively fracturing an oil formation by caving or collapsing will destroy efficient formation sweep mechanisms to recover oil and very little production will result. Before seriously considering caving or collapsing petroleum formations, design engineers would be well-advised to counsel specifically with petroleum engineering technologists about the anticipated recovery results.

Fracture Caving: Fracture caving is an underground mining technique to extensively fracture an oil formation. This increases the overall permeability of the producing zone. The oil is collected in an underlying tunnel where it drains to a collection for transport to the surface. The caving must be done on retreat, which means that fracture caving is initiated at the most remote point of the tunnel workings, sequentially caving on retreat toward the surface openings.

In conventional petroleum production by means of wells, techniques for fracturing the producing zone in the vicinity of the well bore have been used for years. One of two techniques is commonly used. The first method is to lower a measured quantity of explosives down the well bore to a selected point within the producing zone and detonate it. The second method, which is more commonly used, is called hydrofracing. Hydrofracing is a method of fracturing the producing formation by injecting fluids into the production zone under high pressure. In this method, a propping agent such as sand usually is injected to hold the fractures open.

The presumed advantage of fracture caving is that most or all of the producing formation can be fractured rather than only a limited circumference around a well bore. Also, only one fluid pumping system is required rather than having to install a pump in each well.

While the mining techniques used to induce the required degree of fracturing will vary with different reservoirs and reservoir characteristics, the intent in fracture caving remains the same, that is, to increase the permeability of the reservoir and to extract the crude oil from underneath its reservoir.

Modified Fracture Caving: For a reservoir moderately difficult to fracture, explosives probably will be the primary means of inducing fracture and rock breakage. It may prove desirable to use hydrofracing techniques in conjunction with the explosives. The tunnels and workings would be on one primary level underlying the reservoir. A blind hole raise drill would be used to provide access and drilling stations somewhat closer to the reservoir. From each drilling station, fans of holes are drilled up into the reservoir and loaded with explosives. The space or chamber for each drilling station serves as an expansion chamber for the expansion of the fractured rock to create more permeability within the reservoir. After the area has been evacuated and sealed, the explosives are detonated. Limited crude oil then drains from the reservoir into a pipeline in the main tunnel and flows from there to a pumping station for transport to the surface.

Collapsed Slot (Tunnel): After tunnels have been completed beneath the reservoir, the formation could be collapsed into the drift so that the drift becomes the production system. The flow equations for a collapsed slot-producing system would be those indicated as linear, where all the pressure within the system is utilized for production.

The collapsed slot would not permit the selective production of oil. All the reservoir fluids would be produced. This type completion would deplete all the pressure within the formation and, as a result, could subject the area to some degree of subsidence. Once the formation is collapsed, the operation cannot be altered in the future. Access to the drifts would no longer be possible and if by any chance plugging of the drift or flow system occurred, the entire operation within that drift would have to be abandoned.

Conclusions on Gravity Drainage Process — The concept of mining for petroleum in the final analysis must be economic. The process must be competitive with coal, in fill drilling, tertiary recovery of petroleum and alternate (synthetic) sources of liquid energy. The major cost, as would be expected, is the mining program with the cost of drill holes second. Economic analyses detailed by Hutchins et al in the original report indicate gravity drainage is competitive with other alternate energy sources.

As in all petroleum projects, the viability depends upon successfully defining two independent constituent parts, the first being performance of the mechanical factors and the second being the performance of the flow theory involved. In Drip Drainage, the study team concludes that the flow mechanisms are well proven. There are no disqualifying mechanical (mining or equipment design) limitations to oil mining. There is the highest assurance level probability that ultimate recoveries of conventional petroleum can be raised from the current average of 32% to the 80 to 90% range, which represents 150 to 200 billion barrels (5 to 7 times current U.S. reserves) utilizing controlled gravity drainage theory.

Mining of Heavy Oil and Tar Sands

Large quantities of U.S. petroleum reserves are located close to the surface in what is commonly referred to as tar sands or heavy, viscous oil sands. This petroleum is produced, if at all, through conventionally completed oil wells and stimulated with steam. In the case of the Athabasca Tar Sands of Canada, the solid containing the hydrocarbon is being mined and processed at the surface. Recovery efficiencies of these deposits are very low and/or costs of recovery are very high.

The major difficulties in recovering this heavy hydrocarbon through conventional wells are: (a) the formation generally is unconsolidated sand and, hence, the solid particles tend to flow toward the well with the hydrocarbon; (b) the heating material has to be injected into the formation through a well which causes a difficulty of proper placement to gain the greatest efficiency from the injected hot substance; and (c) a large portion of the heating material is produced back with the hydrocarbon, thereby wasting a large amount of the heat.

The customary techniques of completing these oil wells require that special sand control methods (gravel pack) be used to prevent the influx of solids into the producing well. These methods reduce the producing capabilities of the wells. It is the purpose of this section to describe an in situ mining method which is a combination of mining technology and petroleum production technology as a possible solution for the recovery of hydrocarbons from these types of formations.

The Flip-Flop Process: For hydrocarbon formations which are close to the surface, either strip mining of the overburden or trenching can be used to uncover the surface of the heavy oil or tar sand. The formation itself should not be mined. After the surface is exposed, special dams for containing both oil and water in ponds are installed on the surface of the formation. Then, the following oil recovery procedure is applied: (1) The exposed oil-containing surface is covered with hot brine to a height of approximately 3 ft. This brine should contain a surfactant, both anionic and cationic, such as Adafoam, 3M's X-35 or X-37, or any other surface tension reducer which will cause the sand grains to become water-wet.

(2) Steam pipes on the surface of the formation are placed so that the brine soak can be maintained at its maximum temperature. These pipes will also be used to add additional water to maintain an approximate 3 ft height.

(3) The hot brine will conduct its heat to the oil, reducing the oil viscosity and density such that a gravitational head differential will develop, causing the brine to flow down into the formation and force the oil up and out. This interchange of position of brine and oil (Flip-Flop) will be accelerated by a surfactant agent within the brine by creating a differential capillary pressure force between the two liquids. There are thus two forces acting: gravitational and capillary. If the right agent is selected, the capillary force will represent the major component of the two forces for the removal of the oil.

(4) Water (or brine) is continually added to the surface of the sand while oil is removed from the top of the restricting (or skimming) vessels until all the oil has been removed from the top 6 to 10 ft of the formation. This treatment is necessary to stabilize the surface so that oil may flow through this region without moving (or floating) the sand itself. The wetting characteristics of this water will create a minimum connate water-saturation (about 35%) which, in turn, will help to retain the sand grains and keep them from moving when the oil is produced through them to the surface.

(5) Drive pipes with perforated sections at the lead ends can now be driven to the bottom of the hydrocarbon-containing formation. They can be driven at angles underneath an area which has not been trenched or which has not had its overburden removed and can also be driven straight down beneath the section which has been exposed.

(6) A 60 to 70% quality steam with a surfactant is injected into the pipes and may be injected under considerable pressure.

(7) If the formation is thick (over 50 ft), the drive pipes will have to be driven to different depths to increase the amount of heat placed within the formation and also to reduce the distance over which the pressure gradient for moving the hydrocarbon must be applied. Low-quality steam may be injected at different levels alternately so that the oil is constantly being replaced by brine and the sand can stabilize itself during the process of production.

(8) Steam condenses to water and transmits heat into the formation and hydrocarbon. As the the steam condenses to water, it also becomes heavier than the

oil so that it tends to flow toward the bottom of the formation and the oil tends to rise to the top. In addition to this gravitational movement, the wetting agent within the water will create a capillary pressure which will also have a tendency to drive the oil to the exposed surface.

Flow Theory — The initial heat application for stabilizing the top 6 to 10 ft of the section depends on heat being transmitted by conduction as well as mass transport. The surfactant in the water will cause a set of adhesive forces to try to suck the water downward into the sand in the smallest pores. This action (imbition) creates a pressure gradient in the oil system such that the oil is forced to flow out the largest pores. Once water has penetrated into the smallest pores, additional force is applied because of the density difference in the oil and water. At this point, the flow is represented by the equation below:

$$(1) \qquad Q_o = \frac{KA}{\mu_o} \left[\frac{[(\rho_w - \rho_o)(h)]/144 + P_c}{h} \right]$$

where:

Q_o is the volumetric flow rate (ft^3/hr);

K is some proportional function, including permeability. (In the imbition process, this function has not been totally defined, but appears to decrease with time.)

μ_o is the viscosity which decreases as the oil is heated (cp);

ρ_w is the density of the hot water and increases as the water cools (lb/ft^3);

ρ_o is the density of the oil which decreases as the oil is heated (lb/ft^3);

h is the depth from which oil is flowing to the exposed surface (ft);

P_c is the capillary pressure caused by the surfactant (psi); and

A is the cross-sectional area of the flow surface (ft^2).

The process should continue until the top 6 to 10 ft have reached about a 35% water-saturation. The surface bed should then permit oil flow without sand movement.

The second phase of the recovery process is to inject heat in the form of steam into the system, thus generating a bottom water-drive with both imposed potential forces and gravity forcing the oil to the exposed surface. The flow equation for the recovery of the remaining oil is given by equation (2) below:

$$(2) \qquad Q_p = \frac{KA}{\mu_o} \left[\frac{P_i + P_c + [(\rho_w - \rho_o)(h)]/144}{h} \right]$$

where:

P_i is the pressure of steam injection (psi);

h is the depth from the exposed surface of injection (ft); and

the other terms are as previously defined.

The Flip-Flop recovery process as discussed herein may be applied when the surface of the formation safely can be exposed, i.e., the oil does not evolve free gas upon exposure to atmospheric pressure. The process also can be used, with modifications, when free gas will be produced.

Development for Production — It will be unnecessary to expose the total surface of the producing formation. Instead, depending upon the permeability of the formation and the ability to reduce the viscosity of the in-place hydrocarbon, there are several possibilities that exist. Four of these possibilities are illustrated in Figure 5.9

Figure 5.9: Development Modes for Surface Application of Flip-Flop

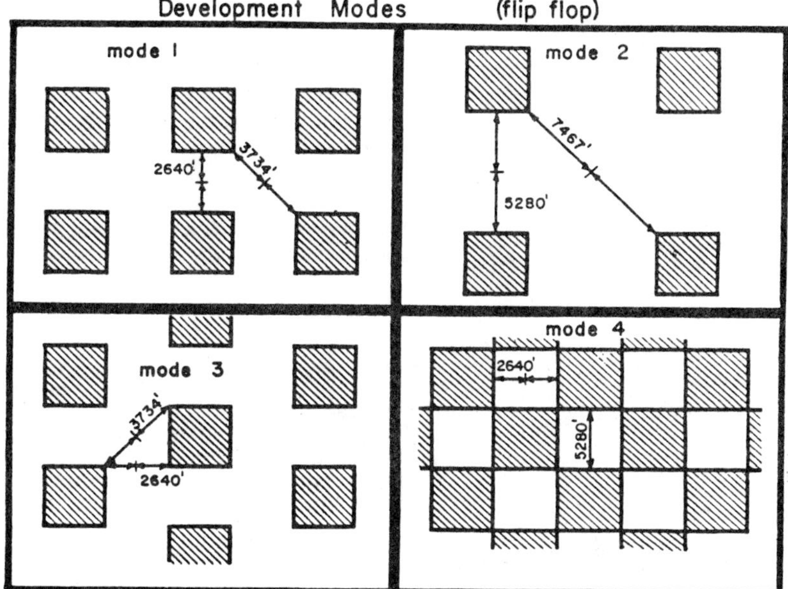

Source: BuMines OFR 56-79

The most probable initial development pattern will be the one illustrated as **mode 2.** This development pattern requires the exposure of one acre of the producing formation for every nine acres from which oil will be extracted. The maximum distance any hydrocarbon has to travel in order to be produced is 7,467 ft. But, over 80% of the oil that will be produced only has to travel a maximum of 5,280 ft. Another mode of operation for initial development is either **mode 1** or **3.** **Mode 1** would probably be preferable because it permits itself to be expanded to **mode 4** in case accelerated production is required. In the case of **mode 1,** the maximum distance any droplet of oil has to travel is 3,734 ft. But, 88% of the recoverable oil has to travel less than 2,640 ft.

These distances of travel and the development intensity are identical for **mode 3.** Both of these modes of operation require the exposure of one acre of the formation for every 4 acres to be produced. **Mode 4** is the most intensely developed of all systems.   100% of the recoverable oil has to travel less than 2,640 ft.

It is unlikely that **mode 4** will ever be used in a normally viscous oil sand. This method may be desirable in the case of a tar sand wherein the viscosity remains relatively high even after the heating process. In all likelihood, either **mode 2** or **mode 3** will be the probable pattern. There is also the possibility of expanding **mode 2** into a **mode 3**-type operation.

Surface Application of Flip-Flop — Once the surface of the formation is exposed, then the actual production operations will be initiated utilizing a Stage 1 and Stage 2. Stage 1 will be the period in which the top 6 to 10 ft of the exposed surface of the formation is stabilized such that as oil flows upward through the zone; the sand particles do not move.

The general operational procedure is illustrated in Figure 5.10, wherein a container is placed on the surface of the formation with steam pipes laid in contact with the surface formation and this container is filled with a hot brine solution. Steam is continually added to maintain the temperature of the water and the oil that floats to the surface is removed from an oil sump within the container and pumped to oil storage at the surface of the ground.

Figure 5.10: Stage 1—Stabilization Configuration
for Surface Application of Flip-Flop

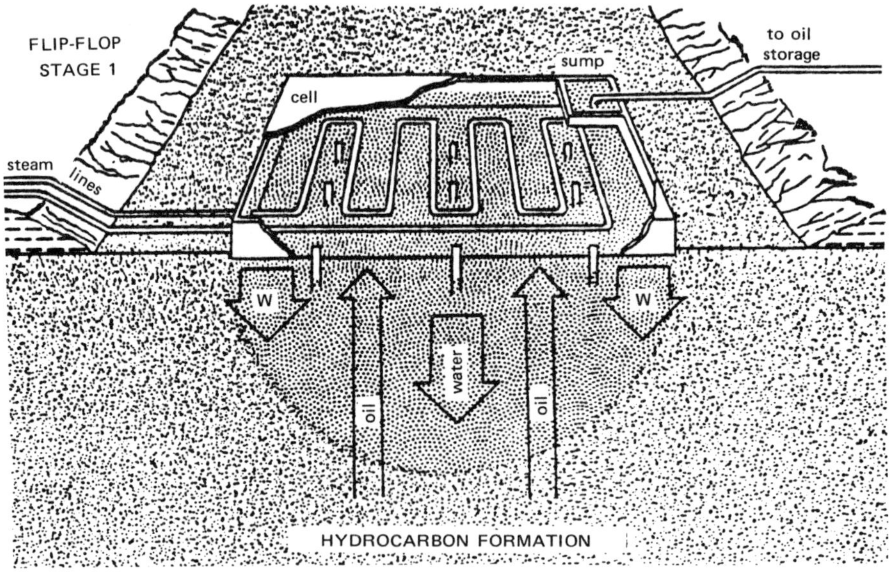

Source: BuMines OFR 56-79

In Stage 2, steam pipes are driven from the surface of the formation down into the formation either vertically or at various angles as shown in Figure 5.11.

Figure 5.11: Stage 2—Steam Injection and Oil Production Phase for Surface Application of the Flip-Flop Process

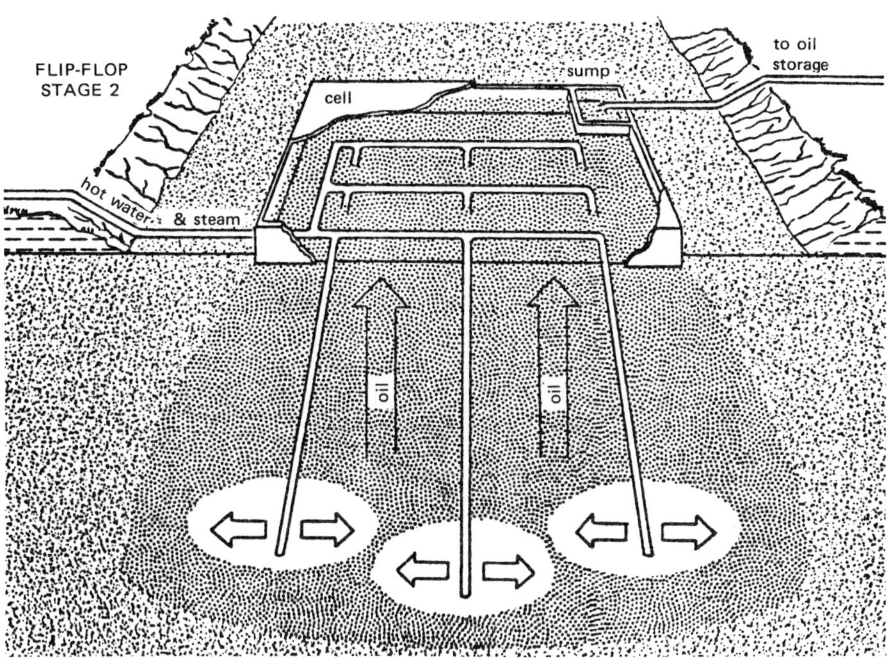

Source: BuMines OFR 56-79

The degree of the angles and the number of drive pipes will be a function of the thickness of the formation and the ability to inject fluid into the formation. It is noted that no attempt is made to maintain the exposed surface of the formation at an elevated temperature. This section of the formation has now been stabilized and only heated oil will be passing through to be collected in the sump.

Figure 5.12 shows in some detail the container that will be placed on top of the formation. This container can be made from fiber glass, utilizing drive pipe or legs to hold the frame in place.

The steam pipes could be built as an integral part of the container, but probably should be built separately. If the steam pipes were built separately, from the container body, they could be removed after the stabilization period and the same container could be used for both Stage 1 and Stage 2 of the operation.

The oil collector indicated in Figure 5.12 should be built as an integral part of the container. The container has not been shown with all the necessary piping hardware required to remove the oil from the system.

Figure 5.12: Sketch of the Surface Dam Used for the
Initial Phase of the Flip-Flop Process

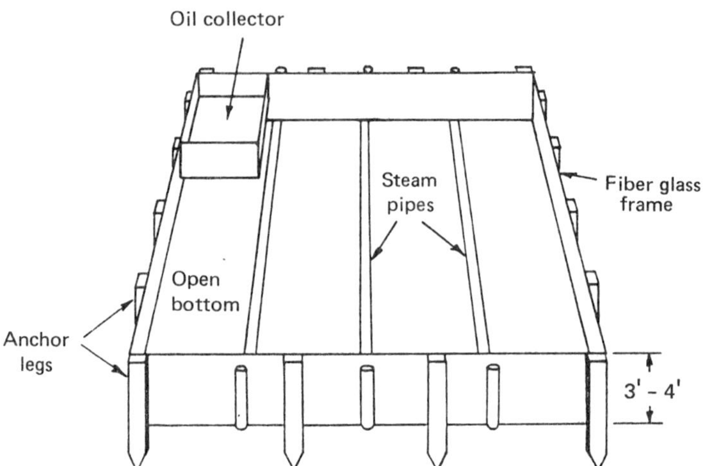

Source: BuMines OFR 56–79

Underground Application of Flip-Flop Process — In many cases, the surface of
the hydrocarbon-containing formation cannot be exposed to the atmosphere
because the fluid contained is under a natural pressure and would flood an excavation. A second reason, less dangerous but more difficult perhaps to control,
is that during the heating and stabilizing process, some hydrocarbons will yield
considerable volumes of flammable hydrocarbon gases, which will make the
operational control procedures more difficult. A third reason for not exposing
the reservoir is that the overburden thickness may be too great to be economical.

When the surface formation cannot be exposed, it becomes necessary to develop
the hydrocarbon-containing formation in a different fashion. Two techniques
have been devised which differ only in the method of obtaining access to the
formation. In the first technique, all access to the formation is from the surface of the ground, whereas in the second technique, access to the formation is
obtained from underground rooms or tunnels which reduce some of the vertical
drilling. As in previous cases, the selection of the technique will be a function
of the depth of the formation and the costs of providing access.

As a general illustration, Figure 5.13 indicates the two possible techniques. The
first technique is conventionally drilling a single hole to the top of the reservoir.
The top of the formation is then jetted to form a cavern approximately 5 ft in
height and 40 ft in diameter. By special drilling techniques, a hole is then
drilled through the producing formation so that steam may be injected into the
bottom of the formation.

The technique of initiating and stabilizing the jetted cavern is slightly different
from that used at the surface. The hole and cavern are both filled with extremely

hot water with the necessary surfactants. Steam is injected in the bottom of the drilled hole and the hot fluids are permitted to migrate up to the cavern to maintain a constant availability of hot water in the region to be stabilized. This process is continued until the region around the cavern has been satisfactorily treated so the oil will flow upward without movement of the sand and destruction of the cavern. The oil released is pumped from the initial well bore to the surface for processing and to reduce the pressure head on the hydrocarbon system.

Figure 5.13: Two Configurations for Underground Application of Flip-Flop

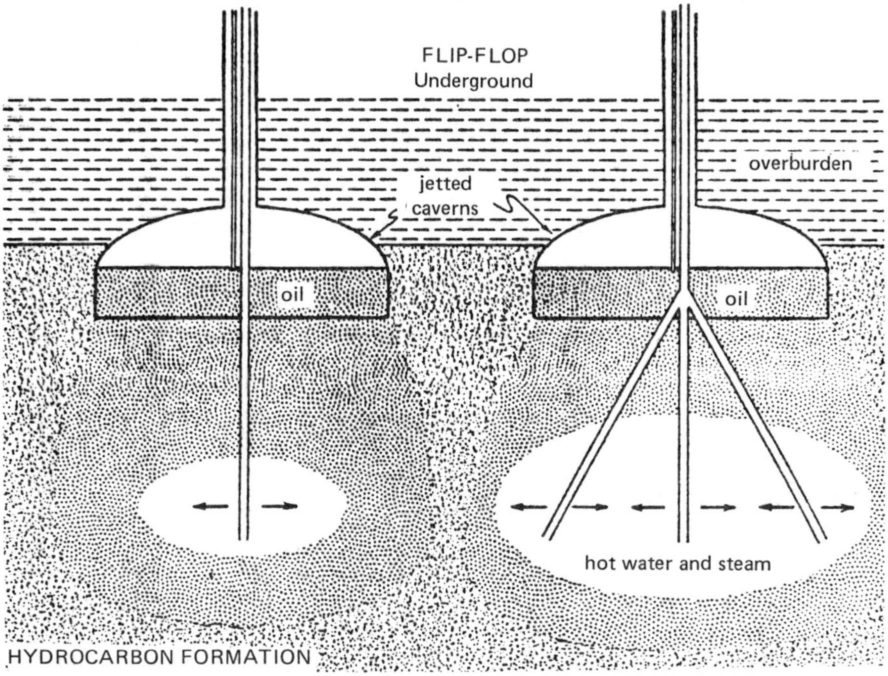

Source: BuMines OFR 56–79

Also illustrated in Figure 5.13 is a similar technique, except that additional steam injection holes have been drilled through the formation. This procedure would increase the area that could be drained with any one drill hole from the surface, but it would require a much larger diameter hole from the surface to the top of the formation. The choice of these two types of completions will be determined by economics and will be a function of the viscosity of the crude oil, formation thickness, formation permeability and depth to the top of the formation.

Conclusions on Flip-Flop Process — The Flip-Flop Process applies to the extraction of conventional, but dead, oil. A conventional petroleum reservoir that only contains liquid petroleum, with no gas or water, and is under little or no over-

burden or other pressure is said to contain dead oil. Usually, little or no oil
can be produced by conventional primary methods.

Since a Flip-Flop Process is primarily used to extract dead oil, i.e., not containing associated gas or water, certain of the environmental problems normally associated with oil production do not exist. Water used in the process can be recycled and, instead of a problem of brine disposal, sufficient makeup water will
be needed to replenish that lost in escaped steam and by evaporative cooling.

Reclamation of mined overburden may not be so difficult since the trenches
probably will have to be back-filled. Thus, spoil piles are generally avoided.

It would appear at this conceptual stage of the Flip-Flop Process that such a
process may offer fewer inherent environmental problems than the more conventional oil mining.

The economics of the Flip-Flop Process depends on many more petroleum engineering variables than the gravity drainage production system. One must spe-
cify the temperature expansibility of the oil, change in viscosity with temperature, the pore size distribution of the sand, the wettability of the sand and
many other factors. Interchange rate of fluids under these variable conditions
have not been totally documented in the literature and for each case some laboratory data will have to be derived. Because of the variables involved and need
for confirming data on specific deposits, economics must be calculated for each
case individually.

However, it is anticipated generally that the capital development cost would be
approximately 10% of the gravity drainage process. The oil recovery factors
should be as good, or nearly so, as the gravity process. Operating cost for the
Flip-Flop Process probably will be higher than for gravity drainage.

In order to address viability of a Flip-Flop Process, an evaluation of flow theory
and the mechanical factors results in different conclusions than those for a Drip
Drainage project. For surface Flip-Flop, the study team concludes that the
mechanical factors pose no unique questions in any of required elements and
that all are shelf technology. The reservoir theory of maintaining a stabilization
bed whose residual water-saturation tension characteristics will allow flow of non-
viscous crude, however, must be demonstrated under laboratory and field conditions and thus poses economic viability risks.

The mechanical factors which are all well-demonstrated are:

   (a) *Exposure of the producing surface* by mining technology;

   (b) *Flooding the surface* with

   (c) *Hot water* for viscosity modification which can easily be con-
        trolled, including

   (d) *Brine* additives to enhance density interchange and also including

   (e) *Surfactants* for capillary improvement;

   (f) *Soaking time* in order for the Stage 1 interchange to take place;

   (g) *Bed depth* of 6 to 10 ft to provide a sufficient stabilization
        thickness; and

(h) *Residual water-saturation* of approximately 35% to provide optimum interfacial tension for maximum bed stability.

The conclusion is that Flip-Flop flow theory has some risk to find a proper
balance of the interrelation of the forces of nature, but that there are no un-
knowns associated with the mechanical factors of project performance.

REFERENCES

(1) Ball Brothers, U.S. Bureau of Mines Monograph 12.
(2) American Petroleum Institute, *Reserves of Crude Oil, Natural Gas Liquids, and Natural Gas in the United States and Canada as of December 31, 1977.* Washington, D.C. American Petroleum Institute, Vol. 32 (June 1978.)

CONCEPTUAL MINING SYSTEMS FOR SELECTED FIELDS

The information in this chapter is based on:

> *Oil Mining. A Technical and Economic Feasibility Study of Oil Production by Mining Methods,* Bureau of Mines Report OFR 55-79, prepared by A. Edey, B.A. Kennedy and L.A. Readdy, of Golder Associates, Inc., for U.S. Bureau of Mines, October 1978.

This chapter describes work performed and results obtained for the technical and economic feasibility study of oil production by mining methods. The characteristics of oil fields which would make them amenable to mining techniques were identified. An inventory of known U.S. oil fields that offer this potential was collated. This was further divided into fields having potential for underground and surface mining systems development. Several candidate fields that fit particular underground and surface mining system types were then selected. Conceptual mining systems were designed for the selected fields. Five underground and four surface methods were chosen.

POTENTIAL TARGETS FOR OIL MINING OPERATIONS

The possible target for oil mining operations is comprised of oil remaining in place in: (1) fields and pools that have reached economic depletion by current petroleum production methods and are not amenable to economic recovery by conventional petroleum industry techniques; (2) oil- or bitumen-saturated rocks that are not normally considered to constitute reservoirs, such as shales and diatomites; and (3) asphalt- and bitumen-impregnated rocks.

Estimates of the oil remaining in place after primary, secondary and nondestructive tertiary recovery vary. However, the evidence is that significantly more than one-half of the oil originally estimated to be in place, remains in the reservoir after all petroleum recovery techniques have been applied (1). A comparison could be made, however, between the sum of the past production and the estimated remaining producible reserves in the U.S. oil fields and the American Petroleum Institute (API) estimate (1975) of total oil in place for the U.S. This comparison indicated that for the U.S. as a whole, 67% of the original oil content will remain in place at the end of the economic recovery process. The corresponding figure for the 100 largest oil fields was 63% remaining after full production. These figures are representative of oil fields that have been, or presently are, under production by conventional petroleum techniques, and do not include poorly known heavy crude reservoirs with limited production and reserves data, tar sands, oil-impregnated diatomite deposits, etc.

The majority (85.7%) of the estimated economically ultimate unrecoverable resource of oil in the U.S. are contained in only eight states, each with a residual resource of more than 10 billion barrels (2). These states, in order of decreasing estimated unrecoverable resource, are: Texas, California, Oklahoma, Louisiana, Alaska, New Mexico, Kansas and Wyoming.

The total estimated unrecoverable resource, subdivided on the basis of geologic age of the reservoir rocks (1) is graphically presented in Figure 6.1. The data indicate that 52.71% of the resource will be contained in Paleozoic, 33.04% in Cenozoic, and 14.24% in Mesozoic age rocks. An insignificant amount with respect to the total resource (i.e., less than $\frac{1}{1,000}$ of 1%) will be contained in Precambrian age rocks. The distribution of the estimated oil originally in place is similar and is 49.73%, 34.27% and 15.99%, respectively. The difference is in part explained by a slight but significant average increase in the percentage of oil left behind in the older rocks as a result of reductions in size of pore space and permeability.

As a generalization, the older the reservoir and the greater its depth of burial the harder and more competent rock can be expected to be, and the higher the API gravity of the oil contained. Conversely, older rocks must be expected to have smaller pore spaces and less oil content per unit volume.

In summary, the oil fields in production or previously in production, contained an original total resource of 446.28 billion barrels of oil. After primary, secondary and tertiary methods have been applied, an estimated resource of 303.5 billion barrels of oil will remain in the ground (1). Not included in these estimates, but potentially of large magnitude, are reservoirs which have never been placed into production normally due to high oil viscosity. An example of this would be the Paris Valley in California.

Since the normal history of petroleum reservoirs is to produce at least some of the gas with the oil, the oil left in place at the end of the primary recovery process, if a significant amount of gas was dissolved in the oil originally, will have a higher viscosity than that previously produced. The residual oil will also have a lower API gravity.

In cases where the oil had lost much or all of its dissolved gases prior to discovery, the oil may be so viscous as to be marginal or uneconomic by conventional, including secondary, petroleum recovery techniques. In such instances a higher unit content of oil may be available for a potential mining project. Example fields include the Paris Valley field in California and other heavy oil pools that have a low cumulative production history.

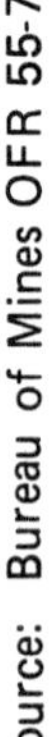

Figure 6.1: Distribution of Estimated Oil Resource Remaining After Full Conventional Production

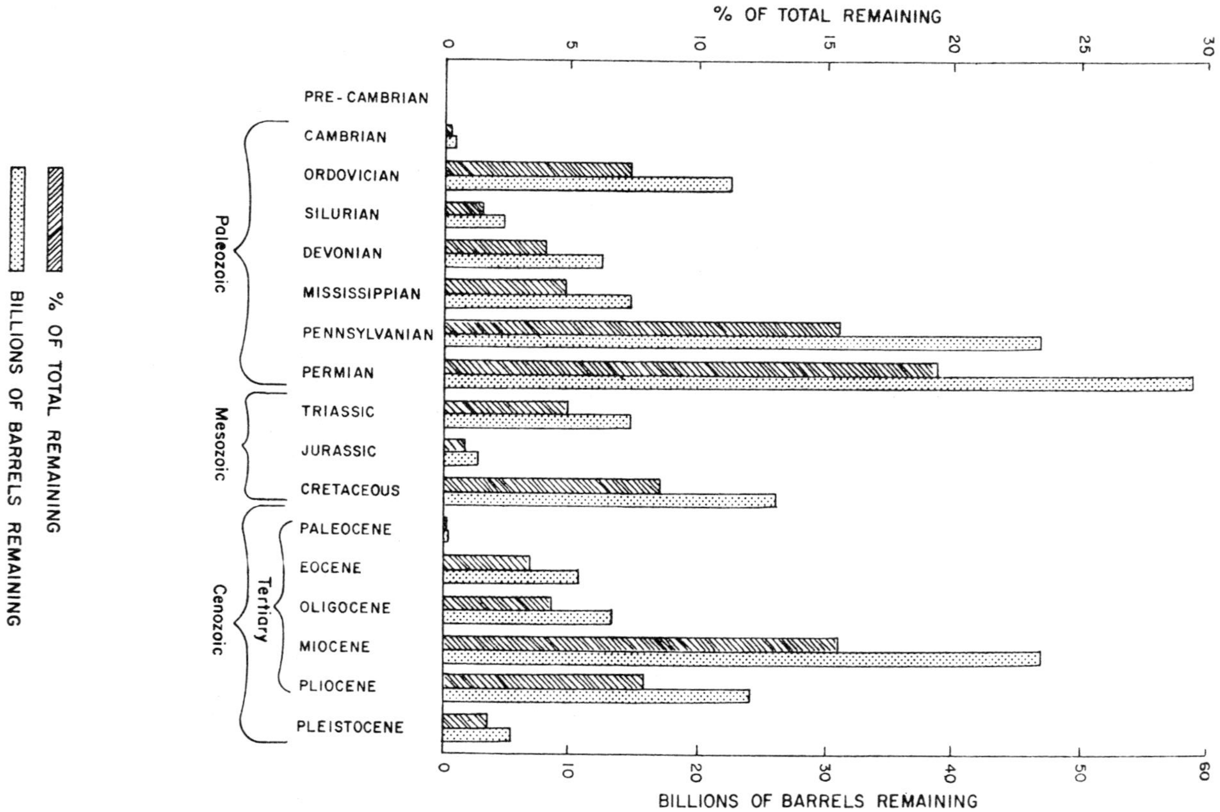

Source: Bureau of Mines OFR 55-79

UNDERGROUND MINING METHODS

Five possible underground oil mining operations were designed and evaluated, two of them involving direct mining, two employing oil drainage, and the fifth using mining techniques to fracture the rock prior to drainage.

Theoretically, the mining engineer is faced with an immense array of potential mining methods from which to choose. In practice, however, the methods which would promise satisfactory results in any particular case are usually few, and in extremely adverse circumstances, may be none at all. Granted a good knowledge of the relevant factors, mainly geological, it is usually relatively simple for an experienced engineer to reduce his options in a specific instance to a manageable number. In this study, however, the problem is complicated by the need to think in generic or typical terms.

This section of the report, therefore, attempts to catalogue all reasonably possible underground methods and to identify the natural circumstances under which they could be applicable. It has been assumed that the reader possesses a general working knowledge of the principles of mining systems and the descriptions have been kept extremely brief.

Oil mining is a term which proves on closer examination to be rather less specific than it at first appears. It has been interpreted, for this study, to include the use of any portion of the array of mining techniques, or their components, to bring about or enhance the production of oil. The possible approaches fall into three categories:

- Extraction Methods, involving physical removal of reservoir material in part or in whole to a treatment plant where the oil would be separated from the host rock. These are also referred to as direct mining methods.

- Drainage of oil from the reservoir rock by methods which are generally similar to conventional production techniques, including such stimulation methods as steam injection. Advantage would be taken, however, of access to the reservoir area through underground mine workings to produce a higher recovery than can be achieved through surface wells. The methods are often called indirect.

- Fracturing and Drainage systems which employ access to reservoirs underground to first fracture the reservoir rock, probably explosively, and then to recover oil by drainage methods.

Extraction Methods

Cut and Fill: In a cut and fill operation the deposit is mined by blasting a series of horizontal slices, normally working from the bottom upwards. Men have to work continuously under the roof of the stope and it is therefore essential that the competency of the oil-bearing zone be high. Within the stope the men work on a floor composed of backfill. The oil-bearing rock would be dumped into stope passes leading to a haulage way, where it would be loaded into trucks, trains, or conveyor belt for transport to the hoisting system.

Depending on the actual competency of the oil-bearing zone, 60 to 90% extraction could be achieved with a series of primary stopes. 60% primary extraction would be obtained using rib pillars large enough to permit later extraction by the same mining method. 90% primary extraction would be obtained using post pillars and secondary extraction would not be attempted.

Preferably, this method should be used in a low potential oil flow situation, because in a high flow situation oil will exude into the stope from the rock roof and sidewalls, and from the broken muckpile. There would then have to be some control of this oil which would create a safety hazard and potentially be lost. In high permeability situations (above 500 md) a further hazard could be caused at the edges of the oil-bearing zone by water entering the stopes.

The most typically normal application of this method is in high value ores where lower production rates and higher costs are acceptable but high recovery is necessary. For oil mining, however, the ore value is low enough to preclude any but the most highly productive, low cost methods. The only form of cut-and-fill worth considering, therefore, is multilift room and pillar or rib and pillar mining using fill as a working platform.

Longwall: In this method, a well established system for mining coal seams, men have to work in very close proximity to the face and face instability could cause unacceptable roof conditions. Reasonably high competency is therefore required in the oil zone.

To have a productive mining system, the oil-bearing rock should be broken by mechanical means. To allow the use of such machines, the rocks should not be above 15,000 psi uniaxial compressive strength. This method would essentially be limited to use in thin oil-bearing zone situations. Preferably, this method should be used in a low flow situation. In the high flow situation, oil would exude from the solid rock into the openings and from the fragmented rocks. This oil would be difficult to control and would create a fire hazard.

Room and Pillar: This is a highly productive method and is widely used in flat-lying, tabular deposits of many mineral types. It is essentially a partial extraction method with major recovery of pillars rarely attempted. To achieve as high a percentage extraction as possible and to allow men to work in safety in the headings, a competent rock is required. The entries should be capable of standing open with little support.

This method is essentially for use in a thin oil-bearing zone situation. The maximum extraction height is directly dependent on rock properties but rarely exceeds 80 feet even in the strongest formations. Preferably, this method should be used in a low flow situation.

Sublevel Open Stoping: In this method, ore is broken by blasting in large open stopes and is withdrawn from the bottom of each stope by front end loaders. While no men have to work in the actual stopes, relatively large openings have to stand open with virtually no support while the ore is extracted. Thus, high competency is essential. This can either be a partial extraction method with the open stopes left empty, or they can be filled after completion with waste deoiled material and pillar extraction attempted.

The ore would be broken in the stopes by drilling and blasting. Oil-bearing rock would be mucked from the stopes using diesel load-haul-dump (LHD) machines which would load into a truck or rail transport system situated at the base of the oil-bearing zone for removal of the material to the hoisting system. This method is essentially suited to thick deposits. While low flow characteristics are preferable, the method could be used in a high flow situation. All development can be located in the oil-bearing zone and competency of the underlying rocks is therefore not essential.

Sublevel Caving: In this method the oil-bearing material would be broken and removed on sublevels while the overlying rock continuously caves in and fills the void that is created. It is a full extraction method and would cause major surface subsidence.

On the sublevels the oil-bearing rock would be broken using fan drilling jumbos, and the broken material would be mucked using diesel LHD machines. These would discharge into ore passes which would take the rock down to the haulage level where it would be loaded via chutes into trucks or trains for transport to the hoisting system. This method is essentially suited to thick deposits. Low flow characteristics are preferable for this method.

Block Caving: In this system, a relatively thin horizontal slice is broken by blasting and drawn from the base of the ore zone. The main volume of the oil-bearing rock which has been undercut in this way is then allowed to break up using only natural forces. Because the bulk of the ore is required to break without mechanical aid and because men are not required to go into stopes, relatively incompetent rock is necessary.

The undercutting operation would probably be performed using long hole fan drilling and blasting from the drawpoints in the underlying rocks. This initially blasted material, followed by naturally broken material, would flow downwards into drawpoints. Here the material would be loaded by diesel LHD machines for transport to ore passes which would deliver the material via chutes to trucks or trains for transport to the hoisting system. A thick oil-bearing zone is a prerequisite for successful application of the method.

In this method no development is required in the oil-bearing zone so that oil exuding from the fragmented rock would be easy to control and thus would not constitute a hazard. In fact, in the high flow situation, a combination method of drainage and extraction could be used in which, after undercutting and the initiation of caving, oil would be allowed to drain. When significant flow ceases, drawing of the broken rock would commence.

While no men are required to work in the oil-bearing zone, the main extraction development will be in the underlying rocks. Thus, it is essential that these rocks be competent so that these tunnels can be excavated and maintained, with support if necessary.

Basis for Method Selection: No study of this type would be complete without including at least one of these direct methods. In practice it was considered that the study should include two systems. In the appropriate conditions, block caving is by far the most economical. It was, therefore, an essential candidate since if block caving could not be shown to be attractive, none of the other mass mining methods would be likely to merit further consideration. The Casmalia deposit in California appeared to possess a large enough thickness of weak oil-bearing rock for block caving. There also seemed to be underlying zones of sufficient competence to support extraction workings and it was therefore selected as the basis for a design of this method.

As a contrast, at least one stable mining method was required for study. Such a method would impose the requirement of operation within the reservoir rock and would also have the potential merit of inducing less environmental disturbance, particularly related to subsidence. Historically, room and pillar mines have been the most productive and economical of such stable operations and this system was therefore selected for study.

The lack of recorded rock strength data made the search for a suitable field very difficult, particularly since a high grade of residual oil was desired. The Circle Ridge field in Wyoming offered the unusual feature of two flat-lying oil-bearing formations of interestingly different thicknesses and was therefore used as a basis for this design.

Fracturing and Drainage Methods

Such methods rely on a mining operation to fragment the oil-bearing zone so that the contained oil is able to drain more freely than is possible from the solid state. Apart from the natural drainage of the oil, stimulation techniques from above or below the oil-bearing zone could be applied. Only enough of the oil-bearing material would be removed from the mine to ensure that proper fragmentation could be achieved. Three possible methods were identified.

Sublevel Shrinkage: Drilling and blasting operations would be carried out in a similar way to sublevel caving, but only sufficient broken material would be removed on each sublevel to enable blasting to continue. Because the method is only geared to breaking and not to full extraction, the sublevel spacing would be considerably wider than with normal sublevel caving.

Because operations have to be carried out mainly from within the oil-bearing rock, its competency must be high, to ensure that the sublevels can be developed and maintained with little or no support. Drilling from these tunnels would be by fan drill jumbos, and the removal of broken material for swell relief purposes would be done by using LHD machines in a similar way to sublevel caving.

Development would be required underneath the broken rock zone, either in the underlying rocks or in the oil-bearing zone itself with boreholes connecting to the broken rock mass to enable oil to be drained out and stimulation techniques to be applied if necessary. This method is essentially applicable to thick ore bodies.

To give the best possibility of safety during fragmentation operations, the reservoir rock should preferably have low flow characteristics at that stage. The eventual success of the system, however, clearly depends on its ability to alter that to a high flow situation.

It is preferable that the underlying rocks be competent so that oil drainage development can be sited underneath the oil-bearing zone. However, such develop-

ment within the oil-bearing zone is possible so the requirement is not critical.
The strength of the overlying formations has no impact on this method.

Block Caving: This method is basically no different from the block caving extraction method except that draw of broken material would cease when the caving action has been propagated to the top of the oil-bearing zone. An incompetent, thick oil-bearing rock would be required, and, as virtually no development would be in the oil-bearing zone, the flow characteristic would not be critical to safety. However, competent underlying rocks would be essential.

Remote Fracturing: Consideration was also given to inducing fracturing of the rock by drilling and blasting fans of holes from an underlying level. This would essentially be the addition of this fracturing operation to one of the drainage concepts described below. In such a system, the competence of the reservoir rock would not be a major factor, provided that a zone could be found underlying the oil rich area in which to locate the development.

Basis for Method Selection: Of the possibilities listed above, sublevel fracturing was considered the most likely to induce consistent additional permeability throughout the reservoir zone of interest.

It was considered probable that the type of porosity and permeability typically found in carbonate reservoirs might be particularly amenable to this type of system. No ideal field could be identified within reasonable time, so the design was based on the Yates field in Texas, although there are many factors that would preclude its specific suitability for oil mining, not least its long conventionally productive future.

Drainage Methods

This group of methods is the furthest removed from a conventional mining approach but also, possibly, is the most promising of the underground options considered. All are variations of the same basic concept, which is that a network of underground tunnels is developed either in or adjacent to the oil-bearing zone. This tunnel system would provide the locations for drilling holes into the oil zone for production purposes and for stimulation if necessary. The availability of such holes at a greatly higher density than would be economically possible from the surface has the potential to produce substantially enhanced recovery in fields of many types.

The systems identified under this heading are only basically different from a mining standpoint by the position in which the tunnel system is located with respect to the oil.

Single Level in Oil Zone: This system, corresponding to that used in past European operations, would be suited to a relatively thin oil zone, all of which could be produced through holes of economical lengths, drilled from a single level. It would require a reasonably competent oil-bearing rock to permit safe and economical development and operation of the system. This must also be coupled with a reasonably high flow potential. The properties of the formations overlying and underlying the oil zone are of no major significance in this system.

Multiple Levels in Oil Zone: In the case of an oil-bearing zone that is too thick to produce economically from a single level, multiple-level operation would be employed. In all other respects, however, such a system would be similar to the previous one.

Remote Drainage: As an alternative to tunnelling within the oil-bearing rock itself, a suitable location could be selected in the rock formations either above or below. This arrangement would permit production from fields which might yield more readily but whose rock competence would make operations within the oil zone unacceptably hazardous or expensive.

The presence of a suitably competent, impermeable formation adjacent to the oil zone is the dominant criterion for this method. It could be either above or below but does not need to be both.

Basis for Method Selection: It was considered essential to include at least one of these concepts in the study. Examination of the available fields quickly showed the difficulty of identifying with enough confidence any field with the necessary high flow potential coupled with the rock strengths necessary for a safe and productive operation. It was therefore decided to concentrate the study on methods which would take advantage of relatively safe ground conditions in formations adjacent to the reservoir and a search was made for fields having such conditions either above or below.

The Irma field in Arkansas possesses a competent caprock and is adjacent to a field with current experience of steam stimulation. It was therefore used as a basis for studying drainage operation from overlying locations. The Wheeler Ridge field in California has, in contrast, a competent underlying formation which appears to be adjacent to a substantial amount of residual oil over an appreciable area. Gravity drainage from below the reservoir was therefore evaluated for this field.

SURFACE MINING SYSTEMS

The number of generic surface mining systems available for oil mining is limited, although the possible variations on them by using different combinations of equipment is quite extensive. All of these are extractive methods in which the oil-bearing rock is removed and processed elsewhere. Any drainage of oil into the pit would be additional recovery over and above that recovered in the main processing plant. Potential surface mining systems can be divided into four categories.

Strip Mining: This involves the removal of the overburden by an excavating machine that directly transports the material across the excavation and then dumps it in the area from which the ore has been removed. Strip mining is suitable for tabular deposits that are either flat or have a very shallow dip, of less than 10 degrees. The maximum overburden depth should be about 150 feet, and the overall pit depth should not exceed approximately 200 feet. The limits of strip mine operations are usually determined more by the capacity of the equipment than by the economics of the deposit.

Terrace Mining: This method involves the removal of the overburden by excavating machines and the transporting of the material around the pit on the level from which it is excavated to be dumped back into the pit in the area from which

the ore has been removed. Terrace mining requires a reasonably tabular deposit
with a large areal extent. Other than these two factors, the system is fairly
flexible. The limits of terrace mining are generally set by the economic limiting
stripping ratio.

Combination of Terrace Mining and Strip Mining: In this method, the upper
overburden material is removed by terrace mining and the last cut to the ore
is made by strip mining. This system is able to achieve greater depths than normal terrace mining due to the relatively low cost of stripping the last 100 feet
of overburden material. It is a flexible system since as the overburden increases
in depth, the terrace mining system handles the extra material that the dragline
cannot remove.

Open Pit Method: This system involves the removal of the waste material up-
grades by truck, conveyor, train, etc., for dumping outside the excavation limits.
If the topography permits, level hauls to waste dumps can be achieved; however,
usually an upgrade haul is encountered. The true open pit method has the greatest flexibility of all four systems, but also, in general, has the highest cost. In
this case the limits of operation will be determined directly by the economic
stripping ratio.

Basis for Method and Field Selection: The surface mining methods described
above are usually far lower in cost per ton of materials extracted than any underground method. It was consequently decided to evaluate, if suitable fields
could be found, each of the four generic surface mining systems.

Strip Mining — As described above, the strip mining system is suitable for shallow, less than 150 feet, flat-lying deposits, since the economic advantages of the
system are limited by the capacity of the equipment to dig to depth. Because
of the lack of detailed data on some of the close-to-surface heavy oil or bitumen deposits, not many candidates were immediately available. One field, however, had the ideal geometry for strip mining, namely the Santa Cruz field in
California. This field is unfortunately small; however, the strip mining system
is quite flexible with regard to size though obviously in a small field the economies of scale cannot be taken advantage of.

Terrace Mining — The terrace mining system is ideal for large flat-lying or gently
dipping deposits where advantage can be taken of level hauls and dumping back
into the mined areas is feasible. This system can be an extremely low-cost producer, providing that the deposit is shallow. Costs increase rapidly with depth.
An ideal candidate field which meets virtually all the criteria for a terrace pit
is the Kern River field in California. This particular field is very large and lends
itself to high tonnages and the mechanized approach of bucket wheel excavators.

Combination Terrace Pit and Open Pit Mining — One of the four generic systems
described is the combination terrace/strip mine. No suitable candidates could be
found for this system; however, a candidate was found for an alternative combination system.

The alternative system is a combination terrace/open pit system, hereafter referred to as a modified terrace pit. In this system the first cut of the overburden, which does not lend itself to terrace mining due to depth and topography,
is removed by an open pit mining system and the last cuts of overburden and

ore are removed by terrace mining. An ideal candidate for this system is the bitumen deposit at Sunnyside, Utah. The same equipment, trucks and shovels, will be used for both the open pit and terrace pit portions of the operation.

Open Pit Mining — The fourth generic surface mining system is the traditional open pit mining system which has the greatest flexibility of all the systems. This system can be applied to any potential surface mining operation. The candidate chosen for the open pit mining system is Edna, California. This ore body has a geometry similar to many metallic and nonmetallic deposits and will necessitate mining to some depth below the topographic surface.

PROCESSING

The processing of oil-bearing materials is probably the least understood area in oil mining technology since the only work in this area has been primarily at a laboratory scale or, where commercial operations exist, is tailored to a unique deposit such as the Athabasca tar sands.

The ultimate objective of any oil mining operation is to produce an oil-rich concentrate which can be marketed directly to a conventional oil refinery. The oil-rich concentrate in certain cases would be the product from underground thermal mining and may contain a high proportion of water, steam condensate, and entrained sand and silt. In other cases, it might be the product from a hot water (flotation) process or it might be mixtures of hydrocarbons distilled and condensed from some thermal or solvent extraction process.

As is the case in hard rock mining for other minerals, each deposit will exhibit certain unique characteristics that may affect the extraction processes. In any of the proposed operations, the ore body will exhibit localized variations in porosity, grain size and silt content. These variations will affect the response of the material to subsequent oil recovery processes.

It is essential that the oil recovery method employed in an oil mining project be specially designed to handle the product from oil mining and to efficiently recover the oil as a marketable product at low cost.

Considerable background experience exists in oil recovery from tar sands, heavy oil sands and oil shales. The extraction of certain naturally occurring waxes in coals has provided some experience in hydrocarbon flotation and solvent extraction. In the past, however, due to the long prevailing low prices for crude oils and the abundant supply from oil wells, there was little incentive to develop and apply oil recovery methods to heavy oil and tar sands.

Oil Recovery Methods

Broadly, there are three basic methods for recovering the oil or hydrocarbons from oil-bearing, naturally occurring materials such as tar sands, bituminous sands, heavy oil sands and related petroleum-bearing reservoir rocks. They are:

- Hot water flotation
- Solvent extraction
- Pyrolysis of retorting and thermal distillation

The simplest and probably the cheapest method is hot water flotation, supplemented by the use of mechanical scrubbing or attrition devices, surface active chemical additives and hydrocarbon solvents for thinners.

Solvent extraction processes can be highly efficient, but involve hazards from explosion and fire because the solvents usually are highly combustible. In addition, solvent loss through absorption in the solid mineral residues has resulted in high operating costs.

Pyrolysis or retorting-thermal distillation can efficiently recover the hydrocarbons and, in effect, partially accomplishes some oil refining steps by stage-removal and condensation of the more volatile constituents. The method also involves hazards similar to those encountered in solvent extraction, such as from explosion and fire. Other special problems include buildup of tars and cokes, dust entrainment in the gas streams, and finally, disposal of the hot residues.

There is active study and research being conducted in the laboratories of certain major oil companies who control, or are interested in, heavy oil deposits. Also, the Department of Energy, Division of Oil, Gas and Shale Technology, headquartered in Bartlesville, Oklahoma has begun to fund programs to evaluate and core drill the shallow heavy oil resources of southeastern Kansas, western Missouri and northeastern Oklahoma. This work should eventually lead to investigations of methods for the extraction of the heavy oils from deposits in those states by methods other than in situ, and thereby stimulate and revive government research work on oil recovery processes.

Water-Flotation Process: The objective of the process is to decrease the viscosity of the bitumen or tar with hot water, and remove it from the sand grains by means of mild attrition scrubbing (aided by sodium hydroxide) and by utilizing the natural froth flotation amenability of oil-sand-and-water mixtures.

A flowsheet of the U.S. Bureau of Mines pilot plant work on the Edna bituminous sandstone is shown in Figure 6.2. Recoveries of from 90 to 95% of the bitumen are reported. Because practically no crushing and grinding are required, which steps are usually the highest cost in metallic ore processing, the operating costs are low. The heating of the pulp is probably the most expensive step in the process.

Where bitumen and clean sand are the major constituents of a low-clay tar sand feed to the process, excellent separations are made. However, when there is a high clay and silt content in the feed, appreciable amounts of the oil become trapped in a middling product and fail to separate into the froth layer.

Most silt and slimes which are dispersed into the ore pulp during processing remain with the sand tailings and because of the dispersed pulp condition (resulting from the use of sodium hydroxide) create difficulties in clarifying the tailings waters in tailings basins. However, parallel problems often occur in the treatment of phosphate rock, taconites and porphyry copper ores, and schemes such as accelerated settlement, flocculation, etc., have been developed to cope with the problem. Silt contamination in the final bitumen product is a more serious problem and must be strictly controlled.

Figure 6.2: Simplified Flowsheet for USBM Oil Extraction System

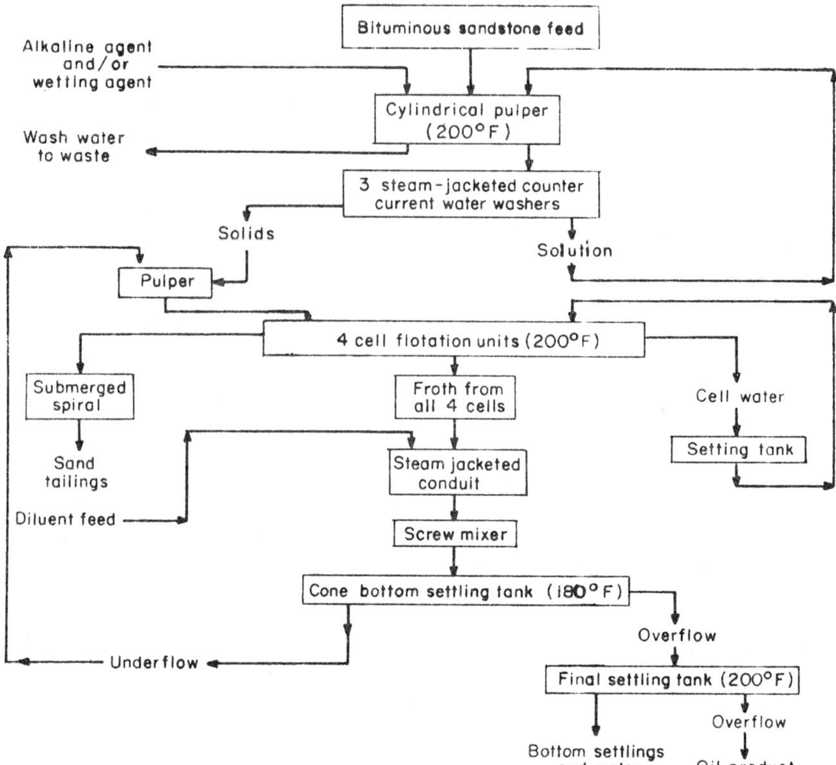

Source: Bureau of Mines OFR 55-79

In comparison with solvent extraction and pyrolysis or retorting processes, the tailings from hot water flotation processes are more readily disposable in impoundment dams and the coarse fractions can be isolated for backfilling of the mine, if desired. Problems associated with disposal of dry, dusty and/or hot residues from oil extraction are thereby avoided.

Due to the extremely fine particle size of the diatoms which make up diatomaceous earths, it is highly unlikely that a hot water process can be employed to recover oils from such materials.

Solvent Extraction Processes: If an oil-bearing sand, shale, limestone or diatomaceous earth is leached with a suitable low viscosity hydrocarbon solvent, the oil can be dissolved in the leaching agent and thus be extracted and washed from the solid material. Subsequently, the oil must be separated from the solvent so that the solvent can be recycled to the leaching stage.

At first glance such processes appear simple, positive and low cost, but there are serious problems associated with the volatility of the solvent and solvent-oil mixtures, and loss of solvent due to absorption in the solid residues. Hazards from explosion and fire are extreme and maintenance of vapor seals throughout the processing steps pose formidable problems to the designer and operator.

Because the leaching step must be carried out on relatively dry material, residue disposal may have to contend with dusts and residual absorbed solvent vapors which will present environmental problems. There are several alternates for the actual leaching step but the washing or rinsing cycle must be followed by vacuum filtering in order to achieve low solvent losses. Filtering costs generally are high both in capital and operation.

As is the case with retorting or thermal distillation, recovery of oil by solvent extraction overlaps the refining of oil because separation of the solvent and oil necessarily must be accomplished by distillation. The capital costs therefore would probably exceed those for a simple oil recovery process plant which yields a crude oil suitable for marketing to a conventional oil refinery, since a partially refined oil would be the end product.

Pyrolysis or Retorting and Thermal Distillation: The application of simple heating can achieve partial oil recovery from high-viscosity crudes contained in sands and related reservoir rocks, and this principle is utilized in the in situ steaming of oil reservoirs. Much of the oil, however, remains on the material and recoveries therefore are low. If the degree of heat application is sufficiently high, virtually all the oil can be removed, some by distillation and some by combustion. Here, as in solvent extraction processes, the oil recoveries by pyrolysis or retorting overlap the field of oil refining because the distillation products must be condensed from gas streams at various temperature ranges.

All pyrolysis processes suffer from the disadvantage of heat requirement to remove moisture from the incoming feed. Diatomaceous earths, for example, may contain from 20 to 30% moisture as well as a high oil content. Recovery of heat from and the disposal of the residues of the final cooking step in pyrolysis is also a problem.

Some interesting results have been reported with processes which incorporate a single, circular traveling grate for the sequential drying, distillation and coking steps. Among the advantages offered by circular traveling grates are high capacity, positive gas seals, sintering of the depleted residues and positive mechanical control of material flow through the process. Most of these features have been proven by many years of operating experience in the pelletizing plants of the taconite iron ore industry.

ENVIRONMENTAL FACTORS

Environmental impacts will obviously vary with the type of mining system used. Surface mining will require the removal and disposal of substantial quantities of overburden. The actual removal of the overburden and oil-bearing material will create substantial quantities of dust during drilling and blasting, where applicable, and in loading and hauling. Water courses that cross the areas to be mined will probably require diversion or, in the worst case, impoundment.

Vegetation will be removed and consequently the erosion potential could be increased. In certain cases, the ground water regime will be altered. Since surface mines are open to the air, the risk of explosion due to flammable gases is reduced when compared to underground mining. The use of backfill dumping in the pit with the consequent deposition of tailings on top of the overburden may increase the potential for leaching of chemicals. These chemicals could come from the extraction process or from natural leaching of minerals. The revegetation of reclaimed areas may be necessary. The operation of equipment will create noise which will impact on the surrounding countryside.

Underground mining will result in less surface air pollution than open pit mines since the majority of the dust and gases will be entrained by the ventilation system, whose discharge can be controlled. The risk of exposure to explosive or toxic gases is higher to underground miners than in the open pit mines. With certain mining systems, the potential subsidence of the natural ground surface could occur. In the worst case this subsidence could be equal in displacement to the mined-out thickness of the ore body. In most cases mine water inflow will have to be pumped out of the mine and will require treatment before disposal or reinjection back into the ground. Some of the proposed operations will be backfilled, which will increase the potential for chemical leaching into the ground water regime. Equipment noise impacts will be great in localized areas of the mine.

Most of the extraction processes suggested utilize dry crushing followed by wet grinding and extraction of the oil by hot water, steam, flotation or solvent extraction. The dust generated during the crushing operation will cause a localized impact; however, dust control equipment will prevent any excessive general impact. The water used in the grinding and extraction circuits will be recycled to minimize losses and makeup water will be added as necessary. The need for a large water supply in water short areas may impact the local ground water table. During the dewatering, desludging and oil/gas separation operations, substantial amounts of odor could be produced. H_2S is the principal contaminant, with other organic sulfides and mercaptans present in smaller amounts. These odors can be detected by humans at very low concentrations (ppb) and thus very small quantities can produce an adverse impact. Odor treatment will be necessary in most cases, especially in those fields close to population centers. Volatile hydrocarbons may evaporate during both mining and processing operation.

CANDIDATE FIELDS AND SELECTED MINING SYSTEMS

The candidate fields chosen for further study in this report and the mining system selected as being the best method for extracting the remaining oil from each field are summarized in Table 6.1.

Table 6.1: Candidate Fields and Selected Mining Systems

. . . . Candidate Field		. . . . Mining System	
Underground	**Location**	**General Type**	**Method**
Casmalia	Southwestern California	extraction	block caving
Circle Range	Wyoming	extraction	room & pillar & cut & fill
Yates	Western Texas	fracturing and drainage	sublevel shrinkage
Irma	Southern Arkansas	drainage	remote drainage (above oil zone)
Wheeler Ridge	Southern California	drainage	remove drainage (below oil zone)

(continued)

Table 6.1: (continued)

| . . . . Candidate Field . . . . | | | . . . . Equipment Type . . . . | |
Surface	Location	General Type	Overburden Removal	Ore Mining
Santa Cruz	Central California	strip	dragline	loaders trucks
Kern River	Southern California	terrace pit	bucketwheel excavator conveyors stackers	bucketwheel excavator conveyors stackers
Sunnyside	Northeastern Utah	modified terrace pit	shovel truck	shovel truck conveyor
Edna	Southern California	open pit	shovel truck	shovel truck

Source: Bureau of Mines OFR 55-79

UNDERGROUND OPERATIONS

This section presents brief descriptions of each of the five underground operations designed in the course of the study, together with comments on their technical feasibility and other significant features.

Throughout the design of all of these examples problems were encountered, with varying degrees of severity, as to the basic practicality of the schemes. This was particularly so where mining would actually take place in reservoir rock. Difficulties were also encountered with either a lack of geological data or an emerging geological picture which diverged from the properties favorable to the system being evaluated.

It was considered essential, however, that all of the potential methods were carried right through to economic evaluation, even if that required that some far-reaching assumptions were made as to their safety during the design stage. Only in this way could the concept be given any ranking at all, and a determination made as to whether there appeared to be a large enough potential profit margin to allow major extra costs to be incurred in offsetting the problems.

Direct Stoping Methods

One of the main direct mining systems which was considered to deserve evaluation in this study was extraction of oil-bearing rock by room-and-pillar or open stoping methods. The Circle Ridge field, in Fremont County, Wyoming appeared to offer suitable conditions in two separate zones for such mining. Key properties of the Circle Ridge field are summarized in Table 6.2. This field produces oil from five pools contained in four formations: the Phosphoria (Embar), Tensleep, Amsden and Madison, all of Paleozoic age. Only two of the four producing formations, the Phosphoria (Permian) and the Tensleep (Pennsylvanian), two relatively flat-lying, oil-bearing formations of differing thickness in reasonably competent rock, have been considered as candidates for mining in this study.

Table 6.2: Circle Ridge Field

| | . . . . Formation . . . . | |
	Phosphoria	Tensleep
Permeability, md	not available	65
Porosity, %	16	14

(continued)

Table 6.2: (continued)

| | . . . . .Formation. . . . . | |
	Phosphoria	Tensleep
Oil gravity, °API	23–24	23–24
In-place grade, bbl/T	0.19	0.18
Depth to top of oil, ft		
Minimum	185	730
Average	850	900–1,200
Thickness of oil zone, ft	110	110
Dip of oil zone	approx. 35°	approx. 35°
Overlying rock		
Type	shale, limestone, sandstone	limestone, dolomite, shale, sandstone
Formation name	Chugwater Gr.	Phosphoria
Age	Triassic	Permian
Underlying rock		
Type	sandstone	limey dolomite, sandstone, siltstone
Formation name	Tensleep	Amsden
Age	Pennsylvanian	Pennsylvanian
Oil-bearing rock		
Type	dolomite	sandstone
Formation name	Phosphoria (Embar)	Tensleep
Age	Permian	Pennsylvanian

Source: Bureau of Mines OFR 55-79

Mining and Processing: The Circle Ridge field offers the opportunity to evaluate the type of noncaving direct mining system expected to prove most favorable, large-scale room and pillar mining. The two oil-bearing formations considered have significantly different thicknesses and are therefore suited to different variants of the system. In the thinner Phosphoria, a simple room and pillar system appears workable but the thickness of more than 100 feet in the Tensleep sandstone would require multiple-pass mining, with placement of fill to provide a working platform. The access to both formations to be mined will be through a common set of vertical shafts located on the boundary of the field.

The actual method of oil extraction can only be definitely determined through bench scale and pilot plant testing, but it is believed that either a hotwater or solvent extraction process could be used with some form of mechanical scrubbing. Considerable cleaning of the oil-rich product will probably be necessary to remove the fines generated in the crushing and grinding steps of the process. Some thinning of the final product will probably also be required utilizing a low viscosity dilutive such as naphtha. Final removal of suspended insoluble material can be accomplished by centrifuging or similar method.

The Circle Ridge processing plant would handle 17.4 million tons of ore per year producing 3,150,000 barrels of oil per year.

Technical Feasibility: Throughout the design of this operation, a consciously optimistic policy was adopted. There were so many obstacles to the generation of a believable design that there was a constant temptation to abandon the entire concept.

The most serious problems arise from the emission of gases into the mine air stream. Even with assured airflow through all working areas at good velocity and without reuse of air, there can be no real assurance that toxic or explosive mixtures will not occur, if only locally. Similarly, there can be very little real likelihood of completely preventing spark generation, if an efficient, productive operation is to be maintained in a siliceous formation such as the Tensleep sandstone.

If the prospect were a bright one economically, or even marginally so, it would be necessary to pursue the potential for improving on its safety through gas drainage, reduced scale operation with enhanced ventilation, remote operation, etc. This is not the situation and the type of operation described has regrettably to be termed unfeasible.

Block Caving

Block caving is, in the correct circumstances, the most economical and simplest of mining systems. It was therefore considered essential to evaluate its potential as a method of extracting poorly consolidated oil reservoir rocks. The Casmalia oil field chosen as the basis for this design was selected due to the expected presence of large vertical thickness of weak, oil-bearing rock underlain by a formation competent and thick enough to accommodate the mine extraction working. The field is located in Santa Barbara County, California. Key properties of Casmalia oilfield are summarized in Table 6.3.

Table 6.3: Casmalia Field

Permeability	>10
Porosity, %	31
Oil gravity, °API	10
In-place grade, bbl/T	0.38
Depth to top of oil, feet	
Maximum	2,000
Average	1,000
Thickness of oil zone, feet	75–700 (main horizon 200)
Dip of oil zone	Varies
Overlying rock	
Type	Diatomaceous mudstone, claystone and shale
Formation name	Sisquoc FM and upper member of Monterey FM
Age	Miocene
Underlying rock	
Type	Siltstone, mudstone, sandstone and shale
Formation name	Point Sal FM and lower member of Monterey FM
Age	Miocene
Oil-bearing rock	
Type	Cherty shale member
Formation name	Monterey FM (middle member)
Age	Miocene

Source: Bureau of Mines OFR 55-79

Mining and Processing: The reservoir rocks are expected to cave readily when undercut and the mine will be of very simple design.   All drifts will be located in the relatively competent lower Monterey, with the undercut fans penetrating to the base of the oil-bearing middle member of the Monterey formation.

The processing method for the Casmalia ore will be similar or identical to that proposed for Circle Ridge.   The plant will handle 10,936,800 tons of ore per year and will produce 3,150,000 barrels per year.

Technical Feasibility: Block caving is a well established and highly successful method of mining, with particular application to weak, poorly consolidated formations or those which have been substantially weakened by alteration or fracturing.   There is little doubt, therefore, that a large proportion of oil fields, if not all, are fundamentally caveable.   There are, however, several severe and possibly fatal problems in introducing this system into an oil field.

Mine ventilation is a major problem area and, even when a comprehensive and expensive system has been provided, the certainty of safe operation must be rated as low.   The loading system specified for the Casmalia field, chosen largely for its minimizing of spark production, could still hardly be considered spark-proof.   In particular, the handling and breaking of oversize boulders in draw-points could pose severe difficulties.   From an explosion/fire point of view, therefore, the system has to be considered unsatisfactory unless an essentially inert, oil-bearing formation can be identified.

The specific Casmalia field has given little cause for optimism as to the suitability of block caving from several other points of view.   As a full caving operation such a mine will induce major subsidence, propagating through all formations from the base of the cave zone to surface.   Any aquifer appearing in this area will therefore drain freely into the mine, if it has not been effectively predrained.

The Casmalia field is overlain by several water-bearing sequences, which would add prohibitive cost to the operation.   Even where there is no overlying water, moreover, there is a basic probability that the reservoir rock itself will contain substantial water, particularly at the lower end, and that such water would be attracted to the caved zone.

This leads to the conclusion that the method is also only likely to find application in a field which is essentially hydrologically inactive.   There may well be a substantial number of such fields, but these restrictions are expected to preclude the acceptance of block caving as a likely major feature of the oil production scene.

Shatter and Drain Method

As a compromise between the methods based on direct mining of the oil-bearing rocks and underground drainage methods, the possibility was investigated of performing an in situ fracturing operation, using mining techniques, and thus inducing additional drainage.   This method and the two that follow in this section share the feature that most or all of the oil production is by drainage through drill holes or underground excavations.

It was, however, necessary to provide some consistent basis for this study and this was done by assuming that each of the three drainage operations would eventually recover 50% of the oil remaining in the reservoir prior to mining. Such a value is very high indeed in relation to the results achieved conventionally in oil field production and may prove to be totally unrealistic.

It was initially considered that a carbonate reservoir might prove amenable to this type of system, and the method was designed around the geology of the Yates field in Pecos County, Texas. Key properties of the Yates field are summarized in Table 6.4. Since this particular field is currently, and will remain for many years, a prolific producer, there is no implication that it is recommended as a potential target for such operations for many years to come.

Table 6.4: Yates Field

Permeability	Wide variability
Porosity, %	31
Oil gravity, °API	20–30
In-place grade, bbl/T	Not applicable
Average depth to top of oil, feet	1,500
Thickness of oil zone, feet	to ~500, average 220
Dip of oil zone	Varies (elliptical anticline)
Overlying rock	
Type	Anhydrite series
Formation name	Seven Rivers and upper part of Queen
Age	Permian
Underlying rock	
Type	Dolomite (?)
Formation name	San Andres (?)
Age	Permian
Oil-bearing rock	
Type	Basal Sands of Anhydrite — Dolomite Limestone — Dolomite Limestone
Formation name	Queen — Grayberg — San Andres
Age	Permian — Permian — Permian

Source: Bureau of Mines OFR 55-79

The Yates field is a crossfolded domal anticline. The oil is trapped in the highly permeable carbonate reservoir by an overlying, thick, impervious anhydrite sequence.

Mining and Processing: The proposed system utilizes conventional drilling and blasting techniques to rubblize the reservoir rock with the intention of inducing improved oil recovery. In order to produce any substantial sustained effect, it is necessary to do more than merely fracture the rock; open fractures must be induced and this will involve some block rotation. It is considered unlikely that such a result will be attainable without the removal of a proportion of the broken rock to permit a degree of expansion during blasting.

The proposed design to achieve this is based on the sublevel caving system of mining and is similar to systems that have been proposed for preparing oil shales for in situ retorting.

The main processing plant feed will be that portion of the reservoir rock which is drawn to aid the blasting process, and will be a hard competent rock with the oil primarily contained in the solution cavities within the rock. Either a hot water or solvent extraction process will be used. The plant will also perform any necessary separation, cleaning and upgrading of the oil produced by drainage from underground. It will process 6,100,000 tons of ore per year and is expected to produce 2,620,000 barrels of oil per year.

Technical Feasibility: The basic objective of this system, enhanced recovery by the use of mining-enhanced permeability, still appears to have important potential; however, the design produced and evaluated here does not contain the answers to all the questions it poses. Three of these unresolved areas are vital: safety, fragmentation efficiency and potential for oil yield.

The safety implications cannot be satisfactorily dealt with at reasonable cost based on present knowledge. It is doubtful that it will prove possible to conduct mining operations on a major production scale and to economic levels of performance, from within an oil reservoir.

The efficiency of any oil recovery system after fragmentation will depend on the size range that can be attained, with large boulders and fines having adverse effects. Under any circumstances this would be a difficult standard to achieve in a blasting operation. It will be made even more difficult, however, by the need to reduce to a minimum, for economic reasons, the proportion of the rock that is drawn as swell. This will ensure that all blasting is performed under choked conditions and make it very difficult for rock to move downwards toward the draw points.

There also remain serious doubts as to whether the process would produce substantially induced oil recovery. The severity of the other technical and economic difficulties facing such a system discouraged further pursuit within the limited resources of this study, but the subject may well deserve additional consideration in the future.

In particular, the concept appears well suited to production from types of reservoirs having locally high but discontinuous porosity and permeability. Such types of trap could well yield a high proportion of their content if connected to the flow system and oil trapped in this way appears more likely to be recoverable than that remaining in intergranular pores. It is also, of course, ironically true that the existence of such trapped oil would be very hard to prove.

Drainage with Steam

One of the most promising practical demonstrations of oil production using mining techniques has been the underground steam injection project at Yarega in the USSR. Results from pilot operations appear encouraging enough to warrant expanded exploitation in that field, and the technique was considered to deserve evaluation in this study. It has been applied to a hypothetical field with properties based on the Irma field in Nevada County, Arkansas (Table 6.5).

Table 6.5: Irma Field

Permeability, md	1,500
Porosity, %	35
Oil gravity, °API	14
In-place grade, bbl/T	0.38

(continued)

Table 6.5: (continued)

Average depth to top of oil, feet	1,150
Thickness of oil zone, feet	20–27
Dip of oil zone	Flat to 5°
Overlying rock	
Type	Marl, shale, sandstone
Formation name	Arkadelphia
Age	Upper Cretaceous
Underlying rock	
Type	Chalk and Marl
Formation name	Saratoga
Age	Upper Cretaceous
Oil-bearing rock	
Type	Calcareous sand, Arenaceous limestone, shales
Formulation name	Nacatoch
Age	Upper Cretaceous

Source: Bureau of Mines OFR 55-79

The main reasons for selecting this field were:

- Low recovery of oil by pumping from surface wells due to the high viscosity of the oil.

- Overlying rocks are relatively strong; reservoir rocks are weak and the underlying rocks are below the water table.

- The field is fairly shallow.

- The field has a reasonably good reserve of oil that cannot be recovered by conventional methods (41.5 million barrels).

- The field has a relatively high permeability.

The specific Irma field is not ideal in all respects. Some of its negative features are:

- It is too small to write off a large mine plant.

- The oil-bearing zone is quite thin, resulting in high unit costs for development (average 20 feet thick for Irma vs 70 feet for the Yarega field).

- The amount of oil per acre foot is rather low—1,200 barrels per acre foot for Irma vs 1,656 barrels per acre foot at Yarega (3).

The field was, however, the most suitable that could be identified within the time available. Particular care was taken to ensure that the eventual evaluation reflected the potential of the method rather than the specific field.

Mining and Processing: The proposed plan involves the driving of a network of galleries in the marlstone caprock overlying the oil-bearing zone. From these galleries wells would be drilled into the reservoir for steam injection and production purposes.

Steam injection will serve several functions. It will lower the viscosity of the oil by heating, drive the oil toward the production wells and provide some lift in the production wells.

Successful steam drives have been performed in the same pay zone in the Smackover field of Arkansas (4). The well density provided by underground drilling of wells on 50-foot centers should increase both the total percent yield and the rate of yield in comparison with steaming operations from surface in which one well may cover acres. Oil production will be 1,750,000 barrels per year.

The oil produced from the steam stimulation method selected for the Irma field will require minimal processing prior to shipment. The proposed plant for Irma will primarily consist of oil/water and oil/gas separation units. Dependent upon the amount of fine clay or silt that is present in the oil, a process such as centrifuging may be required to remove the suspended material before final shipping of the product.

Technical Feasibility: The basic practicality of this operation is probably the highest of any of the underground operations evaluated in this study. The mining operation itself is relatively simple and is to be conducted at modest depths below surface.

The shafts will probably be sunk through unfavorable formations, poorly consolidated materials with considerable water problems. Such circumstances are often encountered, however, in the upper portions of mine shafts and can normally be overcome through the use of grouting and freezing. The horizontal development consists of modest, normal sized tunnels which should pose no unusual difficulties in drifting, support or long-term maintenance.

The only substantial question hanging over these aspects of the scheme concerns the level of geological certainty that can be generated from available data to permit design of a safe mining operation. It is necessary to be able to establish with good confidence that the strata selected will exhibit the required competence and impermeability with good continuity. At this time this appears to be a reasonable expectation.

The drilling of the holes, with adequate safeguards against gas blowouts, steam and oil leaks, etc., is not an established underground technology. All elements of such a system are in common use on surface, however, and there should be no severe problems in developing a rig which is compact and mobile enough for use in a mine.

The certainty of obtaining a satisfactory additional yield of oil from this method is difficult to determine at this time. The availability of lines of holes at the close spacings anticipated creates a very different order of possibility in steam drive efficiency. In particular, the opportunity to employ an essentially linear rather than radial drive could be extremely beneficial.

The generation of reliable predictions of this type, however, will require substantially more specific theoretical work than was possible in this study and will then need validation by field trials. It is felt, however, that the prospects are bright for obtaining good results.

Gravity Drainage

The simplest of all possible systems for underground recovery of oil is the drainage method and it was considered essential to include an example of the technique in this study. If successful in achieving a substantial recovery of the in situ oil, such a system would represent an optimally simple and effective technique. It thus in a real sense represents the lower bound case of all those studied and is, therefore, of great significance in evaluating the future of oil mining. The Wheeler Ridge field, located 25 miles south of Bakersfield, in Kern County, California, was selected for this purpose, mainly based on the following factors:

- The field has relatively high permeability.

- The oil-bearing sands are thin enough to be reached by simple drilling with moderate sized equipment from underground openings.

- The competency of the oil-bearing sands is low and would be very unlikely to permit mining operations within them at reasonable cost.

- The competence of the underlying beds would permit establishment and maintenance of production openings at reasonable cost.

Parameters of the oil field are summarized in Table 6.6.

Table 6.6: Wheeler Ridge Field

Permeability, md	~800
Porosity, %	~32
Oil gravity, °API	22
In-place grade, bbl/T	0.28
Depth to top of oil, feet	
Minimum	1,000
Maximum	1,310
Thickness of oil zone, feet	170
Dip of oil zone	Asymmetrical anticline - dips vary
Overlying rock	
Type	Sand, sandy clay, clay, sandstone, alluvium
Formation name	San Joaquin—Etchegoin, Tulare
Age	Pliocene to Recent
Underlying rock	
Type	Siltstone with interbedded sandstone
Formation name	Fruitville
Age	Upper Miocene
Oil-bearing rock	
Type	Sandstone interbedded with siltstone
Formation name	Santa Margarita (Coal Oil Canyon zone)
Age	Upper Miocene

Source: Bureau of Mines OFR 55-79

Mining and Processing: The method evaluated here is the simplest, and perhaps therefore, the most optimistic, possible. A network of tunnels is mined in a relatively competent formation lying just below the oil-bearing zone. A dense pattern of holes is drilled up into the reservoir and oil recovered by simple gravity drainage. Oil production will be 2.31 million barrels per year.

The processing plant at Wheeler Ridge will only be required to carry out the same function as the plant prepared for Irma, namely, oil/water separation and the removal of any suspended solids.

Technical Feasibility: To a substantial degree, the comments made on the mining aspects of the Irma operation are also applicable here. None of the activities involved are unusual in a mining operation, and the size of all mine openings will be modest enough to make serious problems unlikely in normal circumstances.

The major question, however, is once again that of geological certainty. The comments made with respect to the previous system also apply here, but with the added complication that no natural assumption can be made as to the competence or impermeability of underlying strata.

Indeed, if one seeks a general statement as to the properties of rock immediately underlying a petroleum reservoir, it would be that it is frequently a depleted reservoir material containing residual oil fractions, minor gas and completely saturated with drive water. This type of situation is clearly not the environment for which this system is suited. The necessary favorable circumstances will have to be carefully identified and proven to a good standard by surface drilling.

The oil yield to be expected in total from this operation is, at this stage, more of an assumption than an estimate. As noted previously, there is little or no relevant empirical experience of the behavior of such systems available to serve as the basis for analytical determination of yield.

SURFACE OPERATIONS

This section presents brief descriptions of the four surface mining operations designed as part of this study. Surface mining systems have proved, over many years, to be the cheapest mineral extraction methods available and are usually far cheaper than any underground mining system. The four surface methods considered are modified terrace pit mining, terrace mining, strip mining and the traditional open pit method.

Modified Terrace Pit Mining

The modified terrace pit mining system utilizes a combination of the terrace pit level haul system and the normal open pit system in which the waste is hauled up ramps out of the pit. This waste is then dumped into a previously mined-out portion of the pit. The maximum use is made in this method of the terrace pit hauling system because of its inherent cost advantages.

The geometry of the Sunnyside bituminous sandstone deposit, located 7 miles northeast of the coal mining town of Sunnyside, Carbon County, northeastern Utah, makes it an ideal candidate for this system (see Table 6.7). The system is called a modified terrace pit because the surface and ore body geometry of

Sunnyside will require some haulage upgrade in order to return the pit to the approximate original surface and, more importantly, to build large enough dumps to handle all of the waste. The ore would be crushed in the pit and conveyed to a processing plant located in a flat area in a canyon below the ridge.

Table 6.7: Sunnyside Field

Permeability, md	730
Porosity, %	21
Oil gravity	Not available—asphalt
In-place grade, bbl/T	0.57
Depth to top of oil, feet	
Minimum	0
Maximum	800
Thickness of oil zone, feet	10–350
Dip of oil zone	$3°$ to $10°$ NE
Overlying rock	
Type	Sandstone and shale
Formation name	Green River
Age	Eocene
Underlying rock	
Type	Interbedded sandstone, shale, mudstone and limestone
Formation name	Wasatch
Age	Lower Eocene
Oil-bearing rock	
Type	Bituminous sandstone interbedded with shale, mudstone and limestone
Formation name	Wasatch
Age	Lower Eocene

Source: Bureau of Mines OFR 55-79

Mine Design and Processing: The Sunnyside pit will be mined by the standard drill-blast-shovel loading-truck haulage open pit method. The modified terrace pit approach will permit much larger working areas for the shovels than in a normal open pit and will maximize the use of level hauls. The mine has been designed with 50 foot high benches.

Oil plant feed, delivered by an overland belt conveyor, must be stockpiled in an area with capacity sufficient to supply feed continuously to the separation plant. The stockpile should preferably provide blending capabilities and should be accessible to mechanical reclaiming equipment in case the material tends to bridge over in the reclaiming feeder hoppers.

Processing must provide for control over the top size of material to be treated, and hammer mills or toothed roll crushers probably will suffice. The sized feed can then be delivered to the 3 or 4 required production lines or modules in the plant. The first step of this process would involve a major grinding.

A hot water process is believed to be adequate for oil recovery. Variations in length of time during the pulping-scrubbing steps, pulp temperature and use of chemical additives such as caustic soda, sodium silicate and other wetting agentdispersants must be determined through pilot plant tests prior to design of the production plant.

The degree of cleaning of the oil-rich froth product from the rougher section of the plant will depend on the silt, clay or shale content of the plant feed. Possibly the viscosity of the final oil-rich product from the separation plant must be lowered by dilution with a low-viscosity solvent such as naphtha, which subsequently is recovered by distillation from the plant product (5), after which the suspended insoluble matter can be removed by centrifuging or additional flotation steps. Recirculation of certain middling products may be required to improve oil recoveries and quality of final product. The proposed processing plant would give an annual production of 16,929,000 barrels, for a total of 474,012,000 barrels of oil over the 28-year mine life.

Technical Feasibility: The proposed open pit mining method suggested for the Sunnyside, Utah deposit is completely feasible from a technical standpoint. The open pit method is a well-known and extensively practiced mining method, and the types of equipment used throughout the mining operation are standard items available from known manufacturers. The daily tonnage of materials to be moved is large, but open pits exist elsewhere in the United States which move larger tonnages with the same type of equipment.

The Sunnyside open pit utilizes a modified terrace pit method, whereby hauling of waste and ore is kept as far as possible on the level. This is possible at Sunnyside because of the existing topography and the potential location of the waste dumps. This modified terrace pit approach makes the mining operation a great deal simpler than the normal type open pit where spiral ramps are required for lifting the waste material to the dump sites and the ore to the processing plant. The use of overland conveyors to the processing plant is a commonly used system throughout the world, and the technology has been in existence for many years.

The processing plant design for the Sunnyside material is, as with all processing plants mentioned in this report, one of the major areas of uncertainty. The Sunnyside ore is hard material and would necessitate crushing and grinding prior to liberation of the heavy oils. Since no major pilot plants have been operated or tested with this material, the technical feasibility of the processing method cannot at this time be assessed.

Terrace Mining with Bucket Wheel Excavators

The terrace mining system is well suited to very large deposits which can be mined with level hauls for the overburden transportation and which, by their geometry, permit dumping of the overburden material back into the pit. The Kern River oil field in California, situated 5 miles north of Bakersfield, is an ideal candidate for this method.

The waste material contained in the excavations would be cut and loaded by bucket-wheeled excavators onto belt conveyors. The conveyors would then transport the waste on the terraces around the pit to the dump. The oil-bearing ore would be transported to a processing plant, the oil removed, and the sand tailings returned to the pit and dumped. Key properties of the Kern River field are summarized in Table 6.8.

Table 6.8: Kern River Field

Permeability	Variable (very high)
Porosity, %	34
Oil gravity, °API	12–13
In-place grade, bbl/T	0.62
Depth to top of oil, feet	
Minimum	43
Maximum	900
Average	500
Thickness of oil zone, feet	700
Dip of oil zone	2° to 4° SW
Overlying rock	
Type	Gravel, sand and silt
Formation name	——
Age	Upper Pliocene and Pleistocene
Underlying rock	
Type	Sands
Formation name	Santa Margarita
Age	Miocene
Oil-bearing rock	
Type	Lenticular sands and clays
Formation name	Kern River
Age	Pliocene

Source: Bureau of Mines OFR 55-79

Mining and Processing: For a production of 50,000 barrels per day, operating 360 days per year, 47 million tons of ore would have to be mined each year. This is approximately 130,000 tons of ore per day.

Generally, the maximum overburden thickness in the designed pit at Kern River is 600 feet; consequently, a three-bucket wheel excavator system has been selected with each bucket wheel handling up to 200 feet. The 200 foot high cut would be mined with a 120 foot face, a 40 foot step height and a 40 foot trough cut (Figure 6.3a). The three-bucket wheel excavators used for overburden stripping would each have an effective output of 6,000 bank cubic yards per hour and would feed onto belt wagons, which would transfer the overburden to main trunk conveyors on each bench. The main trunk conveyor system will transport the overburden around the pit to a stacker on each bench elevation, which will dump the material back into the worked-out area of the operation.

Figure 6.3: Terrace Pit Mining System—Kern River, California

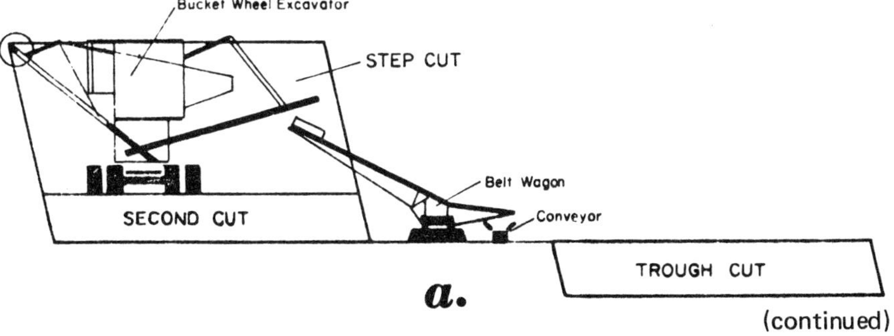

(continued)

Figure 6.3: (continued)

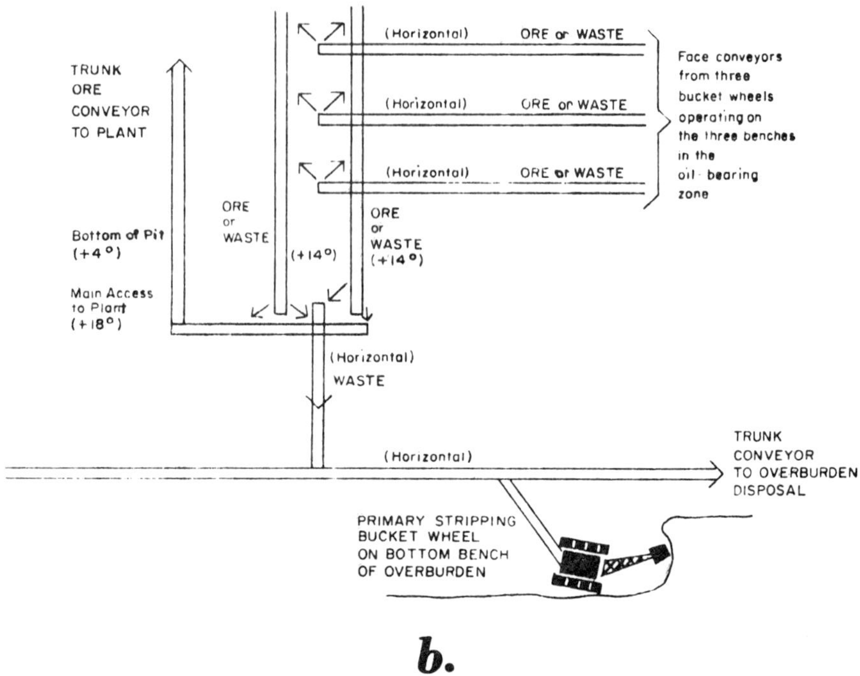

(a)   Bench excavation method
(b)   Conveyor haulage layout

Source: Bureau of Mines OFR 55-79

The method of ore mining permits the system to handle ore and contained waste. Each bucket wheel system in ore would handle up to 100 feet of ore body thickness. The three-bucket wheel excavators for ore mining would each be capable of producing 3,500 bank cubic yards per hour. The excavators would feed onto belt wagons and then onto the bench conveyors.

The following system is considered best. Three-bucket wheel excavator systems would be working in the oil-bearing zone with the middle unit taking the maximum cut permitting the upper and lower bucket wheels to take lower cut heights and consequently concentrating on the greater selectivity necessary to remove included waste from the ore.

These three-bucket wheel excavators would discharge onto their own bench conveyors, which, in turn, would discharge onto a twin belt ramp conveyor system feeding a trunk ore conveyor. The trunk conveyor will convey the ore from the mine area to the plant (Figure 6.3b). The waste material will be discharged onto a further ramp conveyor up to the main overburden conveyor on the lowest stripping bench.

The Kern River process plant feed material will be primarily sands and gravel that are both oil and water wet. The oversize material will have to be separated and crushed. A steam, hot water or solvent extraction technique would probably be used to remove the oil from the feed. The viscosity of the final oil-rich product will have to be lowered by dilution with a low-viscosity solvent such as naphtha. The proposed Kern River process plant will handle 47,000,000 tons of ore per year producing 19,411,920 barrels of oil.

Technical Feasibility: The proposed mining method for the Kern River field is a surface terrace pit utilizing bucket wheel excavators. This method of mining has been used extensively in Europe for many decades in mining for coal. The methods utilized and equipment are well-proven and the Kern River design is predicated upon known conventional equipment. The total material, ore and waste to be moved per day, 600,000 tons, is large by world standards. This daily tonnage is considered, however, to be totally feasible.

The terrace pit conveyor system is a standard method of materials handling, especially in large tonnage operations of this type. The one area of technical feasibility that does have some question is the processing. The processing of oil-bearing sands and gravels is, however, far better understood than hard reservoir rocks, since similar methods are used at the Athabasca tar sands and other places.

Strip Mining

Strip mining is ideally suited to flat-lying deposits overlain with up to 180 feet of preferably medium to soft overburden. If the mineral being mined is valuable enough, the cost of blasting the overburden, if it is hard, can be tolerated. The dragline stripping system digs and directly casts the overburden into a spoil pile, hence requiring no additional overburden transporting system.

An ideal candidate for strip mining is the Santa Cruz deposit in California. Key properties of the Santa Cruz field are summarized in Table 6.9.

Table 6.9: Santa Cruz Field

Permeability	Not available
Porosity, %	Not available
Oil gravity, °API	27
In-place grade, bbl/T	0.57
Depth to top of oil, feet	
Minimum	0
Maximum	100
Average	50
Thickness of oil zone, feet	1–35
Dip of oil zone	3° to 7° W
Overlying zone	
Type	Soil and Marne terrace deposits
Formation name	—
Age	Quaternary

(continued)

Table 6.9: (continued)

Underlying rock		
Type	Sandstone	Quartz Diorite
Formation name	Vaqueros	—
	Sandstone	
Age	Miocene	Pre-Cretaceous
Oil-Bearing rock		
Type	Sandstone interbedded with	
	diatomaceous shale	
Formation name	Monterey Shale	
Age	Miocene	

Source: Bureau of Mines OFR 55-79

The dragline siting on the highwall casts the waste or overburden into the previously mined-out strip. In this manner the bituminous sandstone or ore is uncovered and mined out by loaders and trucks. The trucks would haul ore to a processing plant where the oil would be separated from the sandstone. The spoil piles formed by the dragline would be graded and replanted. The sand tailings from the process plant would be either impounded in a tailings dam or, more likely, would be dewatered and returned to the mine to be dumped either between the peaks of the spoil piles or onto the bottom of the mined-out portion of the working strip.

Mining and Processing: The proposed mining operation at Santa Cruz has been based on assumed ore reserves of 16.7 million tons with an average grade of 10% or 24 gallons of bitumen per ton in the central area, plus 4,000,000 tons of the same grade in the northwest and adjacent area.

4,200 tons of ore must be mined per day. The oil-bearing material is hard and will require drilling and blasting. 6 cubic-yard-capacity hydraulic shovels will match the 15 foot face and have excess capacity for the job. Since the daily ore feed required by the processing plant is so small, it has been assumed that ore mining will be on a five-day-per-week, one-shift-per-day schedule requiring the capability of loading 1,000 tons of ore per hour. Since a 6 cubic-yard hydraulic shovel will produce 500 tons per hour, two such shovels should produce the required 1,000 tons. Six 35-ton trucks will be needed for a mining operation of this size.

As shown in Figure 6.4, the dragline will dig and cast the overburden onto a spoil pile on the other side of the ore mining cut. The hydraulic shovels will trail the dragline in the mining cut, removing the ore and loading it onto the trucks for haulage to the process plant. The spoil piles will be reclaimed as the strip advances.

The feed to the Santa Cruz process plant will be asphaltic sandstone with some oversize material. The feed will probably have to be passed through a toothed-roll crusher to reduce the oversize material prior to treatment. It is believed that a hot water/steam process should be adequate for oil recovery. The proposed oil separation plant would operate three-shifts-per-day, seven-days-a-week, for 340 days a year and would process 4,200 tons per day of ore at a planned recovery rate of 90%. This will result in the production of 378 tons (2,160 barrels) of bitumen per day. This would amount to 730,000 barrels per year for a total of 8,614,000 barrels over the 12-year mine life.

Figure 6.4: Santa Cruz Strip Mining System

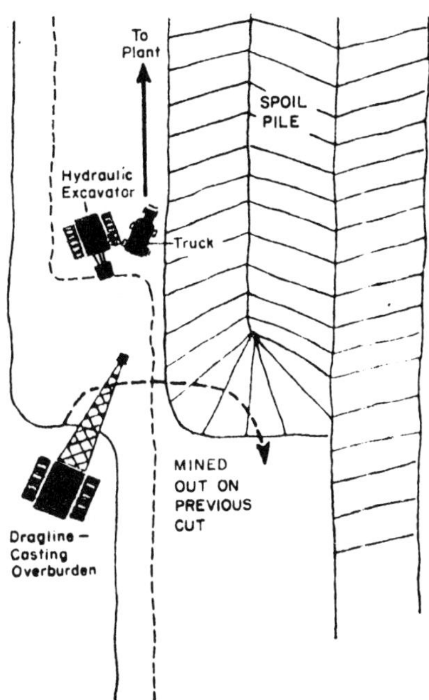

Source: Bureau of Mines OFR 55-79

Technical Feasibility: The strip mining methods selected for the Santa Cruz deposits are well-known and are extensively used. The proposed strip mine at Santa Cruz is very small compared to the majority of strip operations in the coal industry throughout the United States. The equipment is standard and has been available for many years. Once again, the only area of question in the technical feasibility of this proposed operation is the processing.

Open Pit Mining

Ore deposits which come close to the surface but owing either to the topography over the deposit or to extension of the ore body with depth, cannot be mined by systems that backfill the pit. They must be mined by open pit mining systems. In these methods the overburden and ore are hauled out of the pit up haul ramps and the overburden is placed in dumps away from the pit area. An ideal candidate for the open pit method is the Edna bituminous sandstone deposits. The depth of the potential ore body is so great that only after an immense excavation is made (requiring many years of mining) would enough ore be mined out to permit backfilling (with tailings and in-pit waste) without unmined ore being covered.

The Edna deposits cover 410 acres in southwestern San Luis Obispo County, California. Key properties of the Edna field are summarized in Table 6.10.

Table 6.10: Edna Field

Permeability	Not available
Porosity, %	36 (bitumen-free sand)
Oil gravity, °API	15
In-place grade, bbl/T	0.62
Depth to top of oil, feet	
Minimum	0
Maximum	500
Thickness of oil zone, feet	>200
Dip of oil zone	To 40°
Overlying rock	
Type	Unconsolidated alluvial sands and gravels
Formation name	——
Age	Quaternary
Underlying rock	
Type	Shale with interbedded sandstone
Formation name	Monterey Shale
Age	Miocene
Oil-bearing rock	
Type	Arkosic sandstone
Formation name	Pismo
Age	Upper Miocene–lower Pliocene

Source: Bureau of Mines OFR 55-79

Mining and Processing: The stripping ratio at Edna is 1:1 and the postulated ore body lends itself to a truck and shovel open pit. The pit has been designed to produce 50,000 tons of ore per day for 340 days per year. At the 1:1 stripping ratio this will require the daily removal of 100,000 tons of ore and waste.

The ultimate pit would be almost three-quarters of a mile in diameter at the perimeter and 350 feet deep at its deepest point. The major problems that could be encountered with this operation would be grade control due to the known extreme variations in bitumen content that occur at Edna, and the problem of trafficability on the high-bitumen material. The latter problem can be overcome with the use of high flotation tracks on the shovels and bulldozers and high flotation tires on the trucks.

The annual overall production will be 17,000,000 tons of ore and an equivalent amount of waste. The mine would operate with 50-foot benches using standard open pit drilling and blasting techniques as required in the harder areas. The majority of the waste that makes up the 1:1 stripping ratio is included waste within the ore. Only 5,000,000 tons of prestripping would be necessary because of the outcropping nature of the bitumen zone, an extremely favorable situation for early startup. The waste would initially be dumped outside the pit, although later, after detailed mine design and scheduling, in-pit waste dumping would be achieved.

The processing plant for the Edna deposit would utilize the same extraction method as proposed for Santa Cruz. As in that case, the Edna plant will require a crushing circuit for the oversize material, and will handle 17,000,000 tons of ore per year to produce 9,500,000 barrels of oil.

A preliminary report (6) has presented the results of laboratory research into the adaptation of the hot-water separation process to the recovery of hydrocarbons from bituminous sandstones of the Edna deposits. The report discusses the problems involved and the results obtained in treating the Edna bituminous sandstones.

The studies have demonstrated that the hot-water method of separation may be applied satisfactorily to separating hydrocarbons from the Edna bituminous sandstones, although the presence of mineral salts and fine silt in the deposits complicates the operation.

Certain operating problems still remain to be solved but, unquestionably, further study and experimental work will lead to their solution. Although the work has not been advanced to the pilot-plant stage and does not permit estimates to be made of separation costs, the small-scale studies have provided much valuable information of a fundamental nature on the recovery of bitumen from the Edna bituminous sandstones by the hot-water separation method.

Technical Feasibility: The proposed open pit method to be used for the Edna deposit is a conventional open pit mining approach, using conventional readily available equipment.

The tonnages involved would put this operation into the small- to medium-size open pit class. The proposed mining system is totally technically feasible, but as with the other proposed methods in this report, the processing system is the one area of questionable technical feasibility.

ECONOMIC EVALUATION

Based on the nine designs described previously, capital and operating costs were derived. The method of computation is detailed in the original report, together with any significant assumptions that were judged necessary. Each of the proposed oil mining methods was evaluated on a cost basis.

For each method considered, the estimated range of possible production costs, the actual value determined from the specific case studied, and the single value considered most likely to be attained in a favorable situation were determined.

Three methods stand out as offering little hope of viability under any reasonably foreseeable price situation. They are the three methods which involve an element of direct mining of the reservoir rock and have the following major economic disadvantages:

- High capital cost associated with very large mines and processing plants.

- High operating costs compared to surface mines.

- Major problems and cost areas associated with ventilation and water handling.

- Extreme difficulty in ensuring safe working conditions particularly with regard to risks from explosive and/or toxic gases.

- Substantial costs associated with waste disposal and environmental problems.

The worst of the three is the direct stoping operation. One of the main reasons for these high costs is the low grade of rock which is expected to be suitable for such an operation. This method, in common with all the others, is extremely sensitive to reserve grade, particularly in the lower range. It is considered very unlikely indeed, however, that any deposit will be found to combine the needed favorable qualities of rock competence for safe stoping conditions with a high enough grade to produce attractive cost levels.

The Casmalia block caving method appears somewhat better than the stoping method, but still seems unlikely to become of economic interest. The shatter and drain prospect shares all of the problems of the pure direct mining operations, including their sensitivity to ore grade and the efficiency of the drainage system, and the need for basically competent rock.

In summary, therefore, one conclusion of this study is that none of the underground direct mining systems demonstrates the potential of becoming economically feasible. This aptly parallels the conclusion reported earlier that each of them had serious problems of technical feasibility, notably in the safety area.

All of the six other methods considered, however, show some potential for economic success, being particularly attractive when the price of oil exceeds $20.00 per barrel. Four of these are surface mining options, and two are underground drainage methods.

The four surface mining options all suffer from high initial capital cost, particularly on the process plants, and from substantial problems in waste disposal and mitigation of environmental impacts. They all overcome this to a greater or lesser degree, however, due to the inherent economy of the large-scale earth-moving techniques employed.

The most expensive of these is the modified terrace open pit designed for Sunnyside, which is the only one in competent enough rock to require drilling and blasting of all the material, and which has extremely long haul distances for both ore and waste. In spite of these handicaps, however, it would not require too large an increase in the real price of oil for such a mine to become attractive.

The methods used at Santa Cruz and Edna, strip mining and open pit mining, respectively, both give results that border on the attractive. Edna has the best grade of the fields studied and a favorable stripping ratio, whereas Santa Cruz is penalized by its small size, giving a reduced operating life and consequent high amortization cost per barrel.

The Kern River terrace pit operation, employing bucket-wheel excavators, is the best of the surface operations from an economic standpoint, and appears to be viable now if the product is classified as stripper oil, and very nearly so if rated as new oil. It should be noted that any designation of mined oil as old would make such developments a very remote possibility. Kern River is an immense deposit which warrants exploitation on a very large scale, thus giving low unit costs for both mining and processing.

All of the methods studied are very sensitive to low ore grade, and this is the main factor which produces the range of possible costs. Other major factors include stripping ratios and rock hardness for the open pit mines and processing costs for all direct mining systems.

The two underground systems, which rely purely on drainage as the production system, are also in the competitive range of cost. The steam stimulation system used for the Irma field promises costs in the $16.00 per barrel range even in a thin deposit, and could be substantially more economical in a thicker deposit.

The evaluation of gravity drainage at the Wheeler Ridge field was made on very optimistic assumptions—that 50% of the contained oil would be recovered by pure drainage to the drillholes, with no expense incurred for stimulation, steam, etc. If this situation is indeed true, this system would be very attractive economically, in fact, the best of all the systems studied.

It is far more realistic, however, to expect that recoveries would be lower, or that stimulation would be required, and these possibilities are recognized in the preferred cost value. Even a thoroughly pessimistic view of the results to be expected, however, still produces costs which could be of substantial interest.

GENERAL CONCLUSIONS

Conclusions related to underground mining can be summarized as follows:

    (a)   Underground operations involving direct mining cannot be contemplated at this time with adequate confidence in their safety or economic feasibility. There is no obvious way of improving this situation markedly by any readily identifiable research or test project.

    (b)   Underground operations involving drainage of oil offer important attractions. Such projects would have the favorable features of low capital requirements and minimum environmental impact. Considerable further study is needed, however, to establish with confidence the potential yields to be expected from such operations.

Surface mining offers comparable or better economics and greatly superior demonstrable technical feasibility, at least in the mining elements and should, therefore, be pursued with at least as much rigor as underground prospects.

The main areas where additional work is required to further this possibility are in the fields of determination of mineable reserves and the design of oil separation plants. Even with the currently available technology, however, there is no reason why a deposit like Kern River could not now be subjected to a full commercial preliminary feasibility study.

No major mining operation can be undertaken without producing substantial and varied effects on the environment. The methods studied here are no exception to this rule, although the underground drainage methods are substantially less of a problem than the others. The impacts of the surface plants, both mine and processing, and the socioeconomic aspects are, however, all within the range of normal major industrial environmental technology. Disposal of the large quantities of mined waste material can be more of a problem and there may be no reasonable way of making the surface subsidence and hydrological impacts of methods such as block caving acceptable.

Recommendations for Further Studies

During the course of this study, many areas have been identified where additional effort is necessary before mining of oil can become an industry of major importance to the national economy. Suggested categories for such a study are:

- Heavy oils, tar sands, etc. This group would consist of shallow deposits, suitable for surface mining, where conventional production has not been significantly attempted due to problems of viscosity.

- Oils bound in diatomites, etc. This category would consist of oil reserves which have probably never been considered for production and possibly never explored at all, due to unfavorable bonding in the host rock. They also would be shallow fields suited to surface mining.

- Depleted shallow fields. This group includes fields at shallow, surface mineable depths where exploitation by conventional well techniques is either complete to economic limits or can be advantageously replaced by mining.

- Depleted deeper fields. These prospects would lie at depths below the economic range of surface mining and would have been depleted to some extent, and certainly depressurized, by conventional well production techniques. They would have characteristics making improved productive potential by drainage from underground access points a strong possibility, and would have competent impermeable formations either over or underlying them suitable for mine development.

There is also a need for improved technology for evaluating and documenting unrecovered oil remaining in place. The petroleum industry has the technology available to obtain good estimates of the quantity of oil that can be produced conventionally from a reservoir unit.

Estimates of the oil remaining in place after depletion are, however, of a lower order of confidence. This knowledge is based on the limited number of widely spaced core holes where oil content has been measured, augmented by porosity measurements, by geophysical techniques and flow permeability studies. In very uniform reservoir rocks, where there are no radical small scale variations of the porosity, permeability, grain size, etc., these estimates may well be reasonable.

As a basis for design and evaluation of a mining operation, however, very few oil deposits as currently delineated and sampled would be able to demonstrate ore tonnages and grades to satisfy conventional mining standards. It seems possible that, even in the densely drilled Kern River field, the highest ore reserve classification that could be expected would be "indicated," and the typical field would more probably be in the lowest reserve category, "inferred." Methods of study anticipated include the drilling of a series of closely spaced holes within a depleted oil field. These holes would be cored in part using large diameter equipment, to develop good estimates of the oil saturation remaining in the reservoir and the physical nature of the oil contained in the pores. Drilling of a grid of reasonably closed spaced holes, possibly augmented by whipstocked deflections, would allow a detailed nonpetroleum engineering approach to resource evaluation.

Drainage Production Studies: Development of the methodology for making a more accurate estimate of the quantity of oil recoverable in an underground drainage operation, and the rate of its recovery, is essential as a prelude to specific feasibility studies. Progress toward such a situation will include further study of existing reservoir data and the effects of an underground excavation on reservoir properties.

By-Product Considerations: Although not a basic ingredient in the evaluation of oil mining, it is possible that important by-products could be recovered from some such operations. As an example, uranium is known to occur associated with some asphalt deposits and oil fields, and the potential for by-product production uranium from oil mining operations should be investigated. Examples of areas with such potential include fields in California, parts of Texas, and the Joplin tristate area.

Studies of Oil Processing Plant Technology: Processing plants represent a major portion of the cost of the surface mining options, and a large proportion of the technical uncertainty. There is, therefore, a need for a study that would have as its objective the generation of conceptual design and costs of construction and operation for oil processing plants, capable of handling the following types of oil-bearing rock: oil sands, tar sands, and oil-impregnated diatomaceous earth.

REFERENCES

(1) American Petroleum Institute, *Reserves of Crude Oil, Natural Gas Liquids and Natural Gas in the United States and Canada as of December 31, 1976,* Washington, D.C., Vol. 31, May 31, 1977.

(2) American Petroleum Institute, *Reserves of Crude Oil, Natural Gas Liquids and Natural Gas in the United States and Canada as of December 31, 1975,* Washington, D.C., Vol. 30, May 31, 1976.

(3) Teas, L.P., "Irma Oil Field, Nevada County, Arkansas," *Structure of Typical North American Oil Fields,* Vol. 1, Tulsa, Oklahoma, American Association of Petroleum Geologists, 1929.

(4) Fancher, G.H. and Mackay, D.K., *Secondary Recovery of Petroleum in Arkansas—A Survey,* El Dorado, Arkansas, Arkansas Oil and Gas Commission, p 59, 1946.

(5) Oblad, A.G., et al, "Recovery of Bitumen from Oil-Impregnated Sandstone Deposits of Utah" *Oil Shale and Tar Sands,* AIChE Symposium Series No. 155, Vol. 72, pp 69-78.

(6) Taff, J.A., "Physical Properties of Petroleum in California," *Problems of Petroleum Geology,* edited by W.E. Wrather and F.H. Lahre, Tulsa, Oklahoma, American Association of Petroleum Geologists, 1934.

ANOTHER METHOD—OIL RECOVERY
BY NUCLEAR EXPLOSION

The information in this chapter is based on:

"Oil Recovery by Nuclear Explosion," *Nauka i Zhizn* (U.S.S.R.),
No. 2 (1973), by L. Shadrin, translation JPRS #59039, prepared
by U.S. Joint Publications Research Service, May 17, 1973.

INTRODUCTION

One method of stimulating oil recovery during exploitation of oil deposits located in thick reservoirs whose porosity is inhomogeneous and whose permeability is low is the torpedoing of wells, i.e., acting on a reservoir with an explosion. A special charge, i.e., a torpedo, is lowered into the well and is detonated there (by means of a fusehead and wire, leading to the surface). As a result of the formation of deep fissures, the permeability of the cutting face zone of the productive formation increases sharply which makes it possible to increase the daily petroleum yield five to tenfold.

However, this increase of productivity is temporary. In a few days the flow of oil begins to decrease sharply, particularly in formations with low permeability. The reason is that as the result of torpedoing the permeability is increased only in the cutting face zone of the reservoir, limited to a few meters, whereas the oil deposit extends for several hundreds of meters and even kilometers.

It is clear that in order to achieve a stable and long-term effect of stimulation of oil recovery, artificial fissures must be made in as large a volume of the formation as possible, but this requires an explosion that releases incomparably greater energy than can be achieved in charges of the torpedoes that have been used. The obvious conclusion naturally suggests itself: increase torpedo charges. This, however, is unacceptable, since it requires an increase in the size of the torpedoes and a corresponding increase in the drill hole diameter. The additional costs of sinking larger-diameter shafts are not justified.

APPLICATION OF NUCLEAR EXPLOSIONS

The mastery of nuclear energy opened promises for the solution of the problem of stimulating oil recovery. In cases where the use of chemical explosives is not effective nuclear charges may be employed. The energy of nuclear explosion per unit volume of charge is tens, hundreds and even thousands of times greater than the specific energy of chemical explosions. This makes it possible to place nuclear charges of tremendous yield in relatively small containers.

Technology

The very technology of underground nuclear explosion is described in general terms as follows. A charge hole is bored. The container with the charge is lowered into the hole to a given depth, suspended on a reinforced cable. To improve the utilization of the explosion energy the well is carefully sealed, i.e., it is plugged. Then the charge is detonated from a shelter located at a safe distance from the charge hole.

Release of Energy: The nuclear explosion represents a practically instantaneous release of energy during fusion of the nuclei of uranium or plutonium atoms or during synthesis of the nuclei of nonferrous elements.

The energy of a nuclear explosion is expressed in equivalent TNT kilotons (kt). This parameter shows how many kt of the chemical explosive trinitrotoluene is required to produce the same effect as a given nuclear charge. During the detonation of a nuclear device with the yield, for example, of one kt, 4.2×10^{19} ergs of energy are released in 10 billionths of a second. The yield of even such a relatively very "weak" explosion is approximately 2 billion times the rated yield of all Soviet power plants.

One of the main requirements of any method of intensifying oil recovery is that regardless of its effect on oil yield, it must break down the natural conditions of insulation of the formation. Otherwise this may involve rupture of the deposit and escape of the oil into nonproductive strata.

Accuracy of Charge: To prevent this from happening during nuclear stimulation the power of the charge, its depth and its placement in the stratum must be accurately calculated. For this purpose it is assumed that the zone of mechanical action of the explosion on the rock mass will not extend beyond the boundaries of the reservoir. This is why the nuclear explosions used for stimulating oil yield are designed as completely internal explosions, which do not eject broken rocks to the surface. Such explosions are referred to as comouflet.

Effect of Temperature and Pressure: The initial temperature in the nuclear explosion zone exceeds 10 million degrees, and the pressure reaches one billion atmospheres. The powerful shock wave front propagates radially from the charge center. Vaporization, melting, fracturing, displacement and cracking of surrounding rock take place. A spherical cavity is formed in the rock evaporation zone, which expands until the pressure of the gases trapped in it is equal to rock pressure exerted by overlying rocks. The molten rock flows from the roof and walls of the cavity to the bottom, where it collects in a pool and gradually solidifies as an insoluble vitreous mass.

Formation of Collapse Tube: As the gas pressure inside the cavity decreases, the vault and walls collapse (also facilitated by the expansion wave reflected by the earth's surface). This leads to the formation of a vertical collapse tube with a hemispheric base and cupola-shaped dome. The entire volume of the initial spherical cavity is distributed in the spaces between the pieces of broken rock. A collapse tube is formed in any rock as a result of camouflet nuclear explosions. Its size depends on the yield of the nuclear charge, its depth and type of rock.

The permeability of the material in the tube itself is a million times greater than the natural permeability of the productive reservoir, and it increases tens and hundreds of times in the fissure zone. Therefore, the oil or gas escapes from the depth of the formation into the collapse tube at a rate exceeding many times the filtration rate prior to the nuclear explosion, and here the collapse tube behaves as a greatly enlarged well shaft. The oil accumulates here and its recovery essentially acquires the character of sampling from an underground reservoir.

THE SAFETY OF NUCLEAR EXPLOSIONS

Radioactive Contamination

In the nuclear stimulation of oil recovery the basic source of danger is the possibility of contamination of the oil and gas with radioactive by-products. Studies have shown that up to 90% of the radioactive by-products are localized and concentrated in the insoluble vitreous material melt that freezes on the floor of the collapse tube. The rest of the radioactive isotopes, characterized by various lifetimes (from a few seconds to several years) remain in the gas phase of the explosion products, which comes into contact with the oil and is mixed with natural gas.

The obvious conclusion is not to begin the recovery of oil from the nuclear stimulation zone until completion of the natural decay of all radioactive isotopes. But then it would be necessary to wait many years, which is unacceptable from the technical and economic points of view. What is the answer? In practice the following approach is taken. After waiting for the time required for decomposition of short-lived isotopes (from a few days to a few months), the collapse tube is opened by means of a drill hole.

Then the concentration of residual long-lived isotopes decreases to a safe level. For this purpose the contaminated gas is diluted many times or is displaced with air (or clean natural gas). As an additional safety measure the gas removed from the nuclear stimulation zone is used for a certain amount of time only for industrial purposes, as boiler fuel or raw material in the petrochemical industry. Data from numerous studies indicate that the oil in the formation does not pick up any residual radioactivity during nuclear stimulation and may be used as petroleum recovered by any other method.

Seismic Considerations

Nuclear explosions produce elastic vibrations of the rock mass and earth's surface closest to the stimulation zone. Therefore, for a given charge depth, the unit charge yield is limited to that which insures absolute seismic safety for underground installations in the vicinity of the operations. Underground nuclear explosions may be conducted only at oil fields located near sparsely populated regions. All safety measures must be employed to insure complete safety for personnel participating in the operations, as well as for those who inhabit the seismic danger zone. The experience of numerous underground nuclear explosions indicates that the diligent implementation of modern radiation safety measures completely solves the problem of the safety of people and environment during the conduct of nuclear stimulation of oil recovery.

U.S.S.R. INDUSTRIAL EXPERIENCE IN NUCLEAR STIMULATION

Experimental industrial development of nuclear stimulation of oil recovery was first initiated in the Soviet Union. A report of the results of the operations was presented by a group of Soviet scientists and engineers at the 8th World Petroleum Congress (Moscow, 1971). Underground nuclear explosions were conducted in two oil fields, where the expected final oil recovery by other methods could not have exceeded 30%. At the first oil field three nuclear charges with a total yield of about 13 kt were detonated. They were placed at a depth of 1,350 m in the same part of the deposit that contained most of the residual petroleum. After the explosions it was established that the fissure zone extended to a radius of 300 to 400 m from the charge holes, and individual cracks were noted at a distance of up to 800 m. As a result of the nuclear explosions the productivity of the 20 closest wells increased, and the summer oil yield of the deposit increased by more than one-third.

The second target of nuclear stimulation was a field in which the oil reservoir consists of limestones and dolomites, which have extremely nonuniform porosity and permeability. This interfered with oil recovery, so that most of the oil remained in the depths. Two nuclear charges with a yield of 8 kt each were detonated in succession at this oil field. As a result, the productivity of the 7 operational wells located within a radius of 800 m from the explosion epicenters increased approximately one and one-half times. During nuclear stimulation complete safety of servicing personnel and of the local population was insured at both fields. Neither radioactive contamination of the oil nor any change in its physical-chemical properties was noted. It was, therefore, possible to resume normal operation just a few days after the explosions.

UNDERGROUND RESERVOIR CONSTRUCTION

Underground nuclear explosions may also be used for constructing oil reservoirs similar to reservoirs used for storing natural gas (which may also be recovered by nuclear explosion). Such underground warehouses, capable of accommodating millions of tons of petroleum, are needed near large seaports, for development of offshore oil fields on the continental shelf and in Arctic Regions, where oil is stored until the arrival of tankers or completion of pipeline construction.

The first nuclear explosions conducted deep within oil and gas deposits opened a new chapter in the history of the utilization of atomic energy for peaceful purposes. The development of this technology may serve as a virtual fountain of youth for old but no longer rich deposits with vanishing formation energy. Peaceful nuclear explosions will help to unseal and put into operation enormous deposits.

APPLICATION OF EOR TECHNIQUES TO ACHIEVE PRODUCTION POTENTIAL

The information in this chapter is based on:

An Investigation of Primary Factors Affecting Federal Participation in R&D Pertaining to the Accelerated Production of Crude Oil, prepared by J.M. Sharp, of Gulf Universities Research Consortium, for National Science Foundation, under Contract NSF-C-942 (GURC Report 140), September 15, 1974.

Assessment of Enhanced Recovery Technology as a Means for Increasing Total Crude Oil Recovery in Texas, NSF Report RA-N-74-251, prepared by E.A. Lohse, of Texas Governor's Energy Advisory Council, for National Science Foundation, January 6, 1975.

Review of Secondary and Tertiary Recovery of Crude Oil, FEA Report G-75/482, prepared by Lewin and Associates, Inc., for Federal Energy Administration, June 1975.

The Potential and Economics of Enhanced Oil Recovery, FEA Report B-76/221, prepared by Lewin and Associates, Inc., for Federal Energy Administration, April 1976.

Potential Environmental Consequences of Tertiary Oil Recovery, prepared by C. Braxton, R. Stephens, C. Muller, J. White, J. Post, J. Norton, M. Goldberg and P. Stevenson, of Energy Resources Co., Inc., for U.S. Environmental Protection Agency, under Contract 68-01-1912, July 1976.

Enhanced Oil Recovery Potential in the United States, OTA-E-59, prepared by Lewin and Associates, Inc., National Energy Law and Policy Institute, and Wright Water Engineers, Inc., for U.S. Office of Technology Assessment, January 1978.

It has been shown in the preceding chapters that tertiary recovery of crude oil refers to additional recovery from oil fields wherein primary and secondary production has been, or could have been, completed. An important determinant for realizing this additional production from enhanced oil recovery centers on the selection and sequence of the appropriate enhanced recovery techniques. For example, while the use of water injection (waterflooding) is the least costly and most widely used technique of enhanced recovery, it is considerably less productive in the long run than the most costly miscible-augmented water injection.

However, once a field is under waterflooding, it becomes less attractive economically to undertake miscible waterflooding. When the waterflooding is completed with the wells shut down and plugged, the costs of redrilling the fields and cleaning the well of water prior to oil production create an outlook which is economically prohibitive in most cases. Thus, an operator of a field which is being considered for enhanced recovery faces a series of decision points early in the productive life of the field, as diagrammed in Figure 8.1.

Figure 8.1:   Decision Points for Enhanced Recovery

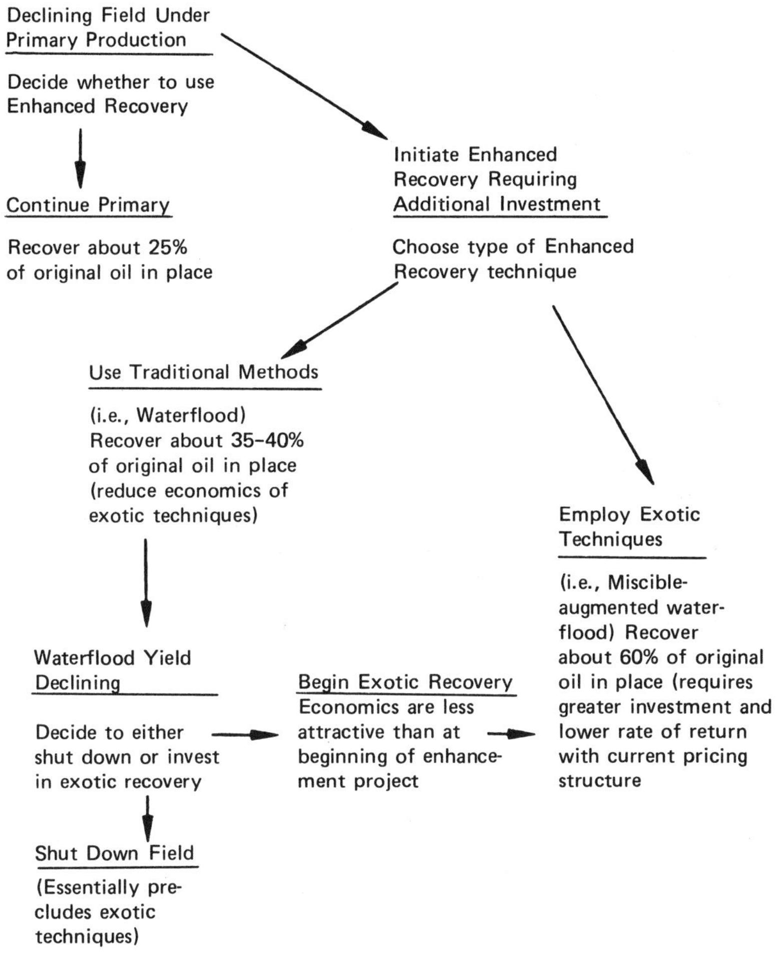

Source: FEA/G-75/482

TERTIARY PRODUCTION FORECASTS

The promising new tertiary or exotic enhanced processes are not fully developed. Little production is now being obtained from these processes and extensive field testing will be necessary before these methods are widely used. With an accelerated program assisted by federal government incentives, the tertiary or exotic processes might produce 423,000 (1) to about 1,500,000 (2) barrels per day by 1985.

Relative Importance of Enhanced Recovery Methods

It is the consensus of the twenty oil companies providing data to the Gulf Universities Research Consortium that four EOR methods will dominate the national scene with respect to the additional amount of crude oil produced by EOR. The relative amounts that should be realized by the application of these four methods, based on their estimates, are given below:

	Percent
Micellar [water-driven miscible slug (surfactant) or microemulsion flooding]	58
Thermal	29
Carbon dioxide	8
Hydrocarbon miscible	5

The percentages for micellar and thermal methods were quite uniform among all participants. Those for CO_2 and hydrocarbon miscible varied widely in terms of percent deviation (ranging from nearly zero to as high as 12%, e.g., for CO_2), but there was complete agreement that the ultimate recovery from these two methods is much smaller than that to be expected from either thermal or micellar methods.

The twenty-company consensus regarding the potential recovery to be realized from EOR methods ranged from an almost assured value of 18 billion barrels to a reasonable value of 36 billion barrels. This amounts roughly to 6 to 12% of the remaining oil in place.

According to Lewin and Associates, Inc., the contribution of five EOR techniques over time (from the three oil-producing states of California, Louisiana and Texas) is projected as shown in Table 8.1

The upper bound case shown in Table 8.1 reflects the maximum anticipated response of releasing tertiary oil to upper tier prices (maintained in real terms). This case assumes a concerted and creative level of federal initiative coupled with a major research/development effort; it assumes that by 1990 tertiary technology will become a proven conventional means of oil recovery.

The timing of tertiary reserves and production are displayed in Figure 8.2, showing the interaction of tertiary reserves and production over time. It should be noted that Lewin predicted that, under the upper bound case, by 1985, 17.5 billion barrels of proved reserves (in these three states) would be added to the nation's oil supply, with the total 30.5 billion barrels being proved by the year 2000.

Table 8.1: Projected Annual Tertiary Crude Oil Production for Three States—Upper Bound Case (Net Production, MM bpd)

	1976	1980	1985	1990	1995	2000
1. Steam Drive	0.1	0.3	1.0	1.3	0.8	0.7
2. In Situ Combustion	*	0.1	0.2	0.1	0.1	0.1
3. CO_2 Miscible and Hydrocarbon Miscible**	0.1	0.3	0.7	1.0	1.5	1.5
4. Surfactant/Polymer	*	*	0.1	0.6	1.4	1.3
5. Polymer Augmented Waterflooding	*	*	*	*	*	*
TOTAL	0.2	0.7	2.0	3.0	3.8	3.6

*Less than 0.1

**The study assumes that current hydrocarbon miscible projects will be converted to CO_2 miscible projects in the future.

Source: NSF-C-942

Figure 8.2: EOR Production and Proved Reserves for Three States—Upper Bound Case

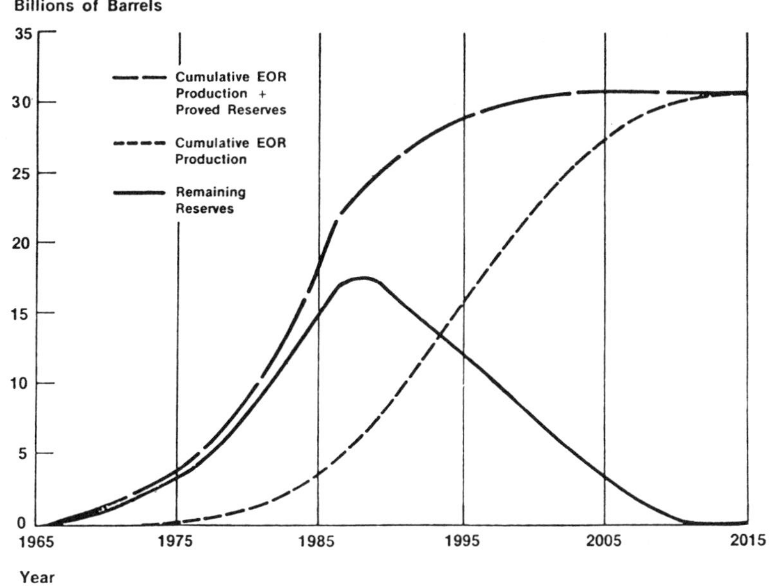

Source: FEA/B-76/221

The above projections were based on data from 245 large onshore reservoirs in California, Louisiana and Texas. For a later assessment (OTA-E-59), however, additional data were collected for onshore reservoirs in those states, for reservoirs in other producing areas of the United States, and for reservoirs in offshore areas (primarily the Gulf of Mexico).

The expanded data base for this assessment was acquired from federal, state and private sources. After all available data were examined and cataloged, they were edited for volumetric consistency. These data were reviewed by the Office of Technology Assessment (OTA) as other sources of information became available. Additional data led to reductions in estimates of remaining oil in place (ROIP) in California reservoirs which contained oil with gravities above 25°API.

The resulting data base for the assessment includes 385 fields from 19 states (Table 8.2). These 385 fields (835 reservoirs) contain 52% of the known ROIP in the United States. The reservoir data in the OTA data base are representative of the known oil reservoirs in the United States.

Referring to Table 8.2, the total U.S. ROIP is 300,338 MM bbl. This value includes 3.3 billion barrels of oil which are included in API indicated reserves as recoverable by secondary methods. It does not include 1.0 billion barrels of enhanced oil in the API proven reserves.

Uncertainty in the Oil Resource

Two EOR processes, surfactant/polymer and CO_2 miscible, are generally applied after a reservoir has been waterflooded. A large portion of the resource for these processes will be located in the reservoir volume which was contacted by water. The oil remaining in this region is termed the residual oil saturation.

There is uncertainty in the estimates of residual oil saturation and hence in the oil which is potentially recoverable with surfactant/polymer and CO_2 miscible processes. A review of the technical literature and discussions with knowledgeable personnel in the oil industry led to the following observations:

(a) There are few reservoirs whose estimates of residual oil have been confirmed by independent measurement.

(b) The uncertainty in the aggregate estimate is due to a lack of confidence in measurement techniques compounded by a limited application of those methods.

(c) The estimates of residual oil saturation may be off by as much as 15 to 25% (or more).

(d) Estimates of the oil recoverable by surfactant/polymer and CO_2 miscible processes will have a large range of uncertainty because of the uncertainties in the estimates of residual oil saturation.

Estimates of the amount of oil that can be recovered by the different EOR processes were based upon an individual analysis of each reservoir in the analysts' data file using reservoir-by-reservoir simulations. Results from the individual reservoir calculations were then compiled and extrapolated to a national total. The accuracy of this approach depends on the extent, representativeness and precision of the reservoir data file.

Table 8.2: Extent of Reservoir Data Base Used to Assess Enhanced Oil Recovery Potential

State	Remaining oil in place (MMB)	OTA data base *			
		Fields	Reservoirs	Remaining oil in place (MMB)	Percent of State
Alabama	519	2	2	354	68
Alaska	14,827	6	9	14,601	98
Arkansas	2,768	3	3	1,328	48
California	62,926	41	67	45,125	72
Colorado	3,002	21	21	1,490	50
Florida	556	3	3	465	84
Illinois	5,726	8	9	2,421	42
Kansas	10,403	28	28	3,345	32
Louisiana					
Onshore	13,696	24	47	6,731	49
Offshore	7,349	24	372	2,983	41
Mississippi	2,988	11	12	1,187	40
Montana	3,796	15	15	1,443	38
New Mexico	11,241	15	18	4,960	44
North Dakota	1,849	6	7	548	30
Oklahoma	25,406	33	35	6,548	26
Pennsylvania	5,344	5	6	1,077	20
Texas	100,591	111	146	54,221	54
Utah	2,725	6	6	1,734	64
West Virginia	2,064	2	2	194	9
Wyoming	10,628	21	27	4,543	43
Total States covered (MMB)	288,404	385	835	115,298	54

*OTA data base is 52% of remaining oil in place in the United States.

Source: OTA-E-59

SELECTING AN APPROPRIATE TECHNIQUE

Before the production potential of EOR can be realized, however, each of the various techniques must be applied in the field according to its own unique suitability. Two major questions face a field operator embarking on a program of enhanced recovery. They are: (1) Are the technical characteristics of the reservoir conducive to enhanced property? and (2) Which, if any, of the enhanced recovery techniques will prove to be economically favorable?

The answers to these two questions vary dramatically from field to field and reservoir formation to reservoir formation, depending on considerations such as:

(a)   the volume of oil remaining in the ground;

(b)   the geological characteristics of the producing formation, including its porosity, depth, permeability and continuity;

(c)   the physical properties of the oil, especially its ability to flow (viscosity);

(d)   the location and condition of existing wells and the need for drilling new wells;

(e)   the costs and availability of equipment, materials and labor;

(f)   the price for the oil and the cost of capital; and

(g)   the timing and extent of the investments and production increases.

Table 8.3 displays the recovery techniques which would likely be selected for the different types of crudes and reservoir formations.

Table 8.3: Selection of Recovery Techniques

Crude Type	Reservoir Characteristics	Suitable Techniques*
Heavy, High viscosity	Shallow formation	
	Good permeability	Steam flood
	Poor permeability	Cyclic steam injection
	Deep formation	
	Good permeability	In situ combustion
	Poor permeability	Enhanced recovery feasible only for large reservoirs
Light, Low viscosity	New, undepleted	
	Good permeability	Primary Water/Gas injection Miscible water/Gas injection
	Poor permeability	Primary Water injection
	25% depleted (post-primary)	
	Good permeability	Water/Gas injection Miscible water/Gas injection
	Poor permeability	Water injection Stripper production

(continued)

Table 8.3:   (continued)

Crude Type	Reservoir Characteristics	Suitable Techniques*
	50% depleted (watered out)	
	Good permeability	Miscible water/Gas injection
	Poor permeability	Enhanced recovery feasible only for large reservoirs

*Under 1975 economic conditions.

Source: FEA/G-75/482

Development of the Screening Guide

The major tertiary methods differ from one another substantially in their applicability to various reservoir and crude oil characteristics. In order to assess the applicability of specific methods to specific reservoirs, an explicit and systematic screening mechanism was developed by Lewin and Associates, Inc. (FEA/B-76-221) for the following five major EOR methods:

    (1)   steam drive;
    (2)   in situ combustion;
    (3)   carbon dioxide miscible flooding;
    (4)   surfactant/polymer flooding; and
    (5)   polymer/augmented waterflooding.

An appendix at the end of this chapter lists the EOR projects, located in California, Louisiana and Texas, which were used in developing the screening guide.

The pioneering efforts of Ted M. Geffen, of the Amoco Production Company, provided the starting point for their development of the screening guide. He has published numerous articles (3)(4) comparing the reservoir and crude oil characteristics most favorable for several EOR techniques. An adaptation of Geffen's criteria for the application of some enhanced recovery methods (5) is presented in Table 8.4.

Other studies such as that by the IOCC (6) have analyzed the conditions of applicability for single EOR techniques. These were put into a comparable format (in the form of a matrix of techniques by characteristics) and used in consultation with various experts. In addition, an analysis of more than 100 EOR projects was conducted.

Based on this analysis, a series of profiles for each technique was generated. The screening matrix was compared to these profiles to see whether the entries in the matrix accurately reflected actual field experience. Where they did not, the matrix entries were modified. Figures 8.3a, 8.3b and 8.3c display these profiles for, respectively, the key reservoir characteristics of gravity, depth and residual oil saturation at project start.

The resultant screening guide (Table 8.5) is a compilation of technical and broad economic factors which approximate the decision parameters used by the oil industry in selecting prospective sites for EOR projects.

Table 8.4: Criteria for Application of Enhanced Recovery Methods

Process	Oil grav., °API	Reservoir temp., °F.	Operating press., psia	Oil visc., cp	Perm., md	Thickness, ft	Favorable factors	Factors which increase risk
High-pressure gas drive*	40	–	3,500	1	–	–	1 Deep formation (see pressure) 2 High dip 3 Undersaturated crude with high C_2-C_6 concentration	1 Extensive fractures 2 Strong water drive 3 Gas cap 4 High vertical permeability in a horizontal reservoir 5 High permeability contrast 6 Natural gas supply limitations
Enriched gas drive* (condensing gas drive)	30	–	1,300 (900-1,000)	3	–	–	1 At low pressures gas required with high C_2 and low N_2 2 High dip 3 Thin pay, low k	1-6 same as above 7 LP-gas supply limitations
LPG slug	30	–	1,300 (800-900 with modif.)	5	–	–	1 Distress (cheap) LP-gas available 2 High dip 3 Thin pay, low k 4 C_2 rich fluid for low pressure reservoir	Same as above
CO_2 (miscible)* (water- or gas-driven)	30	–	1,100	3	–	–	1 CO_2 available 2 High dip 3 Thin pay, low k	1-5 same as above 6 CO_2 supply and transportation requires high initial investment

(continued)

Table 8.4: (continued)

Process	Oil grav., °API	Reservoir temp., °F.	Operating press., psia	Oil visc., cp	Perm., md	Thickness, ft	Favorable factors	Factors which increase risk
Micellar flooding*	–	200	–	10	20-50 in contacted portion of reservoir	–	1 It is ESSENTIAL that good water be available (5,000 pps total dissolved solids; 500 pps Ca^{++} and Mg^{++}) 2 Waterflood sweep 50%	1 Extensive fractures 2 Strong water drive 3 Gas cap 4 High permeability contrast 5 Highly saline (30,000 pps TDS) connate or flood match
Steam drive	10	–	2,500		–	20	1 Existing wells adaptable to steam injection 2 Available gas supply for steam generation 3 Available water which is cheap, slightly alkaline, free of H_2S, oil, dissolved iron, and turbidity	1 Strong water drive 2 Gas cap 3 Low net to gross pay fraction 4 Extensive fractures (not as serious as in other injection methods)
COFCAW	45	–	250		–	10	1 Formation temperature 150° F. 2 Low vertical permeability 3 Available water which is cheap and won't precipitate solids in presence of air 4 Existing wells in condition to withstand high pressure 5 Cheap gas supply for compressors	1 Extensive fractures 2 Gas cap 3 Strong water drive 4 Low net to gross pay fraction 5 Serious preexisting emulsion problems

Source: NSF-RA-N-74-251

Figure 8.3: Profiles for Key Reservoir Characteristics

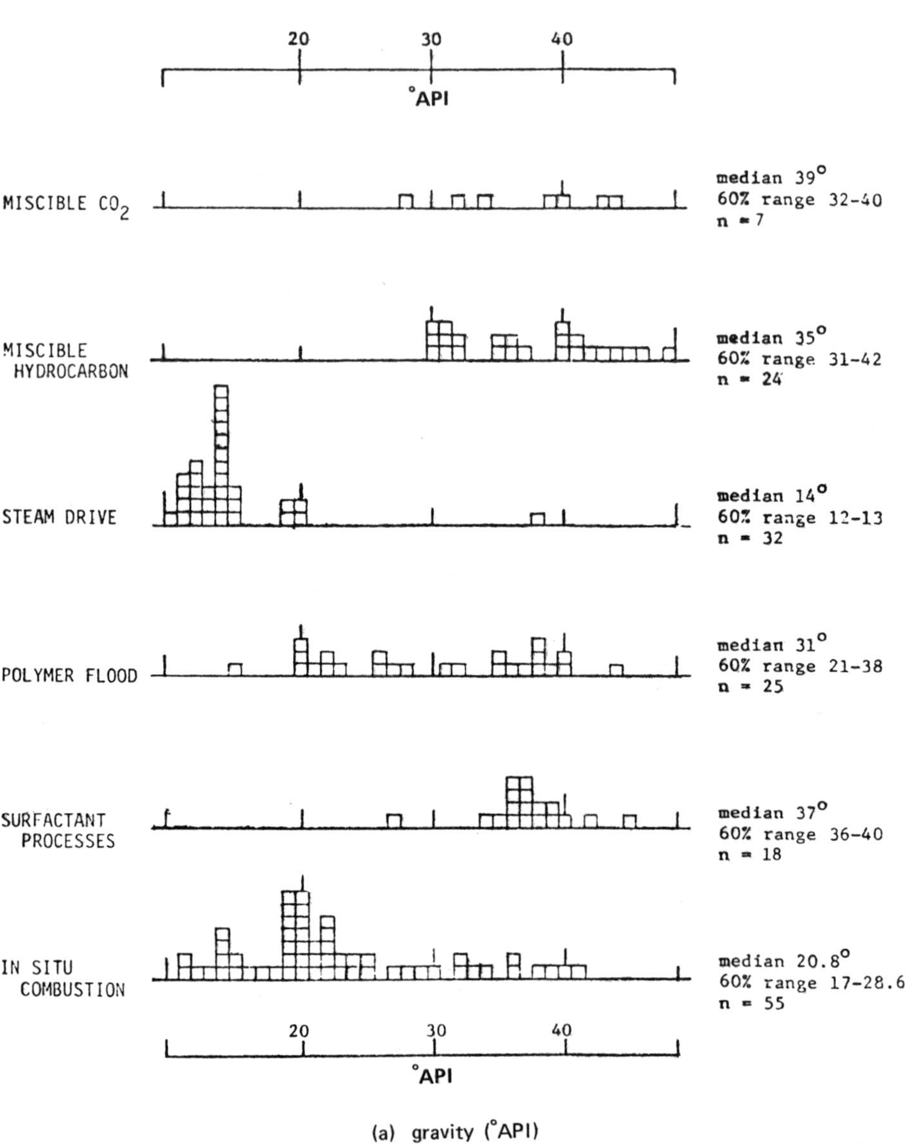

(a) gravity (°API)

(continued)

Figure 8.3: (continued)

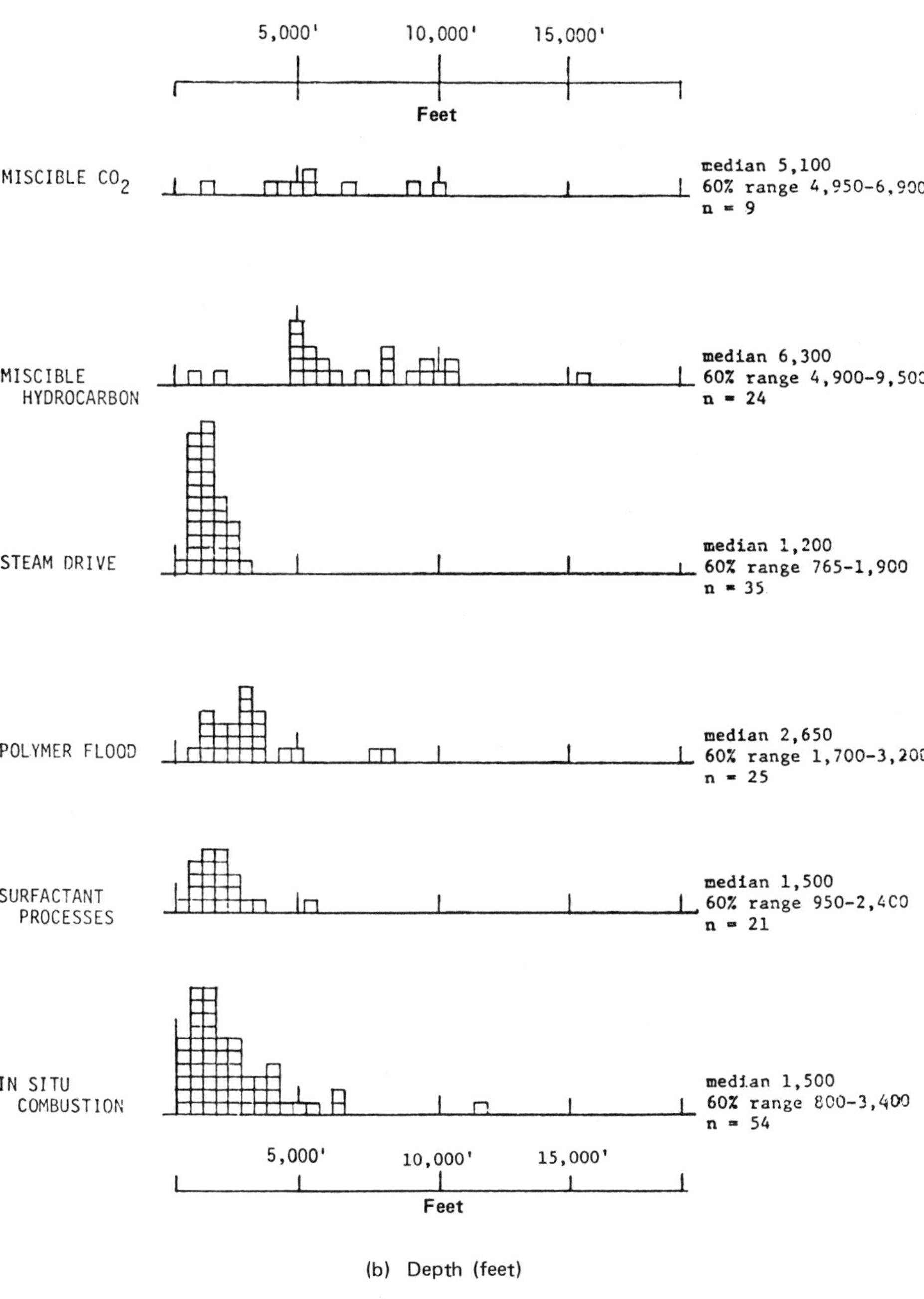

(b) Depth (feet)

(continued)

Figure 8.3:   (continued)

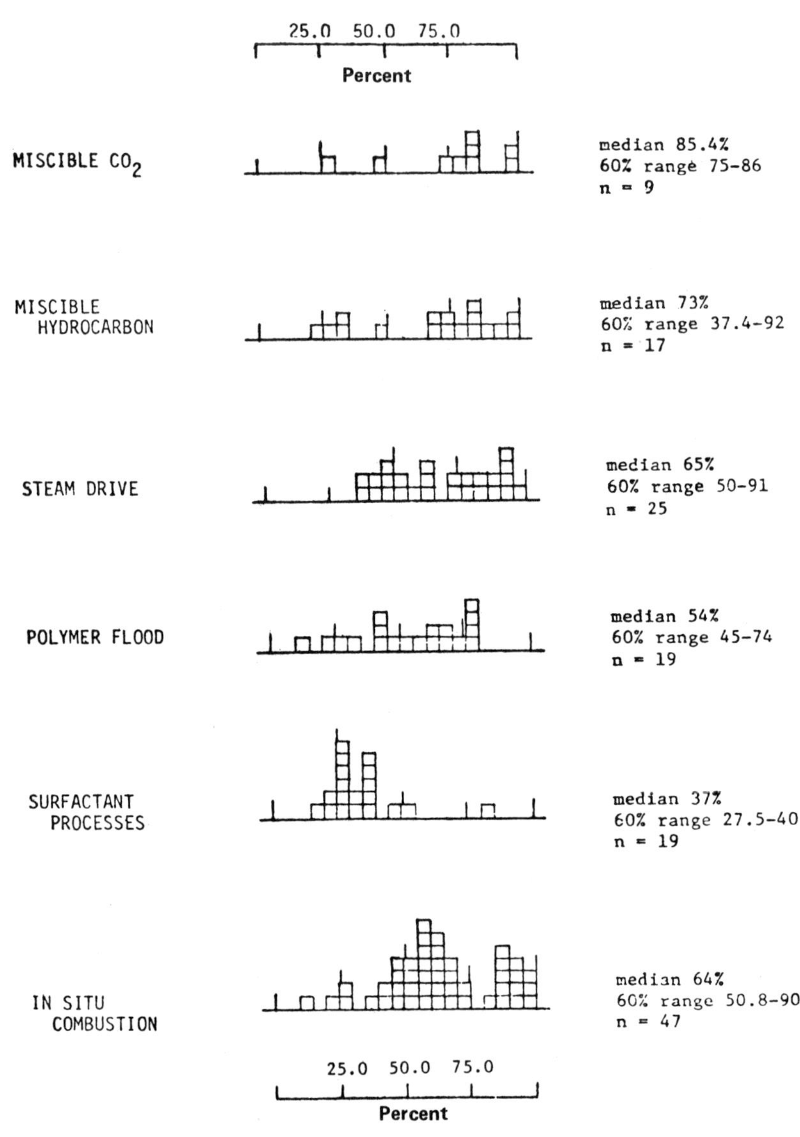

(c)   Residual oil saturation at project start (%)

Source: FEA/B-76/221

Table 8.5: Screening Guide for EOR Site Selection

Screening Parameters	Steam Drive	In Situ Combustion	CO_2 Miscible	Surfactant/ Polymer	Polymer Waterflood
Viscosity, cp at reservoir conditions	NC	NC	<12	<20	<200
Gravity, °API	>10	10–45	>30	>28	>18
(Calif. crudes)	(>10)	(10–45)	(>26)	(>25)	(>16)
Fraction of oil remaining in area to be flooded (before EOR), % PV	50*	50*	25*	25*	50*
Oil concentration, B/AF	>500	>400	NC	NC	NC
Porosity x oil saturation	>0.065	>0.05	NC	NC	NC
Depth, ft	<5,000**	>500	>3,000	NC (8,500)**	NC (8,500)**
Temperature, °F	NC	NC	NC	<200**	<200**
Original bottom hole pressure, psi	NC	NC	>1,500	NC	NC
Net pay thickness, ft	>20	>10	NC	NC	NC
Permeability, md	NC	NC	NC	>20 (with polymer drive)	>20
Transmissibility (permeability x thickness/viscosity)	>100	>20	NC	NC	NC
Natural water drive***	none to weak	none to weak	none to weak	none to weak	none to weak
Gas cap***	none to minor	none to minor	none to minor	none to minor	none to minor
Fractures	NC unless extreme	none to minor	none to minor	none to minor	none to minor
Lithology	NC	NC	NC	sandstone only**	NC
Salinity, ppm TDS	NC	NC	NC	<50,000**	NC
Hardness, ppm calcium and magnesium	NC	NC	NC	<1,000**	NC
Comments	Porosity x thickness (high) 10-acre spacing Economic fresh water available Economic fuel available High net to gross pay Low clay content	High dip preferred Porosity x thickness high 40-acre spacing Low vertical permeability preferred Preferred temperature >150°F High net to gross pay	Thin pay preferred High dip preferred Homogeneous formation preferred Porosity x thickness low Natural CO_2 availability Low vertical permeability in horizontal reservoirs	Homogeneous formation preferred Low clay content Porosity x thickness (high) Prefer waterflood sweep 50%	Use with or prior to waterflood Low calcium and clay content Porosity x thickness (high)

*In portion of field to be flooded. Assuming 80% of area of reservoir contains 95% of remaining oil, the oil saturation for the total field becomes 42% of pore volume.
**Considered a constraint under 1976 technology.
***These criteria apply to reservoirs with substantial remaining primary recovery.
NC = not a critical factor.

Source: FEA/B-76/221

Each of the parameters in the guide lies at or near the outer limits of the range of values that is considered feasible for EOR development. The upper section of the matrix contains parameters usually available in documented form which can be applied on a fairly consistent basis. Below the grid are additional factors which should be considered for each EOR technique, but which are infrequently documented in published sources. The screening guide also notes those limitations that are anticipated to be removed by future technological advances.

Some of the included items refer to technological limitations; others reflect broad economic issues. An example of a technological constraint is the depth limitation for steam drive. Under usual technology, primarily the limitations posed by well casing and pipe, steam cannot be injected deeper than 5,000 feet without losing its heat content or damaging the well equipment.

Economic Limitations: An example of an economic consideration is that residual oil saturation be greater than 25%. This ensures that enough oil will be available to be recovered to pay back the initial investment.

The economic parameters used in the screening guide are not cut-off parameters based on specific oil prices. Rather, they are parameters which recognize that certain reservoir conditions tend to increase substantially the risk of economic failure of a project. These effects are due to the likelihood of low recovery of oil relative to the investment required. Examples of high economic risk parameters are reservoirs with the following:

Active Natural Water Drive — An active natural water drive generally leads to a relatively high recovery rate for the original oil in place, leaving considerably less residual oil for an enhanced recovery project. Thus, reservoirs with a strong natural water drive are screened from consideration since they already produce at high rates and will lead to high natural recovery of the oil in place. Moreover, active water drives will tend to carry injected EOR fluids away from the most productive portions of the reservoir. This criterion is both a technical and an economic screen.

Major Primary or Secondary Gas Cap — A major gas cap has a tendency to allow the injected fluids to flow to the gas cap area causing the fluids to override or bypass the residual oil. Thus, the major gas cap screening criterion eliminates reservoirs with potentially low sweep efficiencies and poor oil recovery.

API Gravity Less Than 10° — A heavy gravity oil is generally too viscous to flow or be moved by natural forces. Such oil is generally classified as nonproducible and is usually not included in the categorization of oil in place. While recovery of such oil is technically feasible, the amount of energy and the investment costs required to recover such oil generally exceed the value of the oil produced.

Residual Oil Saturation Less Than 25% — A low residual oil saturation suggests that there is insufficient oil remaining in the reservoir to cover the investment and operating cost of a tertiary project. The costs of tertiary recovery are generally a function cf the size and pore volume of the project area. Thus, the requirement that residual oil saturation be greater than 25% of pore volume is an economic consideration.

Extensive Fractures — Screening of all reservoirs for EOR with the exception of steam drive depends on the existence and extent of fracturing of the formation.

Reservoirs with extensive fractures have a tendency to channel the injected fluid, air or chemicals, away from the oil-bearing portion of the reservoir. This drastically reduces the effectiveness of the injection and reduces the amount of oil recovered. Except for steam drive, the majority of the EOR techniques are not applicable in reservoirs with extensive fractures.

It should be noted that the extent of fracturing is not normally documented on a reservoir basis. Thus, many reservoirs may pass this screen on the basis of available information, but may be eliminated at a later date by the application of more detailed engineering analysis.

Some factors requiring consideration have been omitted because they are primarily price-dependent. Such factors include a minimum level of recoverable oil, the minimum number of active wells in the reservoir, and the condition of the active wells. Many fields passing the screening guide will not be suitable for enhanced oil recovery due to engineering reasons not found in public documents. On the other hand, some reservoirs that do not appear promising may become feasible in the future due to technological advances. (The screening guide notes those limitations which are anticipated to be removed by future technological advances.)

In order to assign a particular EOR method to a given reservoir, certain criteria must be met, as indicated below.

Applicability of Steam Drive: This depends on the successful screening of four criteria, specifically:

(1)   Reservoir net thickness greater than 20 feet — A net pay thickness of less than 20 feet is generally uneconomic for steam drive. A high proportion of the heat released by the steam would be lost to nonoil-bearing formations above and below the reservoir.

(2)   Residual oil greater than 500 barrels/acre-foot — The practice of using petroleum as the source for generating steam dictates that more than 500 barrels/acre-foot of remaining oil in place (or that residual oil saturation times porosity is greater than 0.065) will be required. Given the relatively fixed rate of steam as input, sufficient oil needs to be in place to provide an economic payout for a steam drive project.

(3)   Reservoir transmissibility greater than 100 — Reservoir oil transmissibility factors less than 100 will slow the rate of oil production, disallowing an economic payout on a steam drive project. (Transmissibility is measured by multiplying permeability times thickness and dividing by the viscosity.)

(4)   Reservoir depth less than 5,000 feet — Reservoir depth is restricted to less than about 5,000 feet due to the prevalence of heat loss, injectivity, and pressure problems which tend to increase with depth. It is anticipated that future technical advances may resolve these depth restrictions.

Applicability of CO_2 Miscible Flooding: This depends on three screening criteria:

(1)   Gravity of 30°API or greater — Reservoir oil with API gravity less than 30° becomes too viscous and dense during injection in relation to the solvent and drive fluids. This leads to the tendency for the injected materials to lose their effectiveness by fingering through or overriding the reservoir oil. The net effect is a poor or minimal sweep efficiency, and thus, lower recovery.

(2)   Reservoir oil viscosity of 12 cp or less — The viscosity constraint is based on empiric observation. The majority of CO_2 projects are only operating in reservoirs with oil viscosities of 12 cp or less.

(3)   Reservoir depth of 3,000 feet or greater — In order for CO_2 to obtain miscibility with reservoir oil, it will need to be injected and consequently to flow at sufficient pressure (estimated at 1,000 to 1,500 psi or greater). A general rule of thumb of two vertical feet per pound of reservoir bottom hole pressure results in the depth requirement of 3,000 feet or greater.

Applicability of Polymer-Augmented Waterflooding: This depends on four criteria:

(1)   Reservoir permeability of 20 md or less — Average reservoir permeability should not be less than about 20 md to allow effective movement through the reservoirs of polymers. Permeability below this level will restrict the rate of flow of the displaced oil, thus providing a slower cash flow against the large initial investment costs.

(2)   Reservoir temperature of 200°F or less — Generally, polymers lose effectiveness at reservoir temperatures greater than about 200°F.

(3)   Reservoir oil viscosity of 200 cp or less — Oil with reservoir viscosity greater than about 200 cp does not flow freely in a manner sufficient to allow an effective polymer flood.

(4)   Gravity greater than 18°API — Similar to the viscosity criterion, crude oils heavier than 18°API will flow too slowly to be moved effectively by a polymer flood.

Applicability of Surfactant/Polymer Flooding: This depends on the criteria for polymer-augmented waterflooding and three additional criteria. The dependence on the polymer criteria is because the surfactant slug must be driven by a substance with a high degree of mobility control. Polymer waterfloods are used to drive surfactant slugs. Thus, a reservoir must pass all the polymer criteria as well as the additional constraints faced by the surfactant process. (Note that developmental work is under way for overcoming these constraints.) These additional criteria are:

(1)   Gravity greater than 28° API — Due to the increased costs of surfactant/polymer processes over simpler polymer waterflood, faster rates of recovery are required to yield an adequate return on investment. Oils heavier than 28° API will flow too slowly for the costly surfactant/polymer processes.

    (2)   Sandstone reservoirs — Surfactant research and field tests have generally been limited to sandstone reservoirs, which comprise almost 70% of the estimated remaining oil in place following primary and secondary recovery. Carbonate reservoirs (which contain much of the estimated remaining oil in place) contain substances (principally calcium and magnesium ions) that are destructive to the usual surfactants.

    (3)   Water hardness of 1,000 ppm or less — The usual surfactants are not compatible with water hardness where the calcium carbonate exceeds 1,000 ppm.

Applicability of In Situ Combustion: This depends on five criteria:

    (1)   API gravity of 45° or less — Effective in situ combustion requires certain burning, coking and deposition characteristics of the reservoir crude oil. These characteristics normally diminish with increases in API gravity, with 45° being the upper limit.

    (2)   Reservoir depth of more than 500 feet — Reservoir depth greater than 500 feet is generally necessary to provide sufficient overburden for effective control of the combustion process.

    (3)   Reservoir net thickness greater than 10 feet — Reservoir formations with thicknesses of less than 10 feet of net pay are normally not economically attractive for in situ projects. Too high a proportion of the generated heat would be lost to nearby nonproductive structures.

    (4)   Residual oil greater than 400 barrels/acre-foot — Sufficient recoverable oil must exist for a project to be economically viable. 400 barrels/acre-foot (or where residual oil saturation times porosity is greater than 0.05) represent the economic limit for in situ projects.

    (5)   Reservoir transmissibility of over 20 — Reservoirs with oil transmissibility factors of less than 20 do not normally allow a sufficiently high rate of oil production to provide an economic payout on in situ projects.

Dominance Rules: Where the decision by logic led to the assignment of more than one EOR method to a reservoir, dominance rules were used to confirm a unique assignment (see Figure 8.4).

These rules express the order of preference among the likely ties that develop, based on the expected crude oil recovery and the degree of economic risk. They reflect assumed company choices in similar situations.

When a reservoir satisfied the criteria for two or more methods, the methods were taken in pairs to the appropriate cell in the exhibit, thus eliminating one of the methods. This was repeated until only one method remained. The reservoir was assigned to the remaining EOR method.

Figure 8.4: Preferred Methods When Two Sets of Screening Criteria are Satisfied (Dominance Rules)

<table>
<tr><td></td><td>Polymer</td><td></td><td></td><td></td></tr>
<tr><td>Steam Drive</td><td>a. Steam Drive
iff* gravity
<25°API
(otherwise
Polymer)</td><td></td><td></td><td></td></tr>
<tr><td></td><td></td><td>Steam Drive</td><td></td><td></td></tr>
<tr><td>In Situ
Combustion</td><td>b. In Situ
iff gravity
<25°API
(otherwise
Polymer)</td><td>c. Steam Drive</td><td></td><td></td></tr>
<tr><td></td><td></td><td></td><td>In Situ Comb.</td><td></td></tr>
<tr><td>Surfactant/
Polymer Flooding</td><td>d. Surf./Poly.</td><td>e. Surf./Poly.</td><td>f. Surf./Poly.</td><td></td></tr>
<tr><td></td><td></td><td></td><td></td><td>Surfactant/Polymer</td></tr>
<tr><td>CO₂ Miscible
Flooding</td><td>g. CO2 Misc.</td><td>h. CO2 Misc.</td><td>i. CO2 Misc.</td><td>j. CO2 Misc.
iff gravity
>35°API
or visc. >6cp.
(otherwise
surf./poly.)</td></tr>
</table>

*iff = "if and only if" (implies reverse dominance if the condition is not met)

Source: FEA/B-76/221

PROFITABILITY CONSIDERATIONS

The widespread popularity of certain processes may in part be explained by examining their respective profitability records. Evaluations of various methods from companies undertaking enhanced recovery have been combined by Energy Resources Co., Inc., in their report prepared for the EPA, as shown in Table 8.6 to yield a set of profitability ratios.

Smaller companies appear to choose the proven processes for enhanced oil recovery, whereas the top twenty are more involved in demonstrating advanced techniques. Independents listed 71% of reporting projects as moneymakers, whereas the comparable figure for programs administered by companies within the top twenty was only 55%. The ratios are likely to improve for many processes as operators learn more about the best prospects upon which to apply them.

Table 8.6: Profitability of Tertiary Processes*

Process	Number Completed or Underway	Number Reporting Financial Status	Number Profitable	Percent Profitable
Combustion	39	30	7	23
Cyclic steam	42	39	39	100
Steam displacement	22	17	9	53
Polymer flood	20	16	6	38
Miscible hydrocarbon	21	18	13	72
Miscible carbon dioxide	9	5	1	20
Micellar solution flood	7	1	1	0
Waterflood alternating gas	4	3	3	100
Surfactant flood	4	2	0	0
Caustic soda flood	2	1	0	0
Hot waterflood	2	2	0	0
Low tension waterflood	1	—	—	—
Soluble oil flood	1	1	0	0
Water/gas injection	1	1	1	100
Water/gas pulsing	1	1	0	0
Total	176	137	80	—

*Based on an analysis of data presented in the March 25, 1974 issue of *The Oil and Gas Journal.*

Source: EPA-68-01-1912

For example, the in situ combustion, having a profitability ratio of only 23%, was designated for six projects scheduled for 1974 or later. Other methods lacking substantial profitability data such as micellar solution flood and combinations of water and gas injection were also slated for the same period.

REFERENCES

(1) Crump, J.R. and Sharp, J.M., *Planning Criteria Relative to a National RDT and E Program Directed to the Enhanced Recovery of Crude Oil and Natural Gas,* Gulf Uni-

versities Research Consortium Rep. 130, Galveston, Texas, Nov. 30, 1973, Rep. ORO-4507-1 (Appendix E of PB-248-951).

(2) National Petroleum Council, *U.S. Energy Outlook,* December 1972.

(3) Geffen, T.M., "Recovery of More Oil from Known Fields," SPE 4948, Society of Petroleum Engineers, American Institute of Mining, Metallurgical and Petroleum Engineers, Inc. (AIME) 1974.

(4) Geffen, T.M., "Improved Oil Recovery Could Help Ease Energy Shortage," *World Oil,* October 1973.

(5) Geffen, T.M., "Oil Production to Expect from Known Technology," *Oil and Gas Journal,* May 7, 1973.

(6) *Secondary and Tertiary Oil Recovery Processes,* Interstate Oil Compact Commission, Oklahoma City, 1974.

APPENDIX—EOR PROJECTS USED IN DEVELOPMENT OF SCREENING GUIDE

Steam Drive

Field	Reservoir/Unit	Operator	Source	Size (acres)
Brea Olinda	First	Chanslor	Company Report*	10
Cat Canyon	Sisquoc/51 Pilot	Continental	Company Report	5
Charco Redondo		Texaco	SPE General**	2.5
Coalinga	Temblor/13 D	Socal	Company Report	400
Cymric	Amnicola/16 Z	Socal	Company Report	70
El Dorado	Eldorado/Hegberg	Cities	SPE General	2
Inglewood	Up. Invest.	Socal	SPE General	3
Kern River	Kern River Sand	Chanslor	Company Report	1
Kern River	Kern River Sand	Chanslor	Company Report	1
Kern River	Getty Oil Portion	Getty	Company Report	1,670
Kern River	A Pilot	Getty	SPE General	133
Kern River	Kern Pilot	Getty	SPE General	228
Kern River	American Naphtha	Socal	Company Report	160
Kern River	Monte Cristo	Socal	Company Report	73
Kern River	Section 3	Socal	Company Report	260
Kern River	Kern River Series	Socal	SPE General	61

(continued)

Field	Reservoir/Unit	Operator	Source	Size (acres)
Midway Sunset	Potter	Chanslor	Company Report	1
Midway Sunset	Potter	Chanslor	Company Report	1
Midway Sunset	Spellacy	Chanslor	Company Report	1
Mt. Poso	Upper Vedder	Shell	SPE-IOR***	110
San Andres	Lombardi	Mobil	Company Report	3,041
Smackover	Nacatoch	Phillips	Company Report	985
Smackover	Nacatoch/Pilot	Phillips	SPE General	22
South Belridge	D&E Zones, Tulare	Mobil	SPE General	204
Toborg	Toborg	Gulf	Company Report	19
Troy	Upper Nacatoch	Mobil	SPE General	15
Winkleman Dome	Nugget	Amoco	Company Report	110
Yorba Linda	Conglomerate and Signet	Shell	Trade Journal	395

*Company submissions to the study.
**General articles and reports published by the Society of Petroleum Engineers (SPE).
***Special projects reported in recent SPE periodical—*Improved Oil Recovery.*

In Situ Combustion

Field	Reservoir/Unit	Operator	Source	Size (acres)
Bellevue	Nacatoch	Getty	Company Report	510
Bellevue	Nacatoch	Cities	SPE-IOR	29
Cedro Hill	Cole "B"/Sec. 132	Mobil	Company Report	240
Cedro Hill	Cole "C"/Sec. 18	Mobil	Company Report	167
Charamousca	Mirando	Mobil	Company Report	110
Charco Redondo		Texaco	SPE General	

(continued)

Field	Reservoir/Unit	Operator	Source	Size (acres)
D.C.R. 79	Cole "C"	Mobil	Company Report	328
Delhi	May/Libby	Sun	Company Report	80
Fosterton Northwest	Roseray	Sun	Trade Journal	257
Fry	Robinson	Marathon	SPE General	
Glen Hummel	Poth A	Sun	SPE-IOR	632
Gloriana	Poth A	Sun	SPE-IOR	800
Government Wells, N.	Government Wells	Mobil	Company Report	600
Hankamer	Miocene	Gulf	Company Report	24
Humboldt	Chanute	Sinclair	SPE General	69
Kermit Yates	Yates	Mobil	Company Report	40
Little Tom	San Miguel	Hanover	SPE-IOR/ERDA	14
Loco	J_s	Conoco	Company Report	24
Lost Hills	300 Zone/2T	Mobil	Company Report	140
Midway Sunset	Moco Zone	Mobil	Company Report	150
Midway Sunset	Potter/Sec. 23	Chanslor	Company Report	20
Midway Sunset	Webster Zone	Mobil	Company Report	390
O'Hern	O'Hern Sand	Mobil	Company Report	260
Salt Creek	Shannon Pool	Amoco	SPE General	1
Sloss	Muddy J1 Sand	Amoco	SPE General	960
Smackover	Nacatoch/Thermal Recovery Unit #1	Gulf	Company Report	6
South Belridge		General	SPE General	3
S.W. Moran	Bartlesville/Iola	Sun		20
Talco	Carr A	Exxon	Company Report	20

(continued)

Field	Reservoir/Unit	Operator	Source	Size (acres)
Trix-Liz	Woodbine C	Sun	Company Report	244
West Casa Blanca	Cole "C"/West Casa Blanca	Mobil	Company Report	269
West Heidelberg	Cotton Valley	Gulf	Company Report	400
West Heidelberg	Christmas 10/ Where	Gulf	Company Report	134
West 76	Cole/D.C.R.C. State Oil	Mobil	Company Report	362

CO_2 Miscible*

Field	Reservoir/Unit	Operator	Source	Size (acres)
Crosset	Devonian/ Northcross	Shell	SPE-IOR	1,155
Kelly Snyder	Canyon Reef/ Sacroc	Socal	Company Report	28,800
Levelland	San Andres	Amoco	SPE-IOR	13
North Cowden	Grayburg	Amoco	Company Report	12
Slaughter	San Andres/ Estate Unit	Amoco	Company Report	12
Wasson	San Andres/ Willard Unit	ARCO	Company Report	415

*Including projects currently undergoing hydrocarbon miscible flooding.

Hydrocarbon Miscible

Field	Reservoir/Unit	Operator	Source	Size (acres)
Block 31	Devonian	ARCO	Company Report	7,840
Clayton	Queen City "B" Sand	Mobil	Company Report	1,087
East Velma	Sims/Middle Block	Amoco	SPE-IOR	180

(continued)

Field	Reservoir/Unit	Operator	Source	Size (acres)
Greeley	Olcese/Olcese Sand	Mobil	Company Report	503
Parks	Penn Bend/Parks	Mobil	Company Report	7,152
Pegasus	San Andres/ Royalty	Mobil	Company Report	239
Salt Creek Light Oil Unit	First Wall Creek	Amoco	SPE-IOR	125
Slaughter	San Andres/ Central Mallet	Amoco	SPE-IOR	12
Spraberry Trend	Spraberry/ Preston	Mobil	Company Report	40
University Waddell	Devonian/ Devonian	Mobil	Company Report	1,280
West Ranch	Glasscock	Mobil	Company Report	480

Surfactant

Field	Reservoir/Unit	Operator	Source	Size (acres)
Big Muddy	Second Wall Creek	Conoco	Company Report	6
Borregos	Zone F4	Exxon	SPE General	1
Bradford	Bradford Third/ Bingham Field Test	Marathon	SPE General	47
Bridgeport	Kirkwood (Cyprus)	Marathon	Company Report	80
Burbank	Burbank/North Burbank Unit	Phillips	Company Report	90
Delaware- Childers	Bartlesville/ Mary Costen Unit	B & N	SPE-IOR	10
El Dorado	650' Shallow Sand	Cities Service	SPE-IOR/ERDA	6

(continued)

Field	Reservoir/Unit	Operator	Source	Size (acres)
Griffin	Maier 50-A	Conoco	Company Report	3
Higgs Unit	Bluff Creek	Union	SPE General	8
Loudon		Esso	SPE General	160
Old Wilmington	Upper Three Zones	Marathon	Company Report	N/A*
Robinson	Robinson/M-1	Marathon	Company Report	330
Robinson	Robinson/119-R	Marathon	Company Report	40
Robinson	Robinson/219-R	Marathon	Company Report	117
Salem	Benoist	Texas	SPE General	5
Siggins	First Siggins	Marathon	Company Report	40
Wichita County Regular	Gunsight	Mobil	Company Report	209

Polymer

Field	Reservoir/Unit	Operator	Source	Size (acres)
C-H	Minnelusa	ARCO	SPE-IOR	1,040
Colby	Colby	Mobil	Company Report	80
East Coalinga	Tremblor Zone II	Shell	ERDA	132
Howard Glasscock	Yates	Sun	Company Report	440
Jordan	San Andres	Shell	SPE General	200
Main Consolidated	Robinson/Swaren T-Flood II Test	Getty	Company Report	1
Morris-Scott Unit	Bartlesville	Betz Lab, Inc.	Trade Journal	20
North Burbank	Burbank	Phillips	SPE General	160
O'Hern	O'Hern	Mobil	Company Report	352

(continued)

Field	Reservoir/Unit	Operator	Source	Size (acres)
Sand Hills	Tubb	Mobil	Company Report	1,680
Sho-Vel-Tum	Penn-Deese/ Tatums	Mobil	Company Report	270
Skull Creek	South Unit	American Petrofina	Company Report	880
South 76	Cole	Mobil	Company Report	230
Stanley Stringer		Kewanee	ERDA	1,550
Stephens Co. Reg.	Caddo Reef/ Curry	Mobil	Company Report	1,680
Taber Mann-ville	D Pool	Chevron	SPE-IOR	520
Vacuum	San Andres	Mobil	Company Report	331
Volpe	Lopez/Benavides Lease	Mobil	Company Report	245
Wilmington	Ranger Zone/ F.B.V.	Mobil	Company Report	302

Caustic Soda

Field	Reservoir/Unit	Operator	Source	Size (acres)
Fruitvale			Bureau of Mines	N/A*
North Ward-Estes	Queen/North	Gulf	Company Report	40
Orcutt Hill	Point Sal	Union	SPE-IOR	420
West Perryton	Morrow	Sun	Company Report	1,344
Wilmington	Tar Zone		Bureau of Mines	N/A*

*Not available

CHEMICAL AND RESEARCH REQUIREMENTS OF EOR

The information in this chapter is based on:

> *An Investigation of Primary Factors Affecting Federal Participation in R&D Pertaining to the Accelerated Production of Crude Oil,* prepared by J.M. Sharp, of Gulf Universities Research Consortium, for National Science Foundation under Contract NSF-C-942 (GURC Report 140), September 15, 1974.

> *Assessment of Enhanced Recovery Technology as a Means for Increasing Crude Oil Recovery in Texas,* NSF Report RA-N-74-251, prepared by E.A. Lohse, of Texas Governor's Advisory Council, for National Science Foundation, January 6, 1975.

> *Potential Environmental Consequences of Tertiary Oil Recovery,* prepared by C. Braxton, R. Stephens, C. Muller, J. White, J. Post, J. Norton, M. Goldberg, and P. Stevenson, of Energy Resources Co., Inc., for U.S. Environmental Protection Agency, under Contract 68-01-1912, July 1976.

> *Enhanced Oil Recovery Potential in the United States,* OTA-E-59, prepared by Lewin & Associates, Inc., National Energy Law and Policy Institute and Wright Water Engineers, Inc., for U.S. Office of Technology Assessment, January, 1978.

The purpose of enhanced recovery is to produce additional amounts of crude oil under economically favorable conditions. Reaching this goal, however, depends not only upon technology but also upon the price of oil, the cost and availability of chemicals or other injected material, costs of injection equipment and procedures and maintenance of the project.

MATERIAL REQUIREMENTS

Tertiary operations are superimposed on, or follow, existing primary or secondary operations. A large percentage of the wells required are already drilled. If secondary is in operation, a large part of the injection equipment is already in place. Tertiary field operations, then avoid the fairly large requirements for steel.

Micellar flood operations, however, require huge amounts of chemicals, well beyond chemical industry capacity if the method is used even moderately. While attempts are being made to manufacture some of these chemicals in the field using field petroleum as the feedstock, there will be a requirement for large chemical plant expansion if 58% of the projected 18 to 36 billion bbl of oil is to be produced by micellar methods which require about 5 to 10 lb of chemical per bbl of oil produced in a successful micellar flood. Enhanced recovery production of a 10 million bbl field may require 80 million lb of polymer, which may have to be tailor made for each field.

CO_2 recovery methods will depend in general on supplies of natural CO_2 in the vicinity if they are to be economical. In such cases, CO_2 wells must be drilled, compression equipment installed, and pipelines constructed to transport the CO_2 to the production field. Under these circumstances, CO_2 methods will require steel in fairly large quantities.

Thermal methods wil also require drilling of some additional wells. Primarily, however, they require the procurement of massive air compression equipment and, possibly, the construction of fuel pipelines (probably natural gas) from nearby sources. Hydrocarbon miscible methods would use nearby natural sources of oil-miscible hydrocarbons; otherwise economics would dictate the sale of LPG, natural gas, etc., rather than its use for crude oil displacement.

Chemicals

Polymers and miscible fluids are not available in sufficient quantities or at economically feasible prices for widespread enhanced recovery operations. Chemical, pharmaceutical and detergent manufacturing companies nationwide, as well as some of the conventional oil field service companies, are however appraising the potential market.

Enhanced oil recovery processes require a variety of chemicals. It appears however that although scores of chemicals have been described in the patent literature for use in EOR applications, only a few will actually have extensive usage.

Chemicals commonly used include broad spectrum petroleum and synthetic petroleum sulfonates; alcohols; polyacrylamide and polysaccharide polymers; sodium dichlorophenol and sodium pentachlorophenol; sodium hydroxide and sodium silicate (1).

Tables 9.1 through 9.5 list some of the compounds in each class which are in use or have been proposed for use in U.S. patents and the literature. The choice of chemicals depends upon: (1) availability of the chemicals, (2) costs of the chemicals, (3) chemical compatibility with the specific reservoir characteristics encountered, and (4) the required concentrations of the chosen chemicals in the reservoir to achieve additional oil recovery.

Chemicals useable for surfactant flooding fall into five general categories: broad spectrum petroleum sulfonates, synthetic petroleum sulfonates, sulfated ethoxylated alcohols, ethoxylated alcohols, and other alcohols. Halogenated compounds, though proposed in the literature, are unlikely to be used in field operations because their possible presence in produced oil streams would poison the catalysts at the refinery.

Table 9.1: Chemicals Proposed for Use as Surfactants

Ditetradecyldimethylammonium chloride	Diethylene glycol sulfate
Dodecyltrimethylammonium chloride	Sodium sulfate oleylethylanilide
Hexadecyltrimethylammonium chloride	Alpha-olefin sulfonate
Alkylphenoxypolyethoxyethanol	Alkyl aryl sulfonate
p-Chloroaniline sulfate laurate	Alkyl aryl naphthenic sulfonate
p-Toluidine sulfate laurate	with monovalent cation
Polyglycerol monolaurate	Hexadecylnaphthalene sulfonate
Glycerol disulfoacetate monomyristate	Sodium lauryl sulfonate
N-Methyltaurine oleamide	Triethanolamine laurate
Monobutylphenylphenol sodium sulfate	Triethanolamine myristate
Polyoxyethylene alkyl phenol	Triethanolamine oleate
Morpholine stearate	n-Dodecyl diethylene glycol sulfate
Pentaerythritol monostearate	Sodium glyceryl monolaurate sulfate
Dihexyl sodium succinate	

Source: EPA 68-01-1912

Table 9.2: Chemicals Proposed for Use as Cosurfactants

Alcoholic liquors	1-Hexanol*
Fusel oil	2-Hexanol*
Alcohols	1-Octanol
Alkaryl alcohols	2-Octanol
Cresol	Isopropanol*
p-Nonylphenol	Aldehydes
Amyl alcohols	Formaldehyde
Isopentanol*	Glutaraldehyde
2-pentanol*	Paraformaldehyde
Phenol	Amides
Decyl alcohols	Amino compounds
Ethanol	Esters
Isobutanol	Sorbitan fatty ester
n-Butanol	Ketones
Cyclohexanol	

*Most commonly used

Source: EPA-68-01-1912

The chemicals most commonly used for mobility control are polyacrylamide, polysaccharides and polyethylene oxide, but others listed in Table 9.3 also are appropriate.

Table 9.3: Materials Proposed for Use as Mobility Buffers

Aldoses, D series, L series	Dextrans
Amines	Deoxyribonucleic acid
Carboxymethylcellulose	Glycerin
Carboxyvinyl polymer	Ketoses, D series, L series

(continued)

Table 9.3: (continued)

Polyacrylamide	Saccharides, conjugated sac-
Polyethylene oxide	charides, mono-, di-, tetra-,
Polyisobutylene in benzene	and polysaccharides
Rubber in benzene	

Source: EPA 68-01-1912

Table 9.4: Hydrocarbons Used as Fraction of Micellar Slug*

Alkylated aryl compounds	Straight-run gasoline
Aryl compounds with monocyclic	Kerosene
compounds	Liquefied petroleum gas
Alkyl phenols	Naphthas
Benzene	Heavy naphthas
Toluene	Refined fraction of crude oil
Aryl compounds with polycyclic	Paraffinic compounds
compounds	Decane
Crude oil**	Dodecane
Partially refined fractions of	Heptane
crude oil	Octane
Overheads from crude	Pentane
columns	Propane
Side cuts from crude	Cycloparaffinic compounds
columns	Cyclohexane
Gas oils	Naphthenic compounds

*Or in Miscible Displacement Processes
**Most commonly used.

Source: EPA-68-01-1912

Electrolytes: Among chemicals usable as electrolytes are inorganic and organic acids, specifically hydrochloric and sulfuric; inorganic and organic bases, notably sodium hydroxide; and organic and inorganic salts, such as sodium sulfate and sodium nitrate (EPA 68-01-1912).

Chemicals to Block Exchange Sites in the Formation (Preflushing): Such chemicals include quaternary ammonium salts, fluoride solutions, potassium permanganate and sodium hydroxide (EPA 68-01-1912).

Chemicals to Initiate Ignition of In Situ Combustion: Hydrazine and hydrogen peroxide serve this purpose on infrequent occasions (EPA 68-01-1912).

Chemicals to Increase Efficiency of Thermal Methods: Quinoline, sodium hydroxide and toluene are used occasionally for this purpose (EPA 68-01-1912).

Chemicals Proposed for Alkaline Flooding: Sodium hydroxide, sodium silicate, ammonium hydroxide, sodium carbonate and potassium hydroxide compose this category (OTA-E-59).

Chemicals Proposed as Oxygen Scavengers: Such chemicals include sodium hydrosulfite, hydrazine and salts of bisulfite (OTA-E-59).

The bactericides and biocides used in tertiary oil recovery are listed in Table 9.5. They are nonspecific and display significant toxicities to organisms other than the target bacteria. Although in many cases chlorination and adjustment in salinity and pH are often sufficient, these compounds may be more widely used to control bacteria which may attack surfactants and polymers used in a micellar-polymer flood.

Table 9.5: Chemicals Used as Bactericides and Biocides

Aldehydes
    Formaldehyde
    Glutaraldehyde
    Paraformaldehyde
Alkyl phosphates
Acetate salts of coco amines
Alkyl amines
Quaternary amines
    Alkyl dimethyl ammonium chloride
    Cocodimethylbenzylammonium chloride
Diamine salts
    Acetate salts of cocodiamines
    Acetate salts of tallow diamines
Calcium sulfate
Sodium hydroxide
Heavy metal salts
Chlorinated phenols
    Alkyl dichlorophenol
    Pentachlorophenol
    Sodium salts of phenols
Substituted phenols

Source: EPA-68-01-1912.

Availability of Chemicals: Table 9.6 shows the chemical availability situation for each of the materials used in bulk for oil recovery. Potential shortages could exist for each chemical and thereby limit the start of new recovery projects.

Polyacrylamide and Polysaccharide — Both polyacrylamide and biopolymers such as polysaccharide, however, can readily be manufactured from various media. The polysaccharide for tertiary oil recovery which is manufactured by a division of Kelco Chemical Company is produced by the microbial action of a strain of bacteria on glucose. The high cost of polysaccharide indicated in Table 9.6 reflects the high prices for sugar and corn during the shortages of the 1970s.

However, use of process materials other than glucose is feasible, which could reduce the cost of polysaccharides. Municipal sewage sludge could serve as a medium for the production of polysaccharides for tertiary oil recovery applications. Though yields of polymer from sludge are expected to be low, its low cost and large, stable supply make sludge an attractive new medium for producing the needed quantities of polymer.

Table 9.6: Availability of Chemicals for Tertiary Oil Recovery

Chemical	Raw Materials	Process	1975 Price ($/lb)	Supply (10^6 lb/yr)	Usage Rate lb/bbl Oil Produced	Estimated Demand (10^6 lb/yr)	Estimated Demand as Percent Supply	Physical Form	Comments on Availability
Polymers Polyacrylamide (partially hydrolyzed)	Acrylonitrile H_2SO_4	Hydration with H_2SO_4, neutralization, ion exchange, polymerization and hydrolysis (partial)	1.50–2.00	50	1	210	>400	Dry powder or liquid	Plant expansions likely to serve other growth markets (e.g., water treatment flocculant). Acrylonitrile feedstock capacity is 1,400 MM lb
Polysaccharide	Glucose *Xanthomonas campestris*	Microbiological action	2.50	40	1	210	500	Powder	New plant capacity or construction should lead to significant price reduction.
Surfactants Linear alkylbenzene sulfonates	Kerosene, (C_{10} to C_{13} or C_{11} to C_{14}), SO_3, oleum, NaOH	Separation of straight chain paraffins, alkylation, sulfonation, neutralization	0.10–0.20	700–730	7–15	1,470–3,150	190–430	Viscous liquid	Possible production at oil refineries and on site to meet needs.
Alcohols Isopropanol	Refinery propylene H_2SO_4	Absorption in concentrated H_2SO_4, hydrolysis of resulting isopropyl ester	0.09	2,500	2–8	420–1,780	16–67	Anhydrous liquid	Common alcohol, short-term shortages possible. Least expensive cosurfactant.
Pentanol	Pentene	Hydration and acid	0.18	NA	2–8	420–1,780	NA	Liquid/low viscosity	
Hexanol	Hexene	Hydration and acid	0.24	NA	2–8	420–1,780	NA	Liquid/low viscosity	
n-Butanol	(1) Crotonaldehyde from ethyl alcohol (2) Propylene, CO, H_2	Reduction by catalytic hydrogenation; Oxo process—propylene to n-butyraldehyde, isobutyraldehyde	0.16	660	2–8	420–1,780	65–250	Liquid	Possible use as substitute during shortages.
Isobutanol	Propylene, CO, H_2	Oxo process (as above)	0.16	170	2–8	420–1,780	250–1,000	Liquid	

Source: EPA-68-01-1912

Surfactants — Surfactants will show a shortfall of 600 to 2,500 million pounds by 1985. Major refiners and specialty chemical manufacturers have stated that they will build the plant capacity needed to serve this demand as soon as it materializes. Certainly, supply and demand will come into closer balance in the long term. However, large shortages may still exist during the transitional period due to the time lags in the construction of new chemical facilities. For tertiary oil production to achieve forecasted levels during the 1985-1990 time frame, alternative sources of sulfonates will be required. An important factor for petroleum sulfonates is transportation cost.

Petroleum sulfonates are highly viscous liquids which must be transported in steam-heated tankers to facilitate loading. This fact discourages long distance multimode movement of large quantities of surfactants. Trucking charges are estimated at 3 to 4 cents per pound per 100 miles. Therefore, imports of sulfonates are likely to be limited. Yet, the typical micellar-polymer project in a field with 5 million bbl of oil remaining in place would require approximately 500 tanker truckloads of surfactants over a few months of time while the slug was being prepared and injected.

To fulfill immediate needs, sulfonation units must be built. The process of sulfonation is basically simple and straightforward and achieves yields of 10 to 15% per treatment in large-scale operations. The simplicity of the process and the high cost and logistical difficulties of transporting the end-product give the integrated petroleum company a number of options for providing its own petroleum sulfonates. As an operator of micellar-polymer flooding projects, the company may choose the location for the sulfonate plant and feedstock inputs. These choices are illustrated in Table 9.7.

The option 1 to produce sulfonates at a refinery will be chosen by operators whose project oil fields are near their refineries and where the capacity is added to serve commercial as well as captive demand and where the unsulfonated crude fraction would be incompatible with the slug. Figure 9.1 indicates the location of existing refineries where sulfonation units might be constructed. Marathon has chosen this approach in Illinois where its oil field is adjacent to the refinery. However, refinery operating plans may not be geared to manage such a unit or may already use the lubricant stock in another part of the process.

Option 2 may be chosen to overcome the possible conflicts between field and refinery operations. The sulfonation unit would be built on-site at the oil field and would be sized to local requirements for sulfonates. If available, lubricating stocks could be transported to this facility from the refinery although savings in transportation costs are then sacrificed. The appeal of this option and option 4 is that on-site manufacturing may reduce transportation costs and provide sulfonates for planned projects quickly and independently of commercial chemical markets.

Either a lubricating stock refined on-site from crude oil or the crude oil itself could be used as a feedstock. Not all oils may prove to be suitably reactive for economic sulfonation, or compatible with the microemulsion slug, but as indicated in Table 9.8, the costs of sulfonating lease crude on-site might compete with market prices.

Cosurfactants — Cosurfactants are not as severe a constraint to the development of micellar-polymer flooding. Though propanol is preferred due to its low cost, there are alternative alcohols available which might serve as substitutes. These include pentanol, hexanol, butanol, phenol, and cresol.

Table 9.7: Tertiary Oil Recovery Options in Manufacturing Sulfonates

```
                                    PLANT LOCATION
                        At Refinery              In the Field
                   ┌──────────────────────┬──────────────────────┐
                   │ 1                    │ 2                    │
                   │  High transporta-    │  Low to High         │
                   │    tion cost         │    transportation    │
                   │                      │    cost              │
   Lubricating     │  High yield          │  Moderate yield      │
     Stocks        │                      │                      │
                   │  (Up to 15% per      │  (Up to 8% per       │
                   │     pass)            │     pass)            │
                   │                      │                      │
 FEEDSTOCK         ├──────────────────────┼──────────────────────┤
                   │ 3                    │ 4                    │
                   │  High transporta-    │  Low transporta-     │
                   │    tion cost         │    tion cost         │
   Crude Oil       │   NOT PRACTICAL      │                      │
                   │                      │                      │
                   │  Low yield           │  Low yield           │
                   │                      │  (5% per pass)       │
                   └──────────────────────┴──────────────────────┘
```

Table 9.8: Oil Field Manufacture of Sulfonates ($/lb)

	COMMERCIAL MARKET PRICE (DOLLARS)	ON-SITE COST (DOLLARS)
Petroleum Sulfonates (Mahogany acid)	0.20	0.03[a]
Raw Materials:		
Sulfur Trioxide		0.10
Isopropanol	0.04	0.04[b]
Sodium Alkaline Salts		0.05
Utilities	0.01	0.02
Transportation (250 mi)	0.10	0.03
Fixed Costs per lb. [c]		
25 MM lb/yr		0.05
150 MM lb/yr		0.02
Cost per lb.	0.35	0.29 - 0.34

[a]"Opportunity Cost" -- One pound of oil at $11 per barrel

[b]Alcohol is used in this process and is available for use as a cosurfactant thus cost included.

[c]Two-year amortization, includes operating costs.

Source: EPA-68-01-1912

Figure 9.1: Sites of Existing Oil Refineries Where Sulfonation Units Could be Constructed to Serve Tertiary Oil Projects

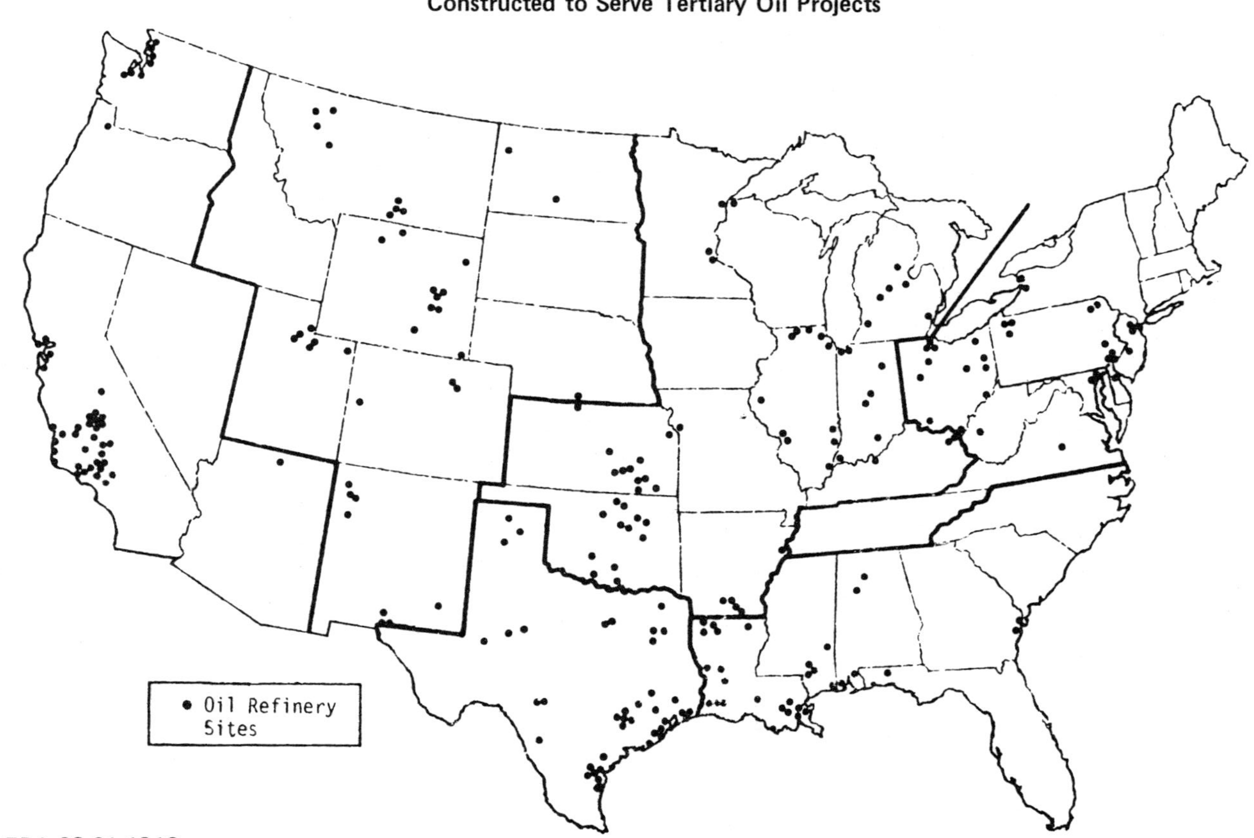

Source: EPA-68-01-1912.

Since sufficient supplies of a particular cosurfactant are required to fulfill the entire "recipe" for the micellar slug at a given project, regional variations in the availability and use of various cosurfactants could develop.

Water Supplies

The availability of suitable water for preflushing preparation of the micellar slug and the mobility buffer and follow-up injection is important to the success of a tertiary oil recovery project. Water with low concentrations of divalent ions and total dissolved solids is required. (Refer to the chapter entitled "Secondary Recovery of Crude Oil" for a discussion of treatment of injection waters for secondary recovery.) Connate water in the reservoirs is rarely suitable for reinjection without treatment to remove high concentrations of divalent ions. Concentrations (mg/ℓ) of calcium and magnesium, respectively, in connate waters average 2,530 and 530 in tertiary formations, 25,800 and 2,500 in Pennsylvania formations (3). These concentrations are 5 to 50 times the acceptable levels for successful micellar flooding.

Supplies of suitable water for a large-scale micellar-polymer flood may not be available. At present, some strains on water resources have occurred with pilot-scale projects. The Cities Service/Energy Research and Development Administration project near El Dorado, Kansas, has built a five-mile-long water pipeline six inches in diameter from the town's lake (raw water supply) to the oil field. The design water flow rate for the pilot-scale project is 6,300 bbl per day (264,600 gpd) (4).

This represents approximately 20% of the estimated daily water consumption by El Dorado's 13,000 residents. Since there are more than 6,200 acres of Admire sand in the El Dorado field while the tests cover only 51.2 acres, there is a potential demand at the field for as much as 800,000 to 1,500,000 gpd of high-quality water from the town's supply, assuming a development rate of 300 to 600 acres per year and a well density one-half as great as the pilot test area. If the field were controlled by a number of operators who undertook simultaneous field-wide development projects, water usage would be 20 to 25 million gpd.

Surface waters such as rivers may be unacceptably high in total dissolved solids and divalent ions, as are shallow unconsolidated aquifers which are subject to evaporative concentration of salts. The difficulty of further defining the effect of water supply availability is the need for detailed data on local consumption trends, availability of undeveloped, uncommitted water resources and the scale and timing of the development of micellar-polymer oil recovery projects. These local water shortages might lead to the need for on-site water treatment and recycling, which would reduce the amount of oil economically recoverable by micellar-polymer flooding, if not making the project economically unattractive.

RESEARCH REQUIREMENTS

Experienced reservoir engineers and petroleum geologists are in short supply. The rate of production of new petroleum engineers and geologists is exceedingly low. Large-scale EOR production will demand larger numbers of such professionals, and probably a more sophisticated level of training for many of them.

University participation is necessary not only to encourage and assist in the training of production engineering and research personnel, but also to provide additional manpower for basic research. Those areas of research needed to support field test programs which are identified as requiring intensive effort are:

(a) Engineering geology—development of methods for synthesizing information from all sources (core analyses, geophysical exploration, outcrop analysis, well logs, primary production history, etc.); the development of maximum information about the characteristics of a reservoir, as necessary to the design of EOR production operations.

(b) Geophysics—improved resolution and definition of reservoir configuration, improved lithological interpretation, definition of reservoir flow patterns, etc., using, for example, improved frequency, pulse, amplified modulation techniques, multiple reflection/refraction techniques, etc.

(c) Surfactant research—generally physical chemistry/chemical physics concerned with the action of promising surfactants under realistic reservoir conditions and with the development of improved surfactant solutions for EOR.

(d) Sweep control fluids and methods—improved performance and economics of mobility control fluids used to maximize sweep efficiency of "miscible" (micellar, CO_2, hydrocarbon) EOR methods; a specific example is the development of polymers that are stable under higher temperatures and wider ranges of ion (especially Ca and Mg) concentration.

(e) Underground combustion—determination of mechanisms of action and means for improved displacement efficiency and mobility control, e.g., effects of oxygen enrichment on combustion rate and efficiency.

(f) Systems analysis—interdependence of enhanced recovery/environment/water supply/manpower/material supply/etc. for overall management and scheduling purposes.

(g) Test method development—improving and simplifying both laboratory and field measurements for field test application, and leading to the development of optimum or standard procedures for use in the cooperative field test program.

REFERENCES

(1) *Enhanced Oil Recovery*, National Petroleum Council, (December 1976).

(2) Robichaux, T.J., "Bactericides Used in Drilling and Completion Operations," U.S. EPA Symposium on Environmental Aspects of Chemical Use in Well Drilling Operations, Houston, Texas, p. 4 (May, 1965).

(3) Collins, A.G., "Chemical Applications in Oil and Gas Well Drilling and Completion Operations" (unpublished).

(4) Cities Service Oil Company, *El Dorado Micellar-Polymer Demonstration*, BERC/TPR-75/1, (October 1975).

ENVIRONMENTAL ASPECTS

The information in the following sections of this chapter is based on:

> *Potential Environmental Consequences of Tertiary Oil Recovery,*
> prepared by C. Braxton, R. Stephens, C. Muller, J. White, J.
> Post, J. Norton, M. Goldberg, and P. Stevenson, of Energy Re-
> sources Co., Inc., for U.S. Environmental Protection Agency,
> under Contract 68-01-1912, July 1976.

IMPACTS OF ENHANCED OIL RECOVERY

While enhanced oil recovery technologies are not generally hazardous or toxic in nature, the implementation of EOR involves certain environmental risks. The processes used for tertiary oil recovery may pose the types of problems to the environment shown in Table 10.1. (Due to the shortage of natural gas and the expectation of curtailed gas availability in the future, miscible hydrocarbon slug processes and miscible flue gas injection processes have not been included.) In addition, the application of each process may result in secondary impacts such as off-site manufacture of chemicals, transportation of bulk chemicals, and re-finery load shifts to handle the higher average sulfur content in lower gravity crudes.

In comparison with exploratory and developmental drilling and conventional oil production, tertiary oil recovery projects pose different environmental problems. Much more is known about the subsurface geology by the time an enhanced recovery project is started. Delineation of the reservoir has resulted from in-fill drilling following the original discovery and additional detailed geologic data are compiled and analyzed before the project. The high pressures which may cause blow-outs in exploratory wells have been reduced during earlier petroleum produc-tion operations so that this risk is essentially eliminated in tertiary recovery. In terms of potential magnitude of environmental problems, the tertiary recovery processes examined in this study can be roughly ranked as follows in descending order of environmental concern: steam displacement, in situ combustion, micellar-polymer flooding, miscible carbon dioxide.

Table 10.1: Potential Environmental Problem Areas

Process	Problem	Causes
Steam displacement	Air quality	SO_x and particulates from steam generators. Hydrocarbons from producing wells and other field sources.
	Water quality	Spills, leaks of chemical foaming agents (if used)
	Water supplies	Process water demand
In situ combustion	Air quality	Hydrocarbons and CO from producing wells
	Water quality	Spills, leaks of low pH water with heavy metals
Micellar-polymer flooding	Water quality	Spills, reservoir and well leaks, disposal of chemically loaded brines
	Air quality	On-site manufacturing of chemicals
	Solid wastes	On-site manufacturing of chemicals
	Water supplies	Process water demand
Carbon dioxide	Air quality	Fugitive emissions of H_2S combined with CO_2
	Water quality	Spills, leaks of low pH water with H_2S

Source: EPA-68-01-1912

The thermal methods of enhanced oil recovery may pose air quality problems if large-scale development occurs in an oil field. A high-density-development oil field steam generator burning 1% sulfur fuel can exceed national ambient air quality standards for sulfur dioxide. Such impacts may represent a stumbling block to the full realization of the resource potential by thermal oil recovery methods. Unanswered questions remain regarding emissions of hydrocarbons from steam displacement and in situ combustion. Potential sources of these emissions may include wellheads and produced water treatment systems, as well as fugitive and area sources. Further data are required to characterize the emissions from these sources in order to identify control technology needs. In some cases, the recovery of casing blow gases containing oil condensate can be accomplished profitably, thus improving the economic return of the project while protecting the environment from these emissions.

Micellar-polymer floods involve large quantities of expensive chemicals, some of which could have an adverse effect on the environment if improperly handled. Economic factors should tend to mitigate environmental risks during the operations phase of a micellar-polymer flood. Well failures or improper injection profiles, which could result in the loss of expensive chemicals or poor reservoir response, must be prevented if the project is to be profitable. Engineering studies are carried out in the early phases of a project to identify and appraise these problems. Since the front-end investment in such projects is large and the returns in many cases are marginal, every effort would be made to minimize foreseeable fluid losses and thus risks to the subsurface environment. Disposal of chemically laden produced water should also be carried out in an environmentally acceptable manner. Field data on the concentration of chemicals in the produced fluid streams are needed to determine the need for effluent control systems.

CHEMICAL HAZARDS OF EOR

Chemicals used in micellar-polymer floods represent the most clearly identifiable group of potential hazards. These are of special concern because many will be left behind in the reservoirs after the recovery project is completed. The primary reservoir remnants of carbon dioxide and thermal methods are low pH brines. The composition of produced gases from fireflooding and steam injection is not well understood, but some initial estimates of the movement of sulfur dioxide and oxides of nitrogen have been made. Fugitive emissions of hydrocarbon vapors from these sources need further study. Gaseous streams from production wells in thermal recovery projects and emissions from steam generator stacks are two sources of potential airborne hazards introduced by tertiary oil recovery.

If, in the recovery process, chemicals escape to the environment in sufficient quantities, their presence can degrade water supply quality and destroy biota. The release of chemicals may occur at the manufacturing plant, on the transportation route, at the oilfield site during preparation of the chemical for use, from the reservoir (during the actual tertiary recovery process), and from the reservoir (after completion of the tertiary recovery process). Many of these activities are within the scope of existing EPA regulations. The potential environmental hazards associated with the handling of these chemicals on the surface are clearly identifiable. However, the presence of concentrations of chemicals in the oil-bearing formations represents a long-term source of pollutants that may be released underground into nearby aquifers at some unknown time in the future. Therefore, an understanding of the potential hazards of chemicals at reservoir conditions is important.

Toxic Effects

Chemicals, which reach the environment, may affect it in a number of ways. Living organisms exercise natural quality controls, both on freshwater and seawater. Some chemicals or their degradation products may eliminate those beneficial organisms in certain areas, or they may decimate only certain species or certain developmental stages. For example, the adult organism may migrate into an area where reproduction is inhibited. The food sources for these organisms may be more sensitive to certain toxic materials than the organisms themselves are; and the beneficial organism populations may become reduced because of a lack of food, rather than by a direct toxic effect upon the organisms.

All of the general types of toxic effects mentioned above and the specific effects to be mentioned later are concentration dependent. Chemicals which may be released from an oil reservoir decrease in concentration because of dilution, filtration, adsorption on both soil and clay particles, precipitation, chemical decomposition or polymerization, and biodegradation with time and distance traveled through an aquifer. Figure 10.1 illustrates the role of these mechanisms in mitigating the effect of chemical leakage or spills.

Figure 10.1: Mechanisms Acting on Chemicals Released from a Reservoir

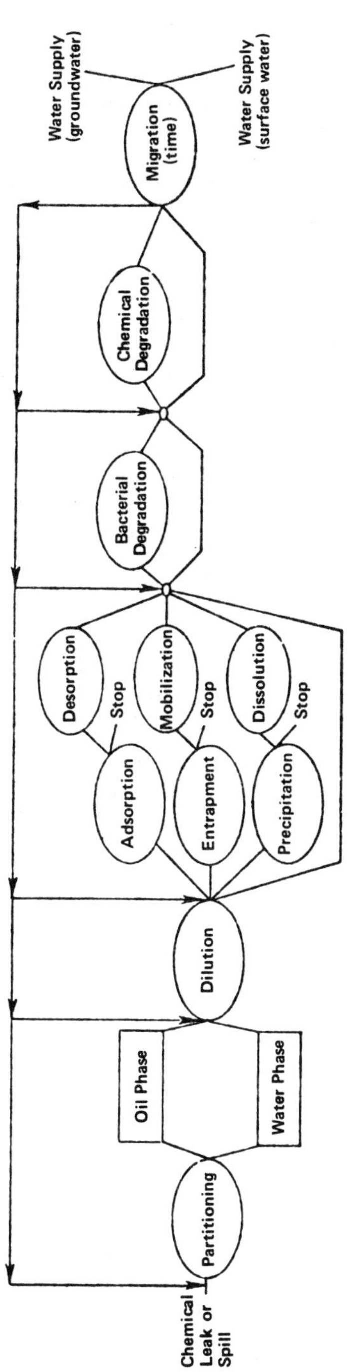

Source: EPA-68-01-1912

The effects of a chemical used in tertiary oil recovery may well be negligible at the levels actually encountered in the vicinity of the recovery site. Furthermore, there is only a possibility that it will ever reach a water body, as shown in the figure. Nevertheless, most of the chemicals used in tertiary oil recovery have definite and well-documented effects on water quality, and many others have documented toxic or carcinogenic effects on animals. It is recognized that the synergistic effects of certain combinations of pollutants portend greater or lesser risks than the simple additive combination of the effects of each pollutant. However, synergistic toxicity studies of chemicals used in micellar-polymer flooding are not available in most instances.

Another important aspect of toxicity is the composition of the ecosystems which are contacted by the pollutants. Definition of these ecosystems and analysis of chemical toxicity data for each ecosystem including questions of biodegradation and bioconcentration are also of concern, but were not undertaken in this study.

Surfactants: The surfactants most commonly used in micellar slugs are long-chain, linear alkyl sulfonates such as lauryl sulfonate and dodecyl sulfonate, and alkyl aryl sulfonates. (Refer to Table 9.1 of the preceding chapter.) These substances are also used as foaming agents and detergents, and their most noticeable water quality effect, at least in high concentrations, is to produce rather ugly foams, both in water bodies and at the tap (if sulfonate-containing water is used as a source of a public water supply). The linear alkyl sulfonates are biodegradable at fairly rapid rates. Surfactants, or surface-active agents, also may impart tastes or odors to water as indicated in Table 10.2.

Table 10.2: Minimum Concentration of Surfactant Required
to Produce a Perceptible Taste and Odor

SURFACE ACTIVE AGENT	MIN. CONC. (mg/l) TASTE	ODOR	TYPE OF TASTE OR ODOR
Alkyl aryl sulfonate	0.6	0.7	Limey, chemical
Alkyl aryl sulfonate	0.4	0.3	Soapy
Alkyl sulfate	3.0	0.2	Solvent
Alkyl sulfonate	1.4	0.3	Bitter, soapy, kerosene
Sulfonated amide	2.5	2.0	Chemical, soapy

Source: EPA-68-01-1912

Toxic effects to aquatic life have been reported for alkyl sulfonates at concentrations which are exceeded in reservoir fluids during micellar-polymer flooding. Reservoir concentrations may be 375 to 1,900 mg/l. Acute lethal concentrations vary from 0.2 to 10.0 mg/l; chronic levels below about 0.63 mg/l produce no appreciable effects on minnow or bluegill populations (1). The 1973 Water Quality Criteria suggest a maximum limit for linear alkyl sulfonates equal to 0.05 times the 96-hour lethal concentration (LC_{50} in the receiving water; such a level would correspond to about 0.2 mg/l. Practically all of the sulfonates (alkyl, alkyl phenyl, and naphthenic) are included as toxic substances on the National Institute for Occupational Safety and Health/Health, Education, and Welfare (NIOSH/HEW) list (2). The toxicities seem to be rather low, however, as a high concentration is required before effects are noted.

Another effect results from the synergistic interaction of sulfonate surfactants with a number of known compounds comprising crude oil, facilitating the solubilization of toxic fractions of the crude oil (3). Presumably, this affects the solubilization of organic toxicants by the sulfonates, with the result that penetration through epithelial barriers (e.g., the intestine) is facilitated (4). Synergistic interactions have been reported with the carcinogen 3,4-benzopyrene (4). Though surfactants based on alkyl benzene sulfonates are not in common use today for tertiary recovery, it should be noted that these compounds are of concern because they do not break down biologically in sewage treatment plants, and can travel long distances in streams and through groundwater without losing their identity. The more commonly used linear alkyl sulfonates are biodegradable; however, they are also two to four times more toxic (3).

Mobility Buffers: The compounds used in mobility control solutions such as polysaccharides and most of the others listed in Table 9.3 of the preceding chapter are nontoxic. If discharged in excessive quantities, they contribute to the degradation of water quality mainly by increasing the biochemical oxygen demand (BOD) and to oxygen depletion in a receiving water body, and by serving as the basis of a rich nutrient medium which will encourage the growth of bacterial species. In typical micellar-polymer floods, very little of the materials will be returned to the surface. However, concentrations of 100 to more than 1,000 mg/l do occur in the produced water from simpler techniques of mobility control flooding with polymers alone.

Hydrocarbon Fractions: Chemical hazards resulting from the use of hydrocarbon fractions other than crude oil in the micellar slug (Table 9.4) may be in the form of emulsified oils, or solutions of the water-soluble fractions of these oils. Due to cost considerations, crude oil has been the most common hydrocarbon constituent since it can easily be recovered from the produced fluids and requires little treatment prior to use. In special cases, other hydrocarbon fractions will be used. These fractions and the water-soluble fractions of crude are of concern here.

The aromatic hydrocarbons are the major group of acutely toxic compounds in oil residues. These also cause problems of fish tainting, and the production of unpleasant tastes and odors (1). In addition, there is a potential carcinogenic problem associated with the introduction of ill-defined petroleum fractions into water supplies. The hydrocarbon compounds can cause tainting of the fish flesh and other aquatic organisms, and may be used in micellar-polymer flooding, thermal methods (quinoline, sodium hydroxide and toluene), and other tertiary recovery processes such as miscible hydrocarbon methods. In general, the hydrocarbons which may have some environmental impact are already utilized as biocides, lubricants, and corrosion inhibitors during primary and secondary oil recovery. Tertiary recovery would be expanding the amount, not changing the kind of chemical used.

Electrolytes and Preflushing Compounds: Some of the electrolytes and compounds used in preflushing are potentially harmful, but the concentrations in micellar projects average less than 10.0 mg/l. Such a concentration is within the criteria for drinking water supplies. Sodium nitrate and the inorganic phosphates are nutrients. If introduced in sufficient quantities, they can encourage the growth of undesirable algal slimes and can accelerate the eutrophication of surface water bodies. Water turbidity can be increased, odors can be created, hydrogen sulfide gases can be produced by bacteria attacking abundant decaying matter, and a swampy taste can be imparted to the water.

If introduced into groundwaters or surface waters which are used as the source of a drinking water supply, sodium nitrate can create further problems, since nitrates in drinking water have been linked rather convincingly to a high incidence of methemoglobinemia among the users (especially infants under three months of age) of the water supply. The magnitude of this effect is, of course, extremely dependent upon the quantities of nitrates actually used. Therefore, the size of the project as well as the concentrations of electrolyte are important. Sodium sulfate is listed as a toxic substance, but only at extremely high concentrations which far exceed predicted levels in tertiary oil recovery processes. Subtoxic effects have been noted for sulfates, including imparting a funny taste and a slight laxative effect; but these, too, seem to be improbable at sulfate concentrations expected as a result of micellar-polymer flood.

Cosurfactants: Of the cosurfactants used in micellar slugs (Table 9.2), isopropanol, isobutanol and amyl alcohol are toxic in high enough concentrations, but at the levels which may result from their escape from an oil reservoir, they should be degradable without too much difficulty by healthy ecosystems. In the normal course of things, they would travel along the usual catabolic pathways, being degraded by bacteria and contributing primarily to biochemical oxygen demand (BOD).

Phenol and Cresol — Two compounds mentioned as possible cosurfactants, phenol and cresol, would create significant water quality problems if introduced into ground or surface waters. Since these alcohols are much more dangerous in the environment, more expensive, and just as effective as the other alcohols, such compounds would appear to be the cosurfactants of last resort in tertiary oil recovery. However, shortages of the other alcohols may induce some substitution of phenol and cresol for micellar-polymer flood projects. The ortho-, meta- and para-forms of cresol, as well as phenol itself, are all listed as toxic substances on the NIOSH/HEW list (2).

The data on that list are limited, for the most part, to toxic effects on mammals (excluding man), so its usefulness to aquatic and human consumption problems might be questioned. Of somewhat more relevance are EPA's aquatic life toxicity data, which include documented toxic effects on a number of first order organisms. Concentrations of 6.5 mg/l have been reported to cause extensive damage to trout reproductive systems, and concentrations of 1.0 mg/l are reported to have no toxic effects on trout (1).

Death to humans exposed to acute phenol poisoning has occurred in periods as short as 30 minutes. Phenol can be absorbed through human intact skin. If death is not forthcoming, damage to the kidneys, liver, pancreas, spleen and lungs may result. Skin absorption may eventually cause gangrene in the afflicted area.

Chronic phenol poisoning is characterized by vomiting, difficulty in swallowing, excessive salivation, diarrhea, loss of appetite, nervous disorders and skin eruption (5).

Effects other than direct toxic ones have been noted for those cosurfactants of last resort. Phenols impart unpleasant tastes and odors to water (with threshold odors being reported at 0.055 mg/l for p-cresol, 0.25 mg/l for m-cresol, 0.26 mg/l for cresol, and 4.2 mg/l for phenol), and furthermore, they are not removed effectively by most conventional water treatment processes (1). They can react with chlorine at public water supply disinfection plants to produce chlorophenolics, whose toxic effects, as well as effects on taste and odor, are much greater than those of phenols and cresols themselves (they are also more environmentally persistent than their parent phenolics). Phenol and cresol are biodegradable, though the rate of degradation is slow.

Biochemical oxygen demand studies have shown that less than one-fifth of the phenols introduced into a seeded water sample are consumed within 5 days. Furthermore, phenol is a tainting substance; that is, it causes distinct unpleasant tastes in fish which have swum in waters containing them. Phenolic compounds, including cresol, have been reported to affect fish taste at the 0.001 mg/l levels (3).

McKee and Wolf concluded, in 1963, that phenols in a concentration of 0.0001 mg/l would not interfere with domestic water supplies; in a concentration of 0.2 mg/l, they would not interfere with fish and aquatic life; at 50.0 mg/l, they would not interfere with irrigation; and at 1,000 mg/l, they would not interfere with livestock watering (3). For most waters, the maximum recommended phenol concentration is 1.0 mg/l.

Bacteriocides and Biocides: Toxic effects of bacteriocides and biocides used in tertiary oil recovery on aquatic life are well documented (1). Aldehydes have been reported to have TL_{50} for fish of 50 to 400 mg/l (3). Chlorinated phenols have a TL_{50} for fish of 0.2 to 1.0 mg/l (6). Quaternary amines have TL_{50} for fish of 0.2 to 5.0 mg/l and other amines have a TL_{50} for fish of 0.4 to 4.0 mg/l (6). All of these compounds can be expected to be toxic to humans.

Chlorinated hydrocarbons are stable in the environment, toxic to some wildlife and other nontarget organisms, and have adverse physiological effects on man. They are stored in fatty tissues rather than being rapidly metabolized. Mild cases of poisoning cause headaches, dizziness, gastrointestinal disturbances, numbness, and weakness of the extremities, apprehension, and hyperirritability and in severe cases, muscular fasciculations and spasms, leading in some cases to convulsions and death (4). Chlorinated hydrocarbons are not completely removed by sewage treatment, nor by water purification plants.

Criteria are needed for many of the chemicals which may be used in enhanced oil recovery so that potential problems may be identified. Furthermore, although the pathways from the oil reservoir to the water user involve low probabilities and large amounts of dilution and a number of coincidental chemical and physical occurrences, the presence of some of these materials at concentrations three orders of magnitude greater than acceptable water quality criteria in abandoned oil fields is a difficult problem in environmental protection for the future. The actual scale of the problem will vary from region to region depending on factors affecting the risk of releasing these chemicals and the proximity of important aquifers or water bodies to oil reservoirs.

Degradation Products

Although many of the chemicals used in enhanced oil recovery processes may form degradation products, such degradation occurs only under extreme conditions of pressure and temperature. For example, typical conditions in a reservoir where micellar flooding is undertaken are estimated to be: pH of 6 to 8; temperatures up to 200°F; pressures up to 3,000 lb/in^2. Degradation occurs most easily at more acidic (pH of 3) or more alkaline (pH of 11) conditions, with temperature in excess of 250°F and pressures in excess of 2,000 lb/in^2.

In addition to the effect of these reservoir conditions on the degradation process, there may be effects due to the other compounds in the formation matrix and their interaction. A chemical used in a micellar-polymer flood may be the limiting factor in a degradation reaction (e.g., it must be present for the reaction to occur), or it may serve as a catalyst for the degradation reaction (e.g., it affects the rate of the reaction).

The surfactants used in tertiary oil recovery processes are a very important group of chemicals because the sulfonation reaction by which most are manufactured is reversible. Ditetradecyldimethylammonium chloride is probably stable, although it may undergo degradation to 1-tetradecene and dimethyltetradecylammonium chloride. Dodecyltrimethylammonium chloride is probably stable, with possible decomposition to 1-dodecene and trimethylammonium chloride.

The details of the toxicity of dodecene and 1-tetradecene are unknown, although they are probable irritants and narcotic in high concentrations. Hexadecyltrimethylammonium chloride is probably stable, with the possible degradation to 1-hexadecene and trimethylammonium chloride. Alkyl phenoxypolyethoxyethanol may hydrolyze to glycol. One hundred milliliters of ethylene glycol is reported to be the lethal dose for man in an acute dose. In general, while most of the surfactants are probably stable, sometimes degradation may occur.

Among the cosurfactants, ketones and p-nonylphenols are both stable, as are isopropanol, n-butanol, isobutanol, 1-hexanol, 2-hexanol, 2-pentanol, and cyclohexanol. Esters may hydrolyze to an acid and an alcohol. Aldehydes are probably stable, although they may be oxidized by strong oxidizing agents. Quaternary amines (used as corrosion inhibitors) may eliminate the longest side chain to form a tertiary amine.

Most of the other chemicals used in tertiary oil recovery are very stable. All mobility buffers appear to be stable. Mercuric chloride (a biocide) is stable, and would stay in solution in all but extreme pH increases. Quinoline used to ignite in situ combustion is stable. However, hydrazine also used to ignite in situ combustion is a very strong reducing agent and will probably be converted to N_2 with a transfer of its hydrogen atoms to other molecules. Of the degradation reactions which may occur, hydrolysis and desulfonation of surfactants such as polyoxyethylene alkyl phenol, diethylene glycol sulfate, alkyl phenoxypolyethoxy ethanol to glycol and p-chloroaniline sulfate laurate to p-chloroaniline are the potential largest hazards.

Carcinogenicity of Chemicals

A number of the chemicals proposed for use in enhanced oil recovery operations have been shown either to induce the formation of tumors in experimental animals or to accelerate the induction of such tumors by other chemical carcinogens. Table 10.3 summarizes the major carcinogenic chemicals which may be used.

Table 10.3: Carcinogenic Chemicals Used in EOR

Surfactants (possibility of synergistic
   interactions with organic carcinogens)
Alkyl aryl sulfonates
Carboxymethylcellulose
Benzene and benzene derivatives
Polycyclic hydrocarbons
Crude oil and crude oil fractions
Formaldehyde and paraformaldehyde
Phenol and phenol derivatives

Source: EPA-68-01-1912

It is almost impossible, however, to make valid quantitative assessments of the human carcinogenic risk associated with the use of various chemicals. In recognition of this fact, the U.S. Environmental Protection Agency has defined the principles to be used in guiding policy making with regard to the regulation of carcinogenic chemicals (7).

The central thrust of these principles is to broaden considerably the definition of carcinogen (e.g., to include materials which have been shown to induce benign tumors) and to assert that in the absence of meaningful procedures to quantitatively determine human carcinogenic dose/response relationships or thresholds, it is necessary to consider as carcinogenic any chemicals which have been shown to induce tumors, at any dosage, in any experimental animal. A positive carcinogenic risk is considered to exist no matter how low the concentration of the substance in the environment. This means that attention should be given to any chemical which has shown carcinogenic effects in some system, no matter what the actual concentrations may be in the vicinity of a tertiary oil recovery project.

Of the chemical surfactants listed in Table 9.1 of the previous chapter, carcinogenic effects are only indicated for the alkyl aryl sulfonates, alkyl benzene sulfonates and some derivatives of alkyl naphthenic sulfonates. However, the synergistic interactions of otherwise noncarcinogenic surfactants with other chemicals will probably prove to be more important than any direct carcinogenic effects they may produce. Alkyl benzene sulfonates, for example, have rather low carcinogenicities when administered alone, but they have been found to increase the tumor-inducing properties of 4-nitroquinoline-1-oxide (8). This effect is presumably due to the solubilization of the carcinogen by the sulfonate, and it may also be produced by other surfactants. Among the mobility buffers, carboxymethylcellulose, polyethylene oxide (polyethylene glycol), dextran, and benzene are suspected carcinogens (9). High long-term doses of glycerin (glycerol) are also indicated as being tumorigenic.

Crude oil (as petroleum) is listed as a suspected carcinogen, and some carcinogenic activity remains in its purified fractions such as oil, which is listed as a carcinogen. Benzene is also a carcinogen, as are certain of the benzene derivatives such as toluene, which is sometimes used in stimulating wells prior to thermal methods of oil recovery. A further danger exists that these compounds may interact with chlorine in drinking water disinfection plants to produce carcinogenic chlorinated aromatics. Extensive data exist documenting the carcinogenicity of polycyclic hydrocarbons (including benzpyrene and some of the naphthenics) (9)(10). Several alkylated aryl compounds, as well as their chlorination products, are included in the references. Pure paraffinics and cycloparaffinics seem to be relatively noncarcinogenic.

The only potential carcinogen listed among electrolytes is the category of organic acids, but inorganic salts are most commonly used. More specific information will be needed about the organic acids used in order to assess the carcinogenicity of these compounds. No known carcinogens are listed for preflushing chemicals.

Definite evidence exists supporting the carcinogenicity of cosurfactants such as formaldehyde, p-formaldehyde, and phenol, but these materials are not most commonly used. No evidence exists for the carcinogenicity of pure preparations of most of the organic alcohols, except for one or two, such as ethanol, which are carcinogenic at extremely high doses administered over long time periods. More information will be needed on which amides, amino compounds or ketones are used.

EFFECTS OF EOR ON GROUNDWATER

Water Quality Conclusions

Operational chemical flooding projects such as micellar-polymer flooding pose few hazards to underground water supplies because numerous physical and chemical processes as shown in Figure 10.1 mitigate chemical leaks or spills. Petroleum sulfonate surfactants are the commonly used materials which could have the most adverse impact on water supplies and water quality. These materials, however, are susceptible to adsorption and precipitation within the oil reservoir as much or more than any other chemical used. Subsurface leaks from miscible carbon dioxide projects using CO_2 containing some H_2S could contaminate groundwater supplies. Indeed, hydrogen sulfide contamination of groundwater supplies has occurred as a result of conventional petroleum production operations.

Future hazards from completed chemical flood projects should be of the same small magnitude as the risks from ongoing recovery projects despite the fact that additional mechanisms associated with well age tend to increase the chances of leaks.

The two most significant potential causes of water quality deterioration during enhanced oil recovery are: (1) spills of chemicals as a result of transportation and handling, and (2) improper disposal of chemically loaded produced water.

Both sources are difficult to control by further regulation. Considerable volumes of freshwater from existing or potential water supplies may be required to carry out large scale projects for steam displacement and to a greater extent for chemical methods of oil recovery such as micellar-polymer flooding. Although this does not appear to be a barrier to early development activities, impacts on water supplies may occur due to local shortages or encroachment of salt water into some aquifers.

Contamination of Groundwater

Primary contamination of groundwater results in large part from leaks in well casings and from seepage through the walls of waste disposal pits. Pollution from lined pits is rare but may have long-lasting effects if brine has escaped to permeable sands outside the lining. Direct contamination of surface water can occur from spills or seeps to the surface of chemicals, water or brines, or oil. Since most water bodies flow eventually into others, pollution usually affects more than the first body it enters. Additionally, contaminants on the land may be washed by rain into surface water systems. In all cases, pollution will continue as long as contaminants remain and water supplies are exposed, perhaps long after production and disposal operations have ceased.

The problems of spills, surface seeps and pipeline breaks have been recognized as a result of conventional oil extraction operations. Because of close supervision, monitoring and modification and upgrading of equipment, it is hoped that the likelihood of leaks and spills during an enhanced oil recovery project will be less than conventional operations. However, during the recovery activity and after it is completed, it is possible for leaks to occur.

Groundwater contamination is particularly bad because it is long-lasting, difficult to trace, and may be very far-reaching. Once a layer of permeable sand has been polluted, it may be years before dilution and percolation clear the contaminant away. Though the actual rate of flow varies among aquifers, 200 ft per year has been suggested as a reasonable rate (11). At this rate of flow, 260 years would be needed to clear an aquifer 10 miles from a river. Some aquifers flow as slowly as 20 ft per year, and would, therefore, require even longer to cleanse naturally.

The most serious problems arise when produced water or brine enters the groundwater system. Pollution may result from a variety of sources. If the brine is used to displace oil in a foreign formation, as in waterflooding, the brine may mix with nearby water supplies. Excess brine disposed into foreign strata may cause contamination of surrounding aquifers. Spills at the wellhead and seepage from pits may ultimately reach groundwater sources. Reinjecting brine into its original stratum is a common and usually safe method of disposal, but it, too, may allow contamination if the disposal well fails.

Bypassing the Natural Filter: An opening in the formation, such as an oil well, may allow pollutants to pass directly into an aquifer without percolating through soil, rock, and organic matter. This poses most of the petroleum-related risks to groundwater supplies. Figure 10.2 shows the manner in which saline water may enter fresh water, polluted water may enter clean water, or substances from the surface may be washed into groundwater.

Though the bypassing of the natural filter may occur in a number of ways, the most common route is well failure. Any permeable stratum penetrated by a well is susceptible to contamination by oil, brine and chemicals passing through the well during both injection and extraction operations. Figure 10.3 shows a producing oil well and the possible routes of pollution into water-bearing strata. The newly drilled wellbore is lined with metal pipe called casing which is cemented to the sides of the hole for rigidity and corrosion resistance. Surface casing is installed near the wellhead to keep the hole clear and to protect the shallow fresh water aquifers. Production casing runs the length of the well. The bottom of the hole is plugged with a shoe, and a packer set inside the casing seals off the well just above the oil-bearing strata. Fluids pass through tubing which is suspended inside the casing and which passes through the packer. The result is a multiple shield between groundwater supplies and oil field fluids.

Yet ruptures do occur. The well provides communication among strata by conducting fluid through the cement-formation annulus, through the casing-cement annulus and then through a crack in the cement, through the tubing-casing annulus and then through cracks in the casing in the cement, or through the tubing and then through holes in tubing, casing, and cement.

Figure 10.2: Contamination of Fresh Water Through Bypass of the Natural Filter System

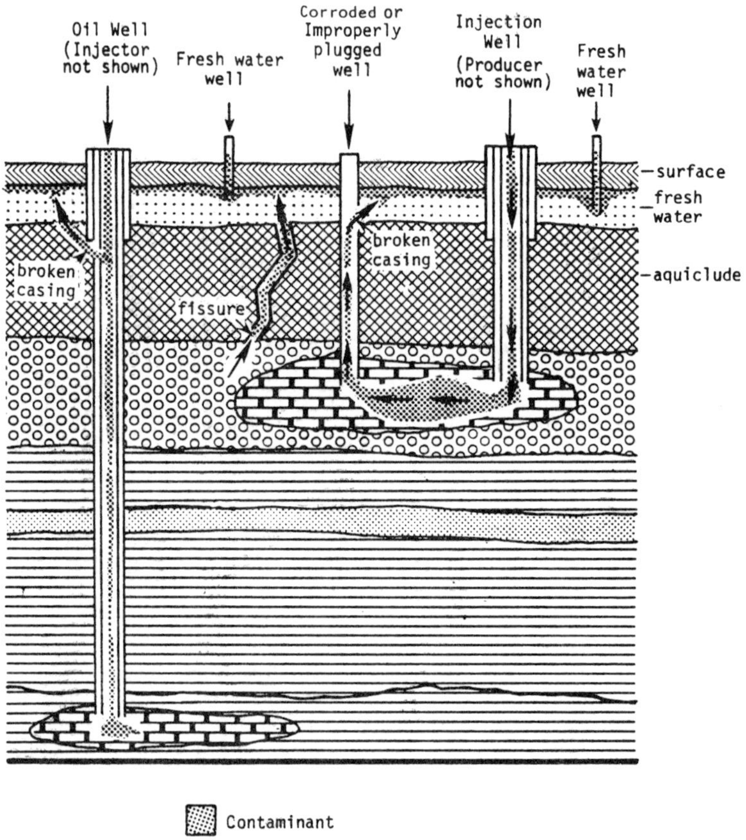

Source: EPA-68-01-1912

**Figure 10.3: Potential Paths of Freshwater Contamination
Resulting from Well Failure**

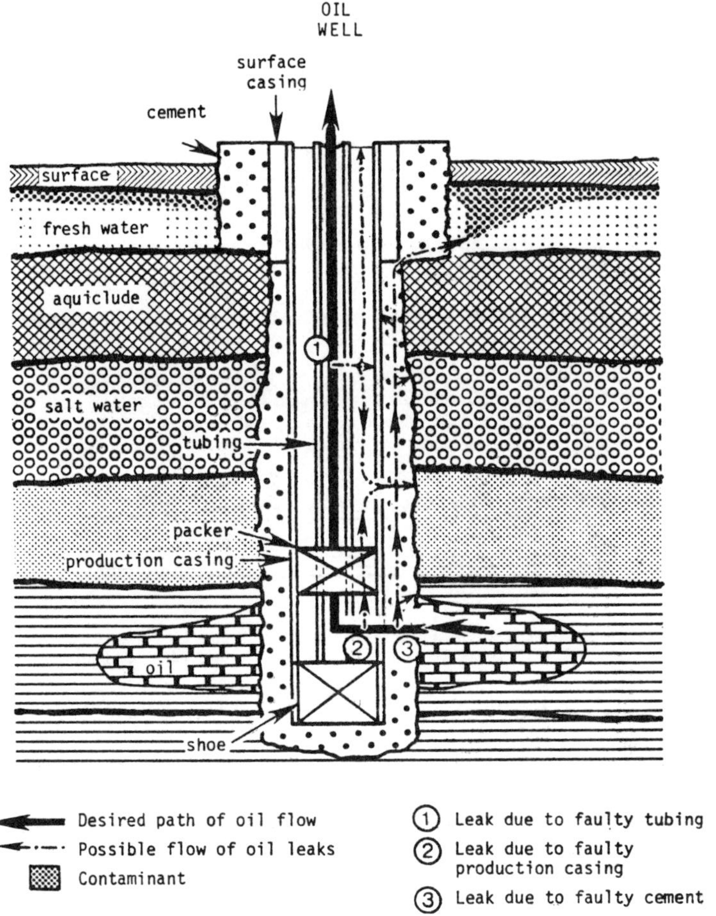

Source: EPA-68-01-1912

Overwhelming the Natural Filter: The natural filtering system is overwhelmed
when the contaminants become so concentrated that percolation cannot render
the solution harmless. Seepage from brine pits pollutes in this manner (Figure
10.4). Most states have recognized this danger and require that all disposal pits
be impervious. The result has been a dramatic reduction in the amount of pollution from this source. The difficulty with brine seeps and chemical spills, as with
groundwater pollution in general, is their longevity. The risk of chemical spills
and seeps is difficult to quantify because many site-specific factors such as operator skill, systems design and construction, modes of shipment, and weather, will
interact at each location.

Figure 10.4: Contamination of Fresh Water by Chemically
Loaded Brine Through (1) Overwhelm of Natural
Filter System, and (2) Pressure

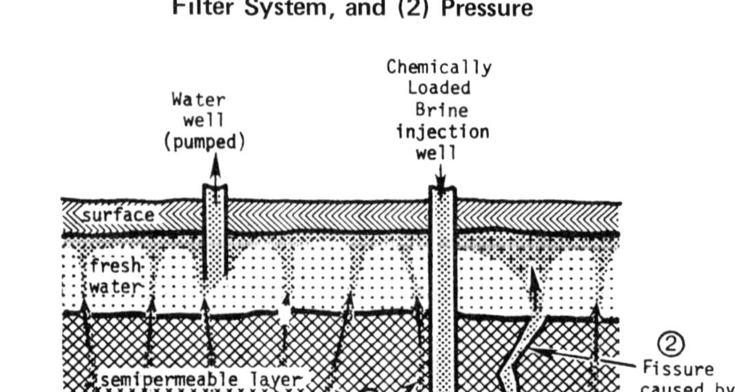

Source: EPA-68-01-1912

Changing the Formation: Differences in pressure, temperature, or chemical concentration cause the flow of groundwater; changes in temperature and acidity
alter the sorptive properties of the formation. Within aquifers or interaquifer
channels (wells or fissures) water flows from higher to lower pressure zones; to
a lesser extent, flows are affected by concentration and temperature differen-
tials. This can result in movement of salt water into freshwater zones in the same
aquifer, into other aquifers containing fresh water, or into surface water (Figure
10.5). If the formation has held toxic substances from another source, a disrup-
tion of the chemical balance may cause the release of such substances into groundwater.

Analysis of Factors Affecting Risk: Risks of contamination of groundwater
within a given oil field are affected by the tertiary oil recovery process chosen
for the field, the ages of the wells drilled into or through the formation in which
the teritary recovery will be carried out, and the seismicity in the area. Corrosion
and pressure are the greatest problems. The tertiary recovery method which is selected may raise the risk of groundwater contamination by further increasing pressure or corrosion rate. Table 10.4 lists some of the typical operating conditions for
recovery methods and compares these conditions with operations during secondary
recovery.

Figure 10.5: Effects of Pressure, Temperature and Chemical Concentrations on Water Flow in Aquifers

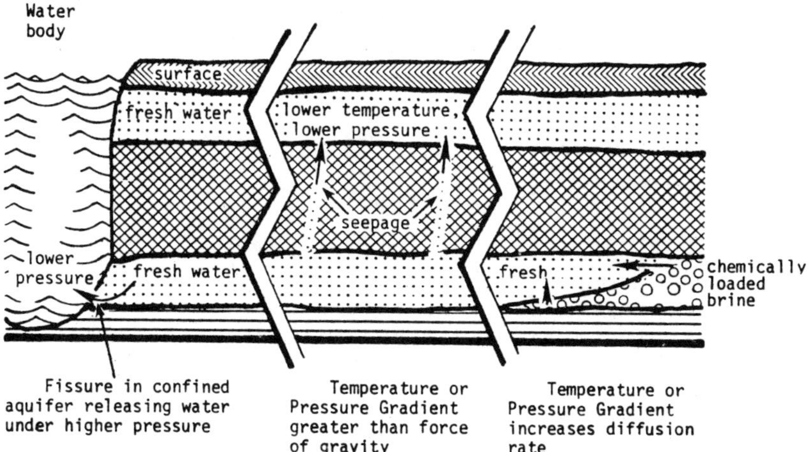

Source: EPA-68-01-1912

Pressure — Increased pressure during injection of the tertiary recovery fluids into the reservoir is unlikely to fracture aquicludes or well bore cement. As shown in Table 10.4, there is a margin of safety between the operating pressures used in tertiary recovery and a conservative estimate of the hydraulic fracture strength of the reservoir rock. The safety factor ratio is smallest for the steam injection reservoir and largest for carbon dioxide processes.

Brine disposal wells present problems for two reasons: first, they are frequently old wells, converted to disposal after a reservoir has been exhausted; secondly, pressure is often used to inject brine. The pressure can fracture aquicludes, allowing migration of brine into fresh water and aquifers. To avoid this, bottom-hole pressure should be less than 1 lb/in^2/ft of well depth. Pressure-induced fractures are usually horizontal in formations as deep as 1,000 ft and are vertical below 1,500 ft (12). Fractures in formations deeper than 1,500 ft might provide communication between strata. Some early steam injection projects blew out, damaging the formation and overburden or fracturing the cement and rupturing or buckling the casings in injection wells. The surface fittings broke on a carbon dioxide/sulfide injection well in Texas resulting in several deaths by H_2S poisoning. The latter accident was caused by a part which failed at an operating pressure below the 10,000 lb/in^2 for which it had been rated.

Though some pressure failures have occurred, they are very unlikely during tertiary recovery operations. There is a margin of safety in the operating pressures of the project and pressures will rarely exceed the original pressure of the fluids confined in the oil-bearing strata. A high pressure is required to initiate in situ combustion, but it is only of brief duration. Average operating pressures in a fireflood are 500 to 600 lb/in^2.

Table 10.4: Maximum Operating Conditions in Some Typical Processes

| | TERTIARY RECOVERY METHOD | | | |
	MICELLAR-POLYMER	STEAM	COM-BUSTION	CARBON DIOXIDE
Assumptions: (Typical Project)				
Reservoir Permeability (millidarcies)	500	500	500	100
Spacing between Wells (feet)	1000	500	1000	2000
Well Depth (feet)	5000	3000	5000	8000
Initial Pressure (lb./in.^{2})	2500	1500	2500	4000
Maximum Pressure during Secondary Recovery[a,b] (lb./in.^{2})	3500	1500	3000	4000[b]
Maximum Pressure during Tertiary Recovery[b] (lb./in.^{2})	3500	2500	4500	4000
Maximum Temperature during Tertiary Recovery (°F)	150	600	1200	200
Hydraulic Fracture Strength of Reservoir Rock[c] (lb./in.^{2})	5000+	3000+	5000+	8000+
Pressure Safety Factor: $\left(\dfrac{\text{Hydraulic Strength}}{\text{Tertiary Recovery}}\right)$	1.4:1	1.2:1	1.2:1	2:1
pH of Residual Fluids	6-8	6-8	2-4	4
Conductivity of Residual Fluids	Moderate	Moderate	High	High

[a] Non-miscible gas injection or waterflooding.

[b] At injector wells bottomhole.

[c] Minimum value for reservoirs.

Source: EPA-68-01-1912

Generally, chances of well failure from applying too much pressure to the reservoir during tertiary recovery are shown during exploratory drilling. When tertiary recovery is begun, a great deal of information is available about the oil field. These data provide the basis for sound engineering decisions in the design and operation of tertiary oil recovery projects. The large front-end investment required in tertiary recovery is an additional incentive for monitoring and controlling operating pressures of the project. In a micellar-polymer flood, the expensive chemicals in the slug could be lost to a fracture before any additional oil had been recovered.

Similar considerations should raise operator awareness of chemical spills at the surface during formulation and injection. The operator of a carbon dioxide project or in situ combustion project must pay for every additional lb/in^2 of injection pressure used. Steam injection methods applied at shallow reservoir depths, still require caution to avoid fracturing the overburden rocks. But problems with injection wells in steam injection have largely been solved by improved cements and completion methods, which compensate for the thermal expansion of the casing.

Corrosion — Corrosion is more of a problem with some tertiary recovery methods. The corrosion rate of steel well casing is affected by the pH and conductivity of the fluids which contact it. Temperature also influences the rate of corrosion. Considering the data in Table 10.4, the corrosiveness of the recovery methods can be evaluated. Table 10.5 ranks the relative corrosiveness of the tertiary recovery methods in relation to primary and secondary recovery. The ranking is based on the corrosiveness of the fluids which remain in the reservoir after the recovery project is completed and on the equilibrium temperature of the oil-bearing strata.

Table 10.5: Relative Corrosiveness of Recovery Methods

Most Corrosive:	Carbon Dioxide Miscible
	In Situ Combustion
	Steam Injection
	Secondary Recovery Waterflood
Least Corrosive:	Micellar-Polymer Flooding
	Primary Recovery

In Table 10.5, it is assumed that a higher conductivity brine is used for the secondary recovery waterflood. Actual corrosiveness will be affected by local geochemical conditions within the oil reservoir and penetrated formations. It is also assumed that the brine from the oil-bearing strata of the primary recovery is of lower conductivity than water used in waterflooding. This is generally true for shallower reservoirs where thermal methods and micellar-polymer floods are feasible. Actual corrosiveness will be affected by local geochemical conditions encountered such as salt sections, high partial pressure of naturally occurring carbon dioxide in gas-condensate produced, or the presence of hydrogen sulfide.

Carbon dioxide processes and the thermal methods of tertiary oil recovery are relatively more corrosive than micellar-polymer flooding. The ranking is based upon the assumption that the process is the only variable. The actual effect of these processes on corrosion of particular wells is dependent on the completion practices, materials used in construction, rock formations, and even the expertise of the operator. Specific data on these factors in each region or field are limited and an analysis of the interrelationships of these variables is beyond the scope of this assessment.

There is a correlation between the ages of the reservoirs and number of wells within the major oil fields where tertiary recovery is feasible. Table 10.6 displays the estimated age profile of wells which may be involved in each tertiary recovery method within each region. For example, the table shows that 75% of all oil wells which may be involved in micellar-polymer flooding in the Pacific West Region are less than 35 years old. These wells were completed with modern materials and practices more oriented to safety and conservation.

Table 10.6: Fraction and Number of Wells in Major Oil Fields of Each Region Where Tertiary Recovery Is Feasible

Tertiary Recovery Method	Date of Wells	Pacific	Rockies	North Central	South Central	South East	North East	Total
Micellar	≤1920	20%	83%	21%	10%	56%	NA	18%
		(512)	(1,656)	(2,580)	(1,421)	(247)		(6,416)
	1921–1940	5%	17%	63%	88%	71%	NA	67%
		(117)	(350)	(7,788)	(12,222)	(3,334)		(23,811)
	≥1941	75%	NA	16%	2%	24%	NA	15%
		(1,893)		(1,920)	(258)	(1,150)		(5,221)
Thermal	≤1920	76%	100%	NA	21%	94%	NA	68%
		(21,820)	(701)		(1,622)	(8,533)		(32,676)
	1921–1946	20%	NA	100%	79%	2%	NA	29%
		(5,593)		(1,920)	(6,090)	(166)		(13,769)
	≥1941	4%	NA	NA	NA	4%	NA	43%
		(1,138)				(383)		(1,521)
Carbon dioxide	≤1920	20%	21%	NA	42%	62%	NA	42%
		(129)	(540)		(11,772)	(8,886)		(20,727)
	1921–1940	14%	79%	98%	34%	31%	NA	42%
		(90)	(1,995)	(5,227)	(9,461)	(4,067)		(20,840)
	≥1941	66%	NA	16%	24%	7%	NA	16%
		(414)		(53)	(6,579)	(912)		(7,958)

Note: NA = none or data not available.

Source: EPA-68-01-1912

Generally, wells or reservoirs that are older were completed in a less sophisticated manner. The divisions of age into three groupings mark distinct changes in the state of the art of oil well drilling, completion and production, as well as differences in the economic and regulatory climate affecting the petroleum extraction industry.

It is not particularly surprising to note the skew toward older ages in the age profile of thermal wells in each region. Many of the shallow, low-gravity oil fields were originally discovered by surface evidence and natural seeps. Although most producing wells in these fields may not date from the turn of the century since they can be redrilled relatively cheaply, using the age profile as a risk indicator is valid since failure of the old plugged and abandoned wells is still possible.

Abandoned wells which have been inadequately plugged are frequent sources of contamination at present. From an environmental standpoint, failures of plugged and abandoned wells are of greater concern than failures of operating wells because there are no monitors of these leaks. Operating wells do fail, but the economic incentives and environmental controls on the well operator should minimize the amount of leakage which occurs. Leakage from an injection well may reduce the effectiveness of tertiary recovery or result in the needless loss of expensive chemicals. Production well leaks may result in loss of oil or higher than necessary pumping expense to lift the additional water which infiltrates the well.

Once production wells have ceased operation and have been plugged, however, no one is concerned with keeping them in good condition. Even when leaks are detected, sometimes no one can be held responsible. The Texas Railroad Commission has about $100,000 allotted annually for replugging operations, but this sum is hardly sufficient since it costs approximately $10,000 to find and plug a single abandoned well.

The problem of leakage of old wells has hindered enhanced recovery operations in parts of New York and Pennsylvania. Wells dating from the late 1800's were often crudely plugged with tree saplings if they failed to reward the early wildcatter sufficiently. Today, leakage from these old-time wells can jeopardize the success of any tertiary recovery project which may be otherwise potentially viable.

Most wells need casing repairs by the time they are 10 to 14 years old. In the Texas Upper Gulf Coast, for example, such repairs were required at a rate of 0.09 repair per well per year, or about 11 years mean time between well failures (13). Older wells are more likely to leak, both because of general aging and corrosion and because earlier technologies of materials, completion, corrosion protection and plugging were less reliable.

Relative Regional Risk Levels: An analysis of the age profile of wells involved in tertiary recovery yields a ranking of the relative risks of well failure in each region. Table 10.7 presents the rankings. A number of factors which are complex and interrelated have been excluded from the rankings for this investigation. Local geochemical conditions may affect well failure rates. Failure incidence is also a function of the depth of the wells involved in each recovery method in each region. The well spacing and number of wells within an oil field also affect the risks presented by each project, but the data have not been developed on these variables.

Table 10.7: Relative Regional Risk of Well Failure*

Micellar Polymer Flooding**

Most likely to fail:	Rocky Mountain
	South Central
	North Central
	Southeast
Least likely to fail:	Pacific West

Carbon Dioxide Miscible Methods***

Most likely to fail:	Southeast
	South Central
	Rocky Mountain
	North Central
Least likely to fail:	Pacific West

Thermal Methods of Oil Recovery†

Most likely to fail:	Pacific West
	Southeast
	South Central
Least likely to fail:	North Central

*Excludes seismic risk and local variations in geochemical corrosiveness.

**Northeast Region not ranked due to lack of data but is probably at a comparable risk level with the Rocky Mountain Region.

***Northeast Region not ranked due to lack of data, but risk level is probably comparable to the South Central Region.

†Northeast and Rocky Mountain Regions not ranked due to lack of data. Risks of well failure in the Northeast and Rocky Mountains are probably comparable to the Pacific West and Southeast Regions, respectively.

Source: EPA-68-01-1912

Well Damage Due to Seismic Activity: Seismic activity can also cause damage to wells in a tertiary recovery project, allowing reservoir fluids to migrate into aquifers.

Figure 10.6 is a seismic risk map of the United States. Earthquakes are virtually unknown, for example, throughout most of Texas. But, in some areas (e.g., Southern California), this risk is high and tremors may contribute to a high localized well failure rate. If localized well failure incidence is known, earthquakes may have contributed their share to that rate, and need not be considered as a separate factor.

Figure 10.6: Seismic Probability Map of the United States

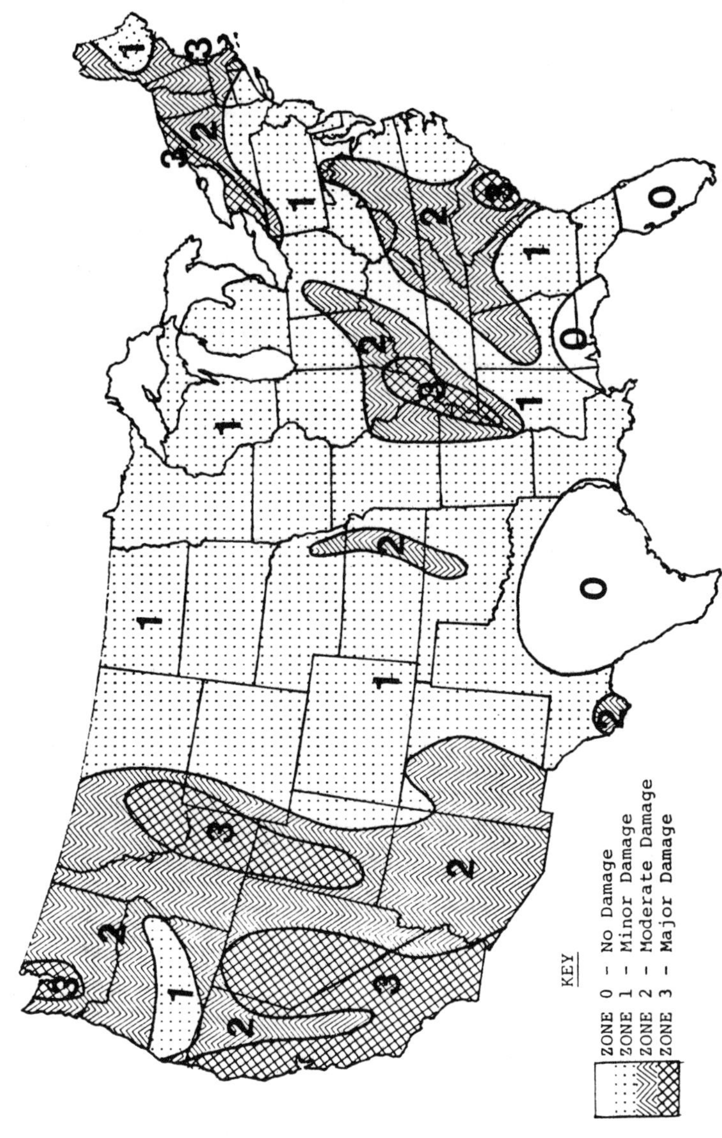

Source: EPA-68-01-1912

EFFECTS OF EOR ON AIR QUALITY

Widespread development of steam displacement projects to recover low gravity crude oils in shallow reservoirs could have a significant impact on air quality. The limits on sulfur content in fuels set forth in many state implementation plans are not adequate to prevent ambient air quality standards for sulfur dioxide from being exceeded. A variety of control strategies are available which may make it possible to keep air quality within the standards despite high-density development of this thermal oil recovery technique. However, the control strategy which is selected or required may limit the development of this petroleum resource and curtail the quantity of oil economically recoverable from the oil field.

Air quality problems may also arise as a result of on-site manufacture of petroleum sulfonates and due to fugitive emissions of vaporous hydrocarbons at thermal oil recovery projects and of hydrogen sulfide at miscible gas displacement projects.

Emissions from many sources in tertiary oil recovery projects have not been adequately characterized. It is important to consider reservoir characteristics and recovery process variables in the development of emission factors and the drafting of environmental policies to control undesirable discharges.

Emissions from Steam Injection

The emissions from oil field steam boilers used in steam injection depend on the size of the project, the oil recovery efficiency of the process, and the heat losses which occur, with more than 100 boilers. Individual boilers range in size from 5 to 500 million Btus per hour. One-third to one-fourth of the oil recovered is generally required for use as fuel and as reservoir production declines an economic limit is reached.

Fuels for boilers may be natural gas, fuel oils, or crude oil from the lease. Though emissions will vary with the content of sulfur and other compounds in the crude, Table 10.8 illustrates the composition which may be expected in typical operations. To these emissions must be added the particulate discharge from boiler cleaning cycles and from trace metals in the crude oil fuel. Hydrocarbon may also be released to the atmosphere at the wells during production. The particular operating and reservoir parameters of oil fields where steam displacement is carried out will determine the actual quantity of emissions.

Emissions from In Situ Combustion

The high temperatures to which the reservoir is subjected in the burning region of a fireflood enable a number of chemical reactions to occur. The compounds created in this manner as well as the combustion residues may be present in the emissions at production wells. Others may be confined in the reservoir after the process is completed.

Possible emissions from an in situ combustion project are varied. Some of the compounds listed in Table 10.9 may appear at the production wells. Table 10.9 is an analysis of the major constituents of the produced gas in a sample survey of 31 in situ combustion projects. These gas streams were measured at near-optimum operating conditions early in the life of the flood. As the flood progresses, the composition of emissions may change to include oxides of sulfur.

No data on oxides of nitrogen or trace contaminants were developed, however. Formation of NO_x compounds is not favored at the typical burning temperatures in a fireflood of 700 to 1200°F. Most of the emissions would probably be nitrogen. Light hydrocarbons, comprising 1% of the produced gas, appear chiefly as propane, methane and, to a lesser extent, ethane.

Table 10.8: Emissions Factors for Fuel Oil Combustion*

Pollutants	Large Sources (1,000 hp or more)	Small Sources (1,000 hp or less)
	. . . . . .lb/1,000 gal fuel oil burned.	
Aldehydes	1	1
Benzo [a] pyrene**	5,000	40,000
Carbon monoxide	3	4
Hydrocarbons	2	3
Nitrogen oxides as NO_2	105	40–80
Sulfur dioxide***	157 S	157 S
Sulfur trioxide***	2.0 S	2 S
Particulates†	8	23

*Reference (14).
**Density of fuel oil equals 8 lb/gal, and 42 gal equals 1 barrel. Values given are for μg/1,000 gal.
***S is percent sulfur in oil (Reference 16).
†Reference (15).

Source: EPA-68-01-1912

Table 10.9: Composition of Produced Gas Stream from In Situ Combustion

Component	Mol Percent
Hydrogen	Trace
Carbon monoxide	2.5–3.0
Carbon dioxide	12–14
Oxygen	1.0
Nitrogen and oxides	78–83
Hydrocarbons*	0–1
pH of produced water	1.6–8.6

*Methane, 0.4 to 0.5; ethane, 0.1; and propane 0.4 to 0.5.

Source: EPA-68-01-1912

Table 10.10 presents an estimate of the volumes of constituent gases produced per barrel of oil produced. It is based on a survey of the gas streams of 31 firefloods (average value) and assumes all sulfur compounds remain in solution with produced water. However, emissions of SO_x may occur in the late phases of a fireflood. As noted for Table 10.9, the formation of NO_x compounds is not favored at the typical burning temperatures in a fireflood of 700° to 1200°F. Most of the emissions would probably be nitrogen.

Regional differences in the operating and reservoir characteristics of in situ combustion projects would affect the emissions produced. The chemical and physical process of combustion within an oil-bearing stratum are complex and the possible environmental effects of combustion products need further study.

Table 10.10: Average Emissions of Gaseous Compounds from In Situ Combustion

Compound	Emissions
Carbon dioxide	435
Carbon monoxide	160
Nitrogen and NO_x	4,220
Oxygen	185
Hydrocarbons	150

*Standard cubic feet per barrel of oil produced.

Source: EPA-68-01-1912

Safety in Handling Gases

Safety in handling carbon dioxide and hydrogen sulfide is also required. On February 2, 1975, a gas leak occurred at a carbon dioxide injection well in Texas, resulting in nine deaths by hydrogen sulfide poisoning. To date this is the only such case recorded, yet it does suggest that such tragic accidents could occur along pipelines or at the well site. Control systems to detect leaks and limit the escape of carbon dioxide to small quantities are available and are required by regulation in some states.

The Office of Technology Assessment has used a different approach to evaluate environmental issues relative to enhanced oil recovery. The information in the following section is based on their work:

> *Enhanced Oil Recovery Potential in the United States*, OTA-E-59, prepared by Lewin & Associates, Inc., National Energy Law and Policy Institute, and Wright Water Engineers, Inc., for U.S. Office of Technology Assessment, January 1978.

HAZARDS OF EOR BY PHYSIOGRAPHIC REGION

The continental United States was divided into four general types of physiographic regions, each of which has certain specific characteristics and vulnerabilities to environmental damage. The four physiographic regions are: (1) the Continental Shelf which includes the broad, shallow gulf coast shelf, the steeper sloping Atlantic shelf, and the narrow steep-edged Pacific coast shelf; (2) the Coastal Plains adjoining the Pacific Ocean, Atlantic Ocean, and the Gulf of Mexico, particularly those of California, Texas, and Louisiana; (3) the Interior Basins such as the Great Plains, Great Lakes, and the central valley of California; and (4) the Rocky Mountains and other mountainous regions.

The Continental Shelf, a shallow, flat, submerged land area at the margin of the continent, slopes gently downward away from the shoreline. The width of the shelf ranges from less than 5 miles along portions of the southern California coastline, to a few hundred miles along parts of the gulf coast. Hazards common to all Continental Shelf oil recovery operations include tidal action, wave action, storm waves, and collisions with ships. In addition, hurricanes in the gulf and Atlantic coasts, landslides and earthquakes in the southern California borderland, difficulty of control, and unstable bottom substrate pose further hazards.

The Coastal Plains along the Atlantic and gulf coasts are as much as 100 to 200 miles wide, and make up nearly 10% of the land in the contiguous 48 states. With minor exceptions, the variance in elevation is less than 500 ft and for more than half of the Coastal Plains is less than 100 ft. Large parts of the nation's Coastal Plains are covered by major population centers. In the arid Southwest, Coastal Plain inhabitants rely heavily on local groundwater supplies. The U.S. Coastal Plains which have the potential for the greatest EOR activity are those of southern California, Louisiana, and Texas.

The Interior Basins include all land areas of the United States except the mountainous areas and the Coastal Plains. Within the interior drainage basins, there are geologic basins which may contain large quantities of oil entrapped beneath the surface. Generally, the geologic formations are older than those in the Coastal Plains. Some EOR activity is expected to take place in the Interior Basins, particularly those of the midcontinent and central California. Typically, the urban centers and farm areas of these basins depend heavily on local groundwater supplies. The groundwater aquifers of these basins are recharged by local rivers and by runoff from bordering mountains.

The mountainous areas are rich in timber and minerals. Some EOR operations are anticipated in the Rocky Mountains, particularly in Wyoming. These mountain areas offer diverse benefits to society since they are prime wildlife and recreational areas; with their relatively high snowpack, they are frequently a major source of groundwater for adjacent plains. These generally are remote unpopulated areas, where direct EOR impacts on the human population are limited but where adverse impacts on the natural environment can be significant.

Causes of Environmental Effects

The following elements and processes are common to all EOR methods: a recovery fluid; an injection system; surface processing; and disposal of spent materials.

The processes and the materials used within the confines of the system pose no environmental threat. Environmental problems result only when the materials are allowed to escape. The following mechanics may be responsible for such escape:

(1) Transit Spills—Spills which may occur when material is being prepared at or transported to the field site.

(2) Onsite Spills—Spills which may occur at the field site from surface lines and/or storage facilities.

(3) Well System Failure—Escape of materials which may occur from failure of the injection or producing well due to casing leaks or channeling.

(4) Reservoir Migration—Fluid may migrate outside of the confining limits of a reservoir through fractures or through a well bore which interconnects reservoirs.

Table 10.11: Matrix Evaluation of Relative Potential for EOR Environmental Impacts
(Impact Model Showing Relative Impacts of Process Effect and Environmental Component)

Process	Process effect	Air				Water				Land				Biota			
		CS	CP	IB	MT	CS	CP	IB	MT	CS	CP	IB	MT	CS	CP	IB	MT
Thermal	Spills	1	1	2	1	1	2	2	2	1	1	1	1	1	2	1	2
Steam	Well Failure . . .	1	2	2	1	1	2	2	2	1	1	1	1	1	1	1	1
In Situ	Reservoir Leaks .	1	1	2	1	1	2	2	2	1	1	1	1	1	1	1	1
Hot Water	Operational . . .	2	4	3	2	2	3	4	3	1	1	1	1	2	4	4	4
Miscible	Spills	1	2	2	1	1	2	1	1	1	1	1	1	1	2	1	1
CO$_2$	Well Failure . . .	1	2	2	1	1	2	2	2	1	1	1	1	1	2	2	2
Hydrocarbons	Reservoir Leaks .	1	1	1	1	1	2	2	2	1	1	1	1	1	1	1	1
	Operational . . .	2	3	3	2	2	2	3	2	1	2	2	3	2	4	3	3
Chemical	Spills	1	1	1	1	2	4	3	4	1	1	1	1	2	4	3	4
Polymer	Well Failure . . .	1	1	1	1	2	3	2	3	1	1	1	1	2	3	2	3
Surfactant/	Reservoir Leaks .	1	1	1	1	2	2	2	2	1	1	1	1	2	2	2	2
Polymer	Operational . . .	2	2	2	2	2	3	4	3	1	3	2	1	2	4	4	4

Scale Units: From 1 negligible or nonexistent, to 4 the most significant.

CS - Continental Shelf; CP - Coastal Plains; IB - Interior Basins; MT - Mountains

REMARKS: Spills - Includes both onsite and transit spills of materials used in process.
Well Failure - Includes leaks from well system of EOR materials.
Reservoir Leaks - Includes the migration from the zone that the EOR process is operative into the external system.
Operational - Includes disposal of spent material, consumption of site natural resources, discharge emissions, and offsite supply of EOR materials.

Source: OTA-E-59

(5) Operations—The effects caused by routine activities and by the support facilities and activities associated with EOR production. To determine environmental problems during operations, the effect of each of the following must be considered: disposal of spent material; consumption of site-associated natural resources; discharge emissions; fugitive emissions; and offsite supply and support efforts.

A simple matrix model was developed to compare the relative significance of environmental impacts from spills, well failure, reservoir leaks, and operations from thermal, miscible, and chemical EOR methods in each of the four physiographic regions. The matrix reflects a subjective assessment and relative ranking of the significance of potential impacts from negligible or nonexistent, 1, to potentially significant, 4. The values assigned on Table 10.11 are comparable only when applied to a specific EOR process and environmental component such as thermal and air.

Table 10.11 relates to potential hazards from each EOR project by physiographic area. To suggest possible total impact of each EOR process, Table 10.12 was developed. This matrix attempts to predict the relative degree of development of the EOR method as a function of the physiographic area. Should time and/or experience indicate different values, they could be substituted without invalidating the matrix presented.

By selecting the appropriate value from Table 10.12 and multiplying it by the value for the same process and physiographic area on Table 10.11, an estimate of the weighted environmental impact of any or all effects can be calculated. Table 10.13 is the sum of each environmental component in a physiographic area times the appropriate value from Table 10.12. After determining the value for each of the physiographic areas for an EOR process, they are totalled and the value transferred to Table 10.13.

The values in Table 10.13 suggest possible relative environmental impacts. For example, chemical EOR projects may have the greatest potential for environmental impacts and thermal the least, or the biota may be the most impacted and land the least. Sweeping conclusions should be drawn with caution, however, because individual sites and production conditions for EOR, and thus possible environmental impacts, vary significantly from setting to setting.

Table 10.12: Potential Distribution of Environmental Impacts for Enhanced Oil Recovery

| | . . . . .Physiographic Area . . . . | | | |
Method	Continental Shelf	Coastal Plain	Interior Basin	Mountains
Thermal	1	4	2	2
Miscible	1	3	3	2
Chemical	1	3	4	2

Scale Units: 1 is improbable; 2 is negligible; 3 is moderate; 4 is significant; and 5 is extensive.

Source: OTA-E-59

Table 10.13: Cross Plot of Environmental Impacts for Enhanced Oil Recovery

| | . . . . Environmental Components. . . . . | | | |
Method	Air	Water	Land	Biota	Total
Thermal	65	79	36	67	247
Miscible	63	67	46	88	264
Chemical	50	112	50	117	329
Total	178	258	132	272	

Source: OTA-E-59

Potential Impacts on the Environment

There are at least seven media in which EOR operations could have environmental impacts: air, surface water, groundwater, land use, seismic disturbances and subsidence, noise, and biological and public health. The matrix described previously attempts to identify the physiographic regions most likely to experience each type of environmental impact.

While each of the four physiographic regions can experience environmental repercussions in these seven media, certain types of impacts will be far more important in some regions than in others. For example, air pollution is a concern primarily in urbanized portions of the Coastal Plains and in Interior Basins where air quality is already in violation of Clean Air Act standards. Similarly, land-use conflicts arise in heavily populated areas where land values tend to be high and multiple-potential uses exist for a given parcel of land.

Groundwater use and pollution are grave concerns in areas where groundwater is a principal component of the water supply, such as in central and coastal California. Surface water pollution is important in areas with high surface runoff and at sites adjacent to surface water bodies.

Noise is a concern in both urban and open areas, although natural ecosystems differ widely in their sensitivity to noise. The compressors and other equipment used in EOR generate high levels of noise, but it is unlikely that this noise will cause any serious environmental impact. The loudest noises, such as those which would accompany preparation for the fracturing of the reservoir or injection of steam in a cyclic steam process, are of short duration. In regions, where the local biota or human population would be adversely affected by noise, maximum muffling and noise abatement procedures will need to be imposed. Occupational Safety and Health Act (OSHA) regulations will serve as a standard for safeguarding humans.

The information in the following section is based on:

Mining for Petroleum: Feasibility Study, BuMines Report OFR 56-79, prepared by J.S. Hutchins, E. Bond, and D.M. Bass, of Energy Development Consultants, Inc., for U.S. Bureau of Mines, July 1978.

ENVIRONMENTAL ISSUES RELATIVE TO MINING

Disposal and Stabilization of Wastes

Excavation of shafts and tunnels during the mining process results in waste rock which must be stored or disposed of at the ground surface. The days are gone when the spoils could be dumped in a heap near the mine shaft as an eyesore for future generations and a source of air and water pollution for untold years. Air, water, and aesthetics must be conserved not only for the present but also against reasonable future contingencies. Mine spoils need to be disposed of in a manner which will assure physical stabilization. This means appropriate slope stability for the pile against gravity and earthquate forces. Since return of the spoils to the mine excavations is seldom economic, the spoil pile must be designed as a permanent structure whose outline will blend into the landscape.

Reclamation

Stabilization of a spoil pile against air and water erosion can be accomplished by proper revegetation of its surfaces. For this to be performed efficiently within a reasonable length of time, top soiling is necessary. Top soil and overburden can be stripped from the surface of the disposal site, stockpiled, and used to provide a plant growth medium on the surface of the mine wastes.

Vegetation must be planned to match the natural landscape, provide a permanent cover compatible with the local ecosystem, and be self-sustaining after the initial period of care. Sufficient topsoil with probable nutrient additives is needed to accomplish this result. Below about 10" of precipitation per year, revegetation becomes more difficult.

Isolation of Toxic and Hazardous Substances

The possibility exists that the mine spoils will contain toxic and hazardous substances which could oxidize and leach into groundwater (e.g., excess boron), or could be taken up by plants to poison grazing animals (e.g., selenium). Analyses of the mine spoils for such materials prior to permanent disposal plans plus burial under sufficient overburden and topsoil normally will solve most problems.

The Federal Surface Mining Control and Reclamation Act of 1977 applies only to coal mines. However, Section 709 authorizes a study of further laws for other types of mining. Rules to be promulgated under the Federal Resource Conservation and Recovery Act of 1976 may, in the future, include mine spoils under the classification of toxic and hazardous wastes.

Few states regulate hard-rock mining and reclamation of lands disturbed by such activities. Colorado is one of the few states which does have a Mined Land Reclamation Act, which covers all kinds of mining operations and which legislates reclamation.

U.S. Forest Service lands and Public Domain lands are covered by rules (or proposed rules) regulating reclamation of spoils from all mining operations.

Mine Tunnels—Subsidence

Although coal mines are well-known for the surface subsidence effects when old underground workings collapse, subsidence is not nearly so common in hard-rock mines. This is because hard-rock tunnels tend to be deeper and to leave less extensive cavities than coal mines. Because subsidence from this cause is extreme, sudden, and local, it can cause extensive and expensive surface damage. Due to the relatively restricted nature of the tunnels proposed for petroleum mining, proper mine abandonment plans should reduce to almost zero the possibility of surface subsidence from tunnel collapse.

Water

Dewatering: The removal of oil and associated waters from geologic formations is known to cause surface subsidence. Examples can be found in the Long Beach area of California and the Texas City area of the Gulf Coast. Such subsidence is usually gradual and evenly distributed on the surface in contrast to the catastrophic failure of old mine workings. Difficulty does occur when such subsidence occurs in coastal areas with consequent flooding of land previously above the high tide mark. Such subsidence can be controlled by reinjection of waters into the formation to replace the liquids withdrawn.

Most states have water laws regulating subsurface aquifers. Before withdrawal of underground water, plans must be approved by the State Engineer.

Water Disposal: All petroleum has some quantity of water associated with it in the geologic formation. These waters usually are quite high in salt content. Separation of oil field brines from the petroleum and its subsequent disposal is a long-standing occurrence to the oil industry and one which has been successfully solved in numerous ways.

Water usually is separated from oil using settling tanks although heat, chemical additions, centrifuges and other means sometimes are employed. It is important to the transportation, processing, and refining of petroleum that as much water as possible be removed at the production site.

Methods — Treatment of oil field brines to the point that they can be discharged into natural bodies of water is prohibitively expensive. Evaporation is too slow and still leaves the problem of disposing of the salt. Normally, for the last 20 to 30 years, the only acceptable water disposal method has been reinjection into the reservoir or into an underlying saline aquifer.

The discharge of foreign substances into the natural waters of the nation is regulated by the Federal Environmental Protection Agency and the individual states under the National Pollution Discharge Elimination System and associated rules by authority of the Federal Water Pollution Control Act as Amended 1977.

Discussion

At this stage of the concept of mining for petroleum, environmental problems are difficult to foresee clearly, but they cannot be too different from those clearly defined in normal oil production and mining operations. It appears that the environmental problems are slightly greater than normal pumped well oil production but far less than area strip coal mining. The single major problem will be underground worker safety.

As noted in the chapter, "Mining of Petroleum," it would appear that the Flip-Flop process presents fewer inherent environmental problems than the more conventional oil mining.

REFERENCES

(1) U.S. Environmental Protection Agency, *Proposed Criteria for Water Quality*, Vol. I (1973).

(2) National Institute for Occupational Safety and Health/Health, Education and Welfare, *Toxic Substances List* (1974).

(3) McKee and Wolf, *Water Quality Criteria* (1963).

(4) National Academy of Sciences, Committee on Water Quality Criteria, *Water Quality Criteria* (1972).

(5) N. Irving Sax, *Dangerous Properties of Industrial Chemicals* (Van Nostrand Reinhold Co.) (1968).

(6) T.J. Robichaux, "Bactericides Used in Drilling and Completion Operations," *U.S. EPA Symposium on Environmental Aspects of Chemical Use in Well Drilling Operations*, Houston, Texas (May 1965).

(7) *Chemical and Engineering News*, p. 17 (November 3, 1975).

(8) Takahashi, *GANN JOURNAL* 61 (1), 27–33 (1970).

(9) U.S. Department of Health, Education and Welfare, "Request for Information on Certain Chemicals," *Federal Register 40* (121): 26391 ff (1975).

(10) U.S. Department of Health, Education and Welfare, *Survey of Compounds Which Have Been Tested for Carcinogenic Activity* (1971).

(11) A.G. Collins, *Chemical Applications in Oil- and Gas-Well Drilling and Completion Operations*, U.S. Bureau of Mines, Bartlesville Petroleum Research Center, U.S. Department of the Interior, Bartlesville, Oklahoma (1975).

(12) A.G. Collins, "Oil and Gas Wells—Potential Polluters of the Environment," *Journal of Water Pollution Control* (December 1971).

(13) J.E. Kastrop, "Downhole Oil Well Maintenance Costs Push Half Billion," *Petroleum Engineer* (July 1975).

(14) W.S. Smith, *Atmospheric Emission from Fuel Oil Combustion*, Public Health Service Publication No. 999-AP-2, R.A. Taft Sanitary Engineering Center, Cincinnati, Ohio (November 1962).

(15) R.P. Hangebrauck, D.J. von Lehmden, and J.E. Meeker, "Emissions of Polynuclear Hydrocarbons and Other Pollutants from Heat-generation and Incineration Processes," *Journal of the Air Pollution Control Association*, 267 (July 1964).

(16) U.S. Environmental Protection Agency, *Compilation of Air Pollutant Emission Factors* (August 1975).

STATUS OF EOR TECHNOLOGY
SUMMARY OF FIELD TESTS
AND PROJECTS

The information in this chapter is based on:

An Investigation of Primary Factors Affecting Federal Participation in R&D Pertaining to the Accelerated Production of Crude Oil, prepared by J.M. Sharp of Gulf Universities Research Consortium for National Science Foundation under Contract NSF–C–942 (GURC Report 140), September 15, 1974.

A Survey of Field Tests of Enhanced Recovery Methods for Crude Oil, prepared by J.M. Sharp of Gulf Universities Research Consortium for National Science Foundation under Contract NSF–C–942 (GURC Report 140–S), December 27, 1974.

Assessment of Enhanced Recovery Technology as a Means for Increasing Crude Oil Recovery in Texas, NSF Report RA–N–74–251, prepared by E.A. Lohse of Texas Governor's Advisory Council for National Science Foundation, January 6, 1975.

The Estimated Recovery Potential of Conventional Source Domestic Crude Oil, prepared by J.W. Devanney III, R. Ciliano and R.J. Stewart of Mathematica, Inc. for U.S. Environmental Protection Agency under Contract 68–01–2445, May 1975.

Potential Environmental Consequences of Tertiary Oil Recovery, prepared by C. Braxton, R. Stephens, C. Muller, J. White, J. Post, J. Norton, M. Goldberg and P. Stevenson of Energy Resources Co., Inc. for U.S. Environmental Protection Agency under Contract 68–01–1912, July 1976.

Enhanced Oil Recovery Potential in the United States, OTA–E–59, prepared by Lewin and Associates, Inc., National Energy Law and Policy Institute and Wright Water Engineers, Inc., for U.S. Office of Technology Assessment, January 1978.

FIELD TESTS

If the upper targets of any of the recent predictions of EOR potential are to be achieved, a significant increase is needed in the rate of technical progress. To achieve this, the level of field tests needs to be significantly increased. It is important that ERDA-sponsored field tests be part of an EOR research strategy designed to complement industry's efforts and to provide information that can be generalized to a variety of classes of reservoirs. The status of field tests as of 1976 is shown in Table 11.1 (1).

Table 11.1: Field Activity in Enhanced Oil Recovery *

	. . . . Number of EOR Projects.						
	Technical . . . .Pilots . . .		Economic . . . .Pilots . . .		Fieldwide . Development .		Acreage Under Current
Technique	Total	Current	Total	Current	Total	Current	Development
Steam drive	17	13	15	14	15	15	15,682
In situ combustion	17	3	6	5	19	10	4,548
CO_2 miscible and nonmiscible	5	4	6	6	2	2	38,618
Surfactant/polymer	12	10	7	7	2	2	1,418
Polymer-augmented waterflooding	3	0	14	9	14	11	14,624
Caustic-augmented waterflooding	5	1	2	0	0	0	63
Hydrocarbon miscible	9	7	6	5	10	8	56,782
Totals	68	44	57	47	62	48	131,735

*From *Research and Development in Enhanced Oil Recovery,* Final Report, Lewin & Associates, Inc., Washington, D.C., ERDA 77-20/1,2,3, December 1976.

Source: OTA E-59

Despite the attractive economic EOR venture that can be portrayed with a hypothetical production profile, past experience with waterflood, fireflood, etc., denies sensible EOR application to any specific field without a field test of results to the entire reservoir. In some cases, this involves many acres and several millions of dollars. The purpose of the field test is to determine whether or not the hypothetical production profile really applies to a large enough portion of a reservoir (which may be many square miles in extent) that it justifies the huge front end investment required.

Application of laboratory-developed EOR technologies to full-field application requires a two-step process, somewhat analogous to the pilot plant procedure for transferring a laboratory chemical process to a full-scale chemical plant commercial operation.

The first level of field test is small in scale (i.e., involving a few acres) in order to assess in the shortest possible time the technological suitability of the EOR method with respect to reservoir chemistry, temperature and pressure. Even this test might require one to three years because of the very slow rate at which fluids move through the reservoir.

The second level of field test, involving a larger amount of acreage, evaluates the economic applicability of the EOR method. Successful completion of the economic test qualifies the EOR method for a specific large-scale or full-field application. Field testing is relatively expensive, having averaged $2.18 MM before 1975.

The total field test procedure can rarely be reduced to less than three to five years, and in some circumstances requires more than ten years. Some minitests are conducted in periods of less than one year in shallow reservoirs where closely spaced wells can be utilized.

Industrial EOR field test activity increased markedly as a result of the 1973 increase in world market price of crude oil. During 1974, one major company operating in Texas increased its national EOR field test program by a factor of five. Field test planning by many companies is in an active but incomplete stage.

Level of Field Test Activity in the United States

To assess the level of field test activity, Gulf Universities Research Consortium requested twenty-two of the crude oil producing companies expected to be most active in EOR technology development and/or testing to submit field test information (NSF–C–942–S). One company did not respond to this survey in 1974 and, of those responding, not all were able to provide detailed information. The level of sampling of total field test activity achieved is summarized as follows.

> Number of companies queried: 22
> Number opting against any response: 1
> Level of response by companies: 21 of 22 (95.5%)
> Those reporting no on-going or planned field test activity: 5 of 21 (24%)
> Those reporting field tests: 16 of 21 (76%)
> Of the 16 providing field test data: Those reporting only descriptive information (method, field, etc.): 4 of 16 (25%)
> Those reporting details (size, duration, cost, etc.): 12 of 16 (75%)
> Total number of active tests reported (1974): 41
> Total number of planned tests reported: 32

The breakdown of the 67 tests (of the reported total of 73) wherein the basic method was identified is as follows:

Basic Method	No. of Active Tests	No. of Planned Tests	Total
Thermal*	11	8	19
Mobility-controlled micellar	10	9	19
Carbon dioxide	8	4	12
Hydrocarbon miscible	4	1	5
Mobility-controlled waterflood	5	2	7
Mobility-controlled caustic flood	3	2	5
Method not specified		6	6
Total field tests reported			73

*In situ combustion with or without water drive but excluding steam drive and huff-and-puff

Figure 11.1: Number of Field Tests Broken Down According to Method

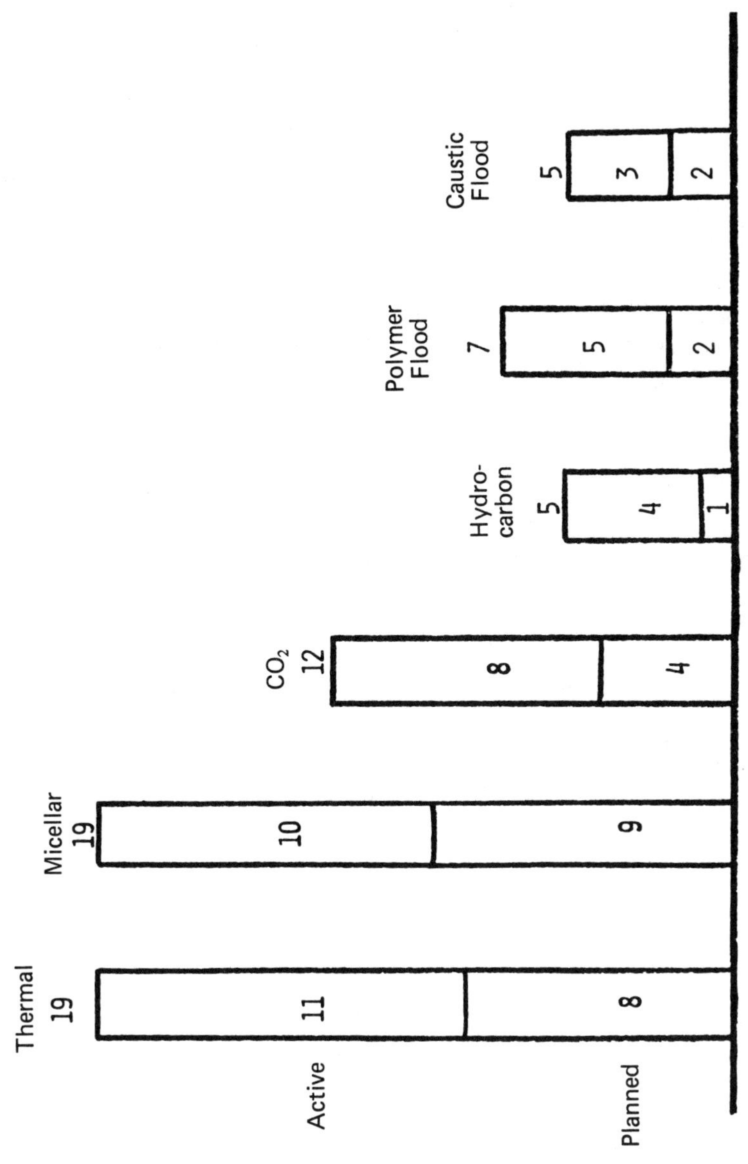

Source: NSF-C-942-S

It should be noted that several of the tests reported to GURC on this survey (steam drive and huff-and-puff tests) are not included in the above totals, since they are (by GURC's definition) considered to be conventional. Also, caustic floods have been identified separately herein from the catch-all chemical flood or micellar category, although sometimes they are not separately identified. This obvious need for standard definitions and nomenclature is one reason why field test inventories are unlikely to be accurate.

The bar chart shown in Figure 11.1 summarizes the distribution of reported tests according to the basic method category being tested. The field test figures for micellar, carbon dioxide and mobility-controlled caustic floods are considered most accurate. There are undoubtedly tests being carried out by companies which involve some type of in situ combustion process, the injection of LPG or other hydrocarbon, and some additives to waterflood operations which have the purpose of controlling reservoir sweep efficiency. Some, perhaps, are being conducted by companies other than the 22 queried in this survey.

On the other hand, micellar and caustic flood processes are few in number, most are patented or the information is proprietary, considerable engineering know-how is required for implementation, and front-end costs are high. These factors minimize the probability of a significant number of tests not being included among the 73 categorized above as micellar or caustic floods. Also, large sources of cheap CO_2 are scarce and are generally well known; hence, the number of large scale CO_2 tests as indicated above should be reasonably accurate. For the remaining three categories, accuracy depends on the definition of large-scale.

Field Test Statistics: The following statistics are descriptive of those field tests reported as being active or in the late planning stages.

Total acreage (38 tests reported)	6,357
Total acreage specified as planned add-ons	
if the initial test increment is successful	16,845
Average acreage/test (38 tests):	
including tests over 500 acres	167
excluding tests over 500 acres	78
Total test cost (31 tests reported), $MM	67.72
Average test cost (of 31 tests), $MM	2.18
Average test cost/acre (of 31 tests):	
including tests over 500 acres, $	19,000
excluding tests over 500 acres, $	25,300
excluding tests over 100 acres, $	39,300
Average years/test (26 tests):	
including tests over 10 years	3.8
excluding tests over 10 years	3.0

Statistics as Function of Recovery Method: Following are statistics for the various types of recovery methods.

Mobility-Controlled Micellar Tests —	
Cost/acre, $:	
High	1,600,000
Low	40,000
Average	88,250

Test acreage range	0.80 to 120
Total acreage	460
Average acreage/test	38

Note: All micellar tests for which data are available report sulfonates as the primary surfactant. Secondary or cosurfactants reported are sulfates and alcohols. Sodium compounds are additives in most of the micellar tests.

In Situ Combustion Tests —
  Cost/acre, $:

High	400,000
Low	1,530
Average:	
including the 720 acre test	11,970
excluding the 720 acre test	45,000
Total test acreage	990
Test acreage range	2.5 to 720
Average acres/test	
including the 720 acre test	99
excluding the 720 acre test	30

Note: All but one test used water drive (either connate, fresh or other); all use natural gas as fuel (except one which uses electric drive also). All those reporting acreage, cost, etc. were in sandstone.

Carbon Dioxide Tests —
  Cost/acre, $:

High	133,000
Low	9,790
Average	17,800
Total test acreage	3,433
Test acreage range (7 tests reporting)	9.4 to 2,200

Note: The seven tests reporting acreage, cost, etc. used CO_2 natural, manufactured, or recovered from natural gas which was pipelined to the test field as a gas. Tests involved both sandstones and carbonates. Purity range of CO_2 was 80 to 100%.

Mobility-Controlled Caustic Flood —Three cases available all report costs of approximately $1,000/acre. Range of test acreage is roughly 100 to 400 acres.

Hydrocarbon Miscible —Two cases reported on same size (12 acres) at a test cost of $100,000/acre.

Mobility-Controlled Waterflood — Three tests reported; two, over 100 acres at approximately $2,000/acre and one, about one acre at more than $1 MM/acre. An error in reporting units appears to be associated with the high cost case.

Geographical Distribution: Distribution of the 60 tests reporting field identification according to geographical region (not petroleum provinces) is as follows.

Gulf Coast States	46%
Midwestern States	22%
Rocky Mountain States	17%
West Coast States	13%
East Coast States	2%

Field Test Activity in Texas

It is expected that one-third of the national enhanced oil recovery production will come from Texas, where in 1975 fifteen active EOR test programs were in progress and four additional tests were known to be planned for the period 1975 to 1978, with breakdown as follows (NSF–RA–N–74–251).

	Active	Planned	Total
CO_2	6	1	7
Thermal	4	2	6
Hydrocarbon miscible	2	–	2
Polymer flood	2	–	2
Caustic flood	1	–	1
Micellar	–	1	1
	15	4	19

Average test cost is $2.18 MM, with 3.8 years duration. Average test cost per acre is $18,709/acre in tests over 500 acres, $20,000/acre in tests of 100 to 499 acres, and $116,000/acre in tests of less than 100 acres. Average duration of all tests is 7.6 years; average for those expected to be completed in less than ten years is 3.7 years.

PROJECT DEVELOPMENT

The potential of additional oil recovery has spurred the development of a number of operations. At least 32 companies have undertaken more than 200 advanced recovery projects in this country. Over 70% have been located in Texas or California, with the remainder in 13 other states. Of those undertaken, approximately 130 are active.

In addition to projects completed or under way, numerous others are being planned or have begun operation. At least 22 enhanced recovery programs were scheduled to commence in 1974, reflecting rapid increases in the price of international crude oil and continued interest in tertiary recovery. Thirteen others were slated for operation before 1980. By 1975 fourteen companies had settled on plans for thirty-five projects in twelve states. Almost half were to be located in California, where thermal recovery methods have proven both feasible and profitable in unlocking low gravity crude oils.

Table 11.2 presents data from a survey by *Oil and Gas Journal* (2) of the planned sophisticated oil recovery projects. Sophisticated applies to enhanced oil recovery methods more complex than conventional waterflooding.

Table 11.2: Sophisticated Oil Recovery Projects

Field	State	Company	Project Type*	Pilot or Fieldwide (FW)	Project Acres	Pay Zone	Depth (ft)	Crude °API	Starting Date
Kern River	Calif.	Chanslor	Comb.	Pilot	10	60 ft	320	14	1974
Tejon	Calif.	Chanslor	Comb.	Pilot	20	70 ft	2,650	16	1975
Carlyle	Kan.	Layton	Comb.	Pilot	20	Bartlesville	—	—	—
Bellevue	La.	Cities	Comb.	Pilot**	10	Nacatoch	400	19	1974
Shoreline	La.	Kirby	Comb.	Pilot***	60	Nacatoch	1,050	22	1973
Heidelberg	Miss.	Gulf	Comb.	Pilot	160	Christmas	5,100	20	1974
Kern River	Calif.	Getty	SS, SD	FW†	450	Kern River	1,000	14	1974
Kern River	Calif.	Getty	SS, SD	FW†	600	Kern River	1,000	14	1975
Kern River	Calif.	Getty	SS, SD	FW†	600	Kern River	1,000	14	1976
Kern River	Calif.	Getty	SS, SD	FW†	600	Kern River	1,000	14	1977
Kern River	Calif.	Getty	SS, SD	FW†	600	Kern River	1,000	14	1978
Kern River	Calif.	Getty	SS, SD	FW†	250	Kern River	1,000	14	1979
Cymric	Calif.	Chanslor	SD	Pilot	10	60 ft	1,500	12	1974
Cymric	Calif.	Std. of Cal.	SD	Pilot	60	Tulare	1,000	13	1974
Kern River	Calif.	Chanslor	SD	FW	298	300 ft	600	13	1975
Midway-Sunset	Calif.	Chanslor	SD	Pilot	10	100 ft	800	11	1974
Midway-Sunset	Calif.	Chanslor	SD	Pilot	10	60 ft	800	11	1974
Midway-Sunset	Calif.	Std. of Cal.	SD	Pilot	15	Monarch	1,000	13	1974
N.E. McKittrick	Calif.	Std. of Cal.	SD	FW	60	Ammicola	1,100	11	1974
W. Coalinga	Calif.	Std. of Cal.	SD	FW	640	Temblor	1,000	13	1974

(continued)

Table 11.2: (continued)

Field	State	Company	Project Type*	Pilot or Fieldwide (FW)	Project Acres	Pay Zone	Depth (ft)	Crude °API	Starting Date
Caddo Pine Island	La.	Texaco	SD	Pilot	30	Nacatoch	-900	19	1975
Port Barre	La.	Texaco	SD	FW	60	1,500 ft	-1,450	19	1974
Poncho	Colo.	Midwest	Poly.	FW	640	"J" sand	6,000	40	1975
Stephens	Texas	Tex.-Pacific	Poly.	Pilot	5	Caddo	3,100	38	1975
Bone Pile	Wyo.	Midwest	Poly.	FW	1,760	Minl "A"	10,800	32	1974
Skull Creek Unit	Wyo.	Texaco	Poly.	FW	920	Newcastle	2,450	32-35	1974
N. Burbank	Okla.	Phillips	MSF	Pilot	—	Bartlesville	—	—	—
S. Okla.	Okla.	Skelly	MSF	Pilot	20	Sims	7,200	28	1974
Bradford	Pa.	Pennzoil	MSF	FW	>100	—	1,900	45	1974
Seeligson Ut.	Texas	Sun	MSF	FW	1,600	13A	4,900	38	1974
LaBarge Unit	Wyo.	Texaco	MSF	Pilot	1.8	Almy	1,000	38	1974
Tinsley	Miss.	Pennzoil	FG	FW	>400	—	4,600	32	1974
Consolidated	Ill.	Texaco	SOF	Pilot	5	Benoist	1,800	39.5	1974
Charlson North Unit	N.D.	Texaco	W&GI	FW	9,100	Madison	8,500	42	1974
Dugout Creek Unit	Wyo.	Texaco	W&GP	Pilot	20	Shannon	2,800	36	1974

*Comb. = combustion; SS = steam soak; SD = steam drive; Poly. = polymer flood; MSF = micellar solution flood; FG = flue gas; SOF = soluble oil flood; W&GI = water and gas injection; W&GP = water and gas pulsing.

**Expansion of 10 acres.

***Expanded 1973.

†Scheduled expansions into new acres of the field.

Source:　EPA 68-01-2445

Dominance by Major Companies

Because of the financial risks involved, major domestic crude oil producers dominate the tertiary recovery industry. The list of projects in Table 11.3 indicates a relatively high degree of industry concentration among the major oil companies. Table 11.3 shows that companies among the Top Twenty initiated 80% of those projects undertaken before 1974; the Big Eight alone accounted for almost two-thirds.

To gauge more accurately the involvement of particular companies in tertiary oil recovery, it is necessary to separate the activities into pilot and field-wide projects. Generally, field-wide projects require a larger investment than pilot projects if research personnel costs are excluded. By this criterion, the larger companies appear more active. Roughly 60% of the projects operated by firms in the Top Twenty are classed as field-wide; for Big Eight firms the corresponding figure is 66%. Relatively fewer projects of the smaller firms, 56%, were dubbed field-wide by their operators. The gap between the Big Eight and the smaller firms was initially nonexistent for projects slated to begin in 1974 or subsequent years; both groups expected roughly 60% of these to be field-wide.

Table 11.3: Tertiary Oil Recovery Projects for Firms Among the Top Twenty

	Overall Domestic Production Rank*	Tertiary Recovery Rank**	Projects Planned	Projects Completed or Under Way	Total Projects
Exxon	1	11***	0	2	2
Texaco	2	3 ***	7	9	16
Gulf	3	6	1	10	11
Shell	4	7	0	10	10
Standard of California	5	2	4	16	20
ARCO	6	8	0	9	9
Amoco	7	5	0	14	14
Mobil	8	1	0	37	37
Getty	9	9***	6	2	8
Union	10	11***	0	2	2
Sun	11	9***	1	7	8
Conoco	12	3 ***	0	16	16
Marathon	13	10***	0	3	3
Phillips	14	10***	1	2	3
Cities	15	11***	1	1	2
Skelley	18	11***	1	1	2
TOTAL			22	141	163

*Based on the Federal Trade Commission Report released on July 18, 1973.

**Determined on the basis of projects planned and projects underway.

***Indicates more than one company in same risk.

Source: EPA 68-01-1912

Table 11.4: Location of Tertiary Oil Recovery Projects

	Calif.	Texas	Wy.	Ok.	La.	Ill.	Miss.	Ark.	Pa.	Col.	N.D.	Ind.	Ky.	Neb.	N. Mex	Kan.	TOTAL
Thermal																	
Combustion	12	15	1	3	7	1	2	2						1		1	45
Steam Soak	48																48
Steam Drive	26	1	1		2		1	2									33
Hot Waterflood				2													2
Subtotal Thermal	86	16	2	5	9	1	3	4						1			127
Chemical																	
Polymer Flood	1	12	5	3						1			1		1		24
Micellar Solution Flood		1	1	2	1	3			3		1				1		13
Surfactant Flood		2	1									1					4
Soluble Oil Flood		1				1											2
Caustic Soda Flood		2															2
Low Tension Water Flood																	
Subtotal Chemical	1	18	7	5	1	4			3	1	1	1	1		2		45
Gas																	
Miscible Hydrocarbon	5	13	2	1													21
Miscible CO$_2$		6					1			1							8
Waterflood/ Alternating Gas			3	1													4
Water/Gas Injection		1									1						2
Water/Gas Pulsing		1	1														2
Flue Gas							1										1
Subtotal Gas	5	21	6	2			2			1	1						38
TOTAL	92	55	15	12	10	5	5	4	3	2	2	1	1	1	1	2	211

Source: EPA-68-01-1912

However, firms among the Top Twenty but outside the Big Eight appear to have escalated their efforts in the area; field-wide activities were slated for 70% of their new projects, indicating an increasing commitment to processes which have been proven on a pilot scale.

While experimenting with a variety of tertiary recovery methods, most firms appear to rely heavily on two or three. Although steam soak, combustion, steam drive and polymer flood are popular individually with many companies, each company is involved to a different extent in each method.

Classification by Location and Method

Cross-classification of tertiary recovery projects by process and location demonstrates an uneven distribution, as seen by the breakdowns of projects by location and method found in Table 11.4. It is seen from Table 11.4 that California has over 80% of all steam drive programs.

Projects in Texas: According to Table 11.4, Texas, with slightly more than one-fourth of all projects has a third of all combustion projects. Likewise, 62% of all miscible hydrocarbon and 75% of all miscible CO_2 sites have been located in Texas, where there is both a relatively large number of carbonate reservoirs favorable to inert gas miscible drives and ready availability of hydrocarbon and CO_2 miscible fluids. Enhanced oil recovery projects in use in Texas for all years through July 1974 are summarized in Table 11.5, as prepared by Gulf Universities Research Consortium for the Texas Governor's Energy Advisory Council. A complete, detailed list of EOR projects in Texas is given in NSF–RA–N–74–251.

Table 11.5: Enhanced Recovery Projects in Use in Texas Through July 1974

Method	Total Number, Past and Present
Thermal:	
In situ combustion	44
Steam drive	<u>19</u>
	63
Miscible:	
Hydrocarbon miscible	17
Carbon dioxide miscible	6
Flue gas miscible	1
Soluble oil flood	1
Low-tension waterflood	6
Surfactant flood	4
Caustic soda flood	2
Gas injection	3
Gas pulsing	<u>1</u>
	41
Polymer flood (mobility control combined with a miscible flood)	11

Source: NSF–RA–N–74–251

REFERENCES

(1) *Research and Development in Enhanced Oil Recovery*, Final Report, Lewin and Associates, Inc., ERDA 77-20/1, 2, 3 (December 1976).

(2) Bleakley, W.B., "Journal Survey Shows Recovery Projects Up," *Oil and Gas Journal,* p 69 (March 25, 1974).

RECOVERY OF HEAVY OIL

HEAVY OIL RESOURCES

Heavy oil resources are assuming greater importance in meeting the nation's energy needs. Production of this low gravity oil, 20°API or less, has been low because of the high viscosities involved. The San Francisco Energy Research Center (SFERC) has taken as a primary goal the development of improved heavy oil production methods. A few statistics indicate how important this goal is to the nation. The heavy oil resource (viscous oil in known reservoirs not economically recoverable) was estimated at 55 billion barrels in 1966 (1). The proved reserve (crude oil economically recoverable) was estimated at 34 billion barrels in 1975 (2). Development of improved heavy oil recovery methods could greatly add to the nation's proved reserves.

RECOVERY METHODS

Thermal processes are generally thought to be the best methods for heavy oil recovery. Heat is applied to the reservoir to lower oil viscosity and increase rate of flow to the producing well. The two most commonly used methods are steam injection and in situ combustion. Over 100 of these projects are under way (3). Many are reported to be successful, while others are not.

One problem which can affect both methods is bypassing of injected fluids in heterogeneous reservoirs. Sands of different permeabilities overlie one another and injected fluids ultimately channel through the high permeability zones. Injected air, or steam, can bypass oil-rich zones and flow directly to the producing well. This leads to poor oil recovery and to high operating costs per barrel of produced oil. Heterogeneous reservoirs are common, and solving this problem is essential for economic heavy oil recovery.

The information in this section is based on: *Solvent Stimulation Tests in Two California Oilfields,* BuMines Report RI-7978, prepared by H.J. Lechtenberg, G.L. Gates, W.H. Caraway and O.C Baptist, of San Francisco Energy Research Laboratory, for U.S. Bureau of Mines, 1974.

SOLVENT STIMULATION IN CALIFORNIA

Field Data

The Wilmington low-gravity oil field is located in the Los Angeles Basin of California. The field is a multilayered reservoir. In some areas the field is known to have a natural water drive. Oil production has exceeded 1.5 billion barrels. In recent years, the field has produced at the highest rate of any field in the United States. Production in 1972 was over 191 million barrels of oil. Several waterfloods and other secondary recovery processes have been utilized to increase recovery. Water injection, in addition to increasing oil recovery, is used to dispose of waste water and to prevent land subsidence.

The uppermost broad oil-producing horizon of the Wilmington field is the Tar zone. This sand initially contained an estimated 2,150 barrels per acre-foot of stock tank oil in place. Primary recovery was estimated to be 12% of the oil in place. The sand is unconsolidated with porosities ranging from 34 to 40%. Air permeabilities range from 1 to 8 darcys. The oil gravity is low, ranging from 14° to 17°API. The bottom hole pressure in the area of the stimulation tests ranges from 400 to 800 psi.

Wilmington crude oil, like most California oils, contains many cyclic compounds and a relatively large amount of nonhydrocarbons. A composite sample of the Wilmington crude oils contained, in weight percent, 1.53 sulfur, 0.65 nitrogen, and 0.44 oxygen. It is an asphaltic oil containing a small amount of gasoline and large amounts of high molecular weight cyclic compounds. The hydrocarbon portion of the oil has been shown to contain primarily naphthenic compounds. Bureau of Mines characterization methods indicate that the naphthenic and aromatic hydrocarbons predominate in the Tar zone crude oil.

The Fruitvale field that also produces from reservoirs having low-gravity oil is located in the San Joaquin Valley of California near the west edge of the city of Bakersfield. Over 99 million barrels of oil have been produced from the field. The 1972 oil production exceeded 1 million barrels. Secondary recovery projects have been started but have been small. The sand is unconsolidated, and silt in the produced liquids is a problem. The bottom hole pressure of the well treated was approximately 1,000 psi.

Well Treatment

Solvent injection tests were run on four wells in the Tar zone of the Wilmington oil field and one well in the Fruitvale oil field by the Bureau of Mines in cooperation with oil companies, to investigate solvent injection as a means of improving oil recovery from these viscous oil reservoirs.

The field tests consisted of injecting solvent, followed or not followed by water, and allowing the well to soak for a few days or, as in one well, by circulating the injected solvent down the casing and up through the tubing for a week. The solvents used were a coker oil in the Wilmington field and a weed oil in the Fruitvale field. Both solvents contained large quantities of aromatic hydrocarbons and were readily available to the operators in the fields at a reasonable price.

The coker oil was injected either down the tubing or the casing and allowed to soak for a period of 3 to 5 days. Lease saltwater was injected after the solvent in three of the wells to displace the coker oil further into the formation. The volume of coker oil injected in the wells ranged from 300 to 1,000 barrels, and the volume of saltwater when used was from 550 to 1,000 barrels. In two of the wells, an additive, n-butylamine, was added to the coker oil and mixed thoroughly prior to injection.

The well in the Fruitvale field was treated in two stages, Initially, the well was treated with a mixture of 180 barrels of weed oil plus 19 gallons of n-butylamine and 19 gallons of a nonionic surfactant. The well was then circulated for 4 days and again treated with 150 barrels of weed oil, 15 gallons of n-butylamine and 15 gallons of nonionic surfactant. The well was circulated an additional 3 days and returned to production. Table 12.1 gives the treatment data for the five wells.

Table 12.1: Well Treatment Data

Solvent	Coker oil				Weed oil,
	Well A	Well B	Well C	Well D	Well E
Volume...................bbl..	1,000	550	300	300	[1] 300
Additive.................pct..	-	-	0.5 n-butylamine	0.5 n-butylamine	0.3 n-butylamine plus 0.3 surfactant.
Chaser, water............bbl..	-	550	1,000	1,000	-
Gravitated through............	Tubing	Casing	Casing	Casing	-
Soak time..............days..	3	5	3	3	-
Circulating time.......days..	-	-	-	-	7
Solvent injected per foot of perforations........bbl..	4.4	2.9	1.4	1.3	1.7

[1] 2-stage treatment: 180 bbl injected and circulated for 4 days and 150 bbl injected and circulated for 3 additional days.

Source: BuMines RI-7978

The early poststimulation rate of production was considerably greater than the prestimulation rate, and the produced mixture contained a high percentage of coker oil. Both the rate of production of the oil-solvent mixture and the percentage of coker oil declined sharply within a short time after the stimulation. Almost all of the 1,000 barrels of coker oil used in well A was recovered after 72 days and about 85% of this was recovered within the first 10 days of production.

The injected solvent was followed by a saltwater chaser in wells B, C, and D. The percentage of coker oil recovered in these three wells ranged from 33 to 49%, which was considerably less than the almost 100% recovered from well A.

Oil Production

Cumulative incremental oil production (the total oil produced in excess of that which would have been produced if the prestimulation rate had been maintained) was approximately 4,000 barrels since the stimulation in well A. The average of the daily oil production, total barrels of oil produced divided by the number of days the well had produced during the time period of the test, has approximated 26 barrels of oil per day compared with prestimulation rate of 19 barrels of oil per day.

The cumulative incremental oil production, the monthly oil production, the average daily oil production, and the prestimulation daily oil rate for well A are shown in Figure 12.1.

Figure 12.1: Oil Production—Well A

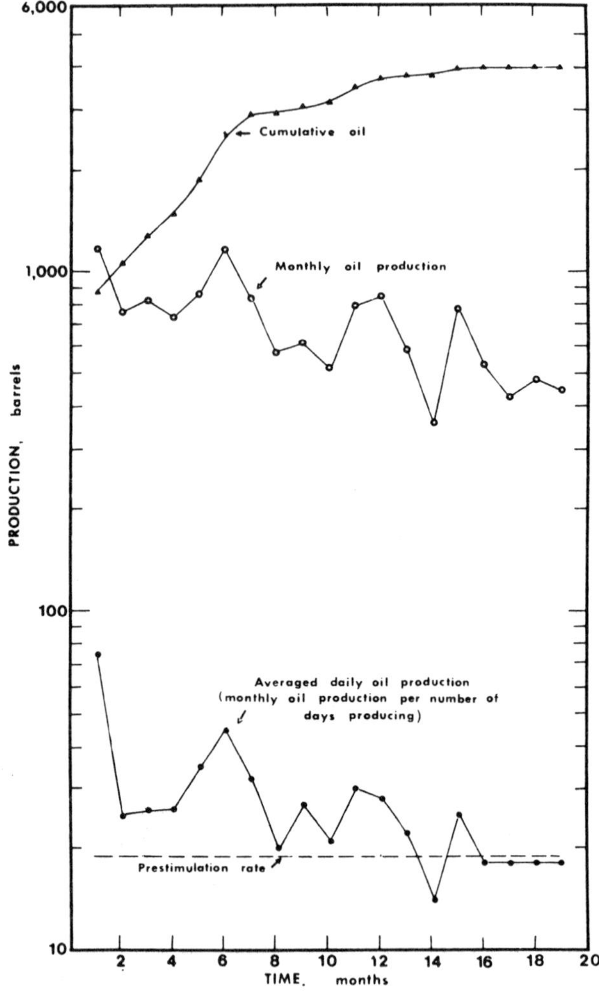

Source: BuMines RI-7978

Wells C and D in the Wilmington field, which were treated with coker oil and saltwater, had a short initial period of flush production. This included the solvent recovered. The daily oil production then returned to or near the prestimulation rate, and this rate was maintained for a period of 3 to 5 months. At this point, oil production began to increase greatly and showed a sizable gain over the prestimulation rate.

Some of the increase in oil production for wells C and D could have been due to a possible miniflood caused by an offset well located 250 to 300 ft away from these wells.

Table 12.2 shows the results of all wells that were stimulated by solvent injection. These results can be compared with the prestimulation history as shown in Table 12.3.

Table 12.2: Solvent Stimulation Results

	Well A	Well B	Well C	Well D	Well E
Additional oil...............bbl..	2,949	249	1,208	4,361	2,760
Solvent recovered............bbl..	1,000	270	(^{1})	(^{1})	330
Average of daily oil production bbl..	26.3	15.7	23.4	26.6	9.6
Water cut.....................pct..	96.6	93.4	93.1	94.6	85.2
Time since treatment......months..	19	13	11	11	13
Ratio of additional oil to solvent injected..........................	2.9	0.45	4.03	14.54	8.36

[1] Uncertain

Source: BuMines RI-7978

Table 12.3: Prestimulation Data for Wells

	Well A	Well B	Well C	Well D	Well E
Gross production.........bbl/day..	244	245	306	559	34
Net oil production.......bbl/day..	19	16	19	13	3
Water cut.....................pct..	92.7	93.9	94.3	97.8	89.0
Gravity....................° API..	15.1	15.2	16.1	15.8	17.0
Perforated interval...............	2,359-2,584	2,444-2,634	2,332-2,558	2,355-2,606	3,700-4,200
Bottom hole pressure.........psi..	400	800	400	400	1,000

Source: BuMines RI-7978

The average of the daily oil production rate over the test period for four of the five wells treated with solvent ranged from 5 to 14 barrels of oil greater than the prestimulation rate. The fifth well showed no improvement. The average of the daily oil production rates of the wells before and after solvent stimulation is shown in Figure 12.2.

Figure 12.2 also shows the number of months used to calculate the poststimulation average daily rate.

Conclusions

The data show that on the average five barrels of additional oil were obtained for each barrel of solvent used in these five wells. This favorable ratio of additional oil to solvent injected indicates that four of the five wells have already recovered sufficient additional oil to be an economic success.

A large part of the increase in oil produced in the treated wells may be due to the cleaning of asphaltenes and other solid deposits from the well bore, casing perforations, gravel pack and nearby formation.

Figure 12.2: Daily Production Rate Before and After
Solvent Stimulation

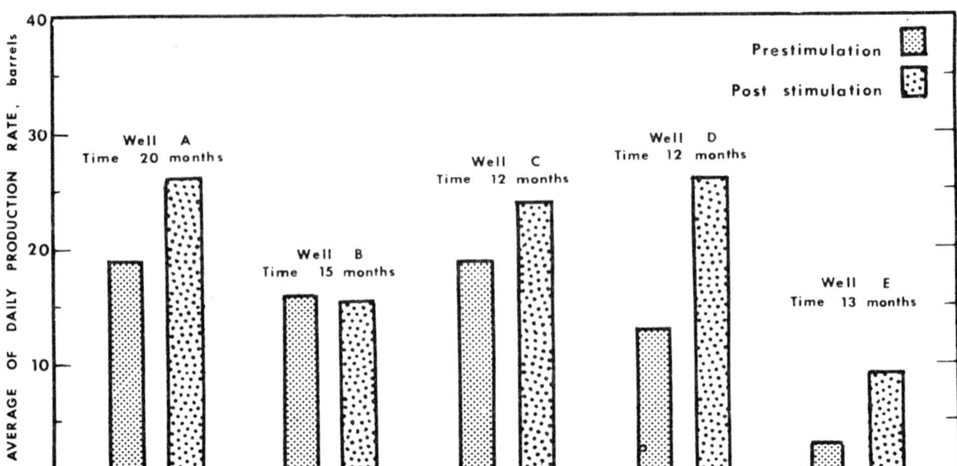

Source: BuMines RI-7978

The well in the Fruitvale field (well E) was producing only about 3 barrels of oil per day before stimulation and was approaching the end of its economic life. The rate of oil production was increased to 8 barrels by the solvent treatment, and this rate has been maintained for a considerable time. Since the well will be operated as long as this rate is maintained, the life of the well has been lengthened, and the ultimate recovery of oil increased.

Monitoring production from treated wells indicated that essentially all of the solvent was recovered with the produced oil in wells in which the solvent was not followed by water. However, solvent was only partially recovered from wells in which solvent was followed by water.

The water injected in three wells apparently caused a temporary water block. A period of high water-oil ratio lasted until essentially all of the injected water had been produced, then the oil production increased until it was considerably above the prestimulation rate and this higher rate continued.

Several combinations of solvents and methods of stimulation were used in these tests. Because of the large number of variables in each test, it is impossible to evaluate the relative effect of each variable on the overall results.

Cumulative poststimulation oil production indicates that cyclic solvent injection can be an economic success. More laboratory research and field experimentation are needed to determine the limits of applicability of the methods.

The information in the following section is based on: *Laboratory Study of Heavy Oil Recovery Using Polymer and Solvent,* SFERC/RI-76-1, prepared by G.D. Peterson, H.J. Lechtenberg, W.H. Caraway and G.L. Gates of San Francisco Energy Research Center, for U.S. Energy Research and Development Administration, February 1976.

SAN FRANCISCO ENERGY RESEARCH CENTER LABORATORY PROCESS

To fully develop heavy oil resources a variety of approaches will probably be necessary. The laboratory process described in this paper used a combination of polymer and solvent. Polymer-thickened water and multivalent ion solution were injected to reduce flow through high permeability sands. Solvent was then injected to mix with the in-place crude oil and decrease oil viscosity. Polymer-thickened water was again injected to provide a mobility control behind the oil phase. The column was then waterflooded to produce mobilized oil.

In summary, polymer and multivalent ion were used to increase sweep efficiency, and solvent to decrease oil viscosity. This process was developed as a result of previous work by SFERC to recover heavy oil using solvent.

Polymer and Multivalent Ion

Small amounts of polyacrylamide polymers can be dissolved in water to provide high-viscosity solutions. A 1,500 ppm solution will have a viscosity of over 800 cp when properly prepared. Care must be taken to avoid agglomeration of dry polymer particles upon contact with water. For field applications, special mixing equipment is required (4).

Polyacrylamide solutions have a dual effect in reducing water mobility, which makes them of special interest in heavy oil recovery. The first effect is the high viscosity of the injected solution which enables a stronger "pushing" of oil to the producing well. The second effect is adsorption of polymer on the surface of reservoir rock. Polyacrylamides are polar compounds with a chemical attraction for both rock surface and water molecules. The polymer adheres to rock surface and partially extends through the pore space because of the great polymer chain length.

Water molecules passing through are attracted to the polymer, resulting in a water phase permeability reduction. Oil molecules, being nonpolar, pass through the pore space relatively unaffected (5). The two effects of high viscosity and permeability reduction combine to produce better reservoir sweep efficiency. Oil drops trapped in pore spaces which could not be moved by waterflooding can be moved by the polymer solution.

Adsorbed polymer will not be permanently retained in the reservoir as subsequent waterflooding causes gradual polymer elution. To provide a permanent and much larger water permeability reduction, multivalent ion solutions are used with the polymer solutions. The multivalent ions, principally aluminum ion, crosslink individual polymers to form a larger, more stable polymer network.

Permeability to water is further reduced over that of polymer solutions alone. The crosslinking is done in situ in three steps. Polymer solution is first injected, followed by aluminum citrate solution, followed again by polymer solution. Needham, Threlkeld and Gall compared the effects of polymer solution alone and polymer with aluminum ion in Repetto sandstone cores. They report a residual resistance factor (the ratio of water mobility before and after treatment) of 3 using polymer solution and about 18 using polymer and aluminum ion (6).

The effectiveness of the polymer and multivalent ion treatment is particularly important with respect to heterogeneous reservoirs. After waterflooding, high permeability zones will receive a large portion of the injected polymer and multivalent ion treatment and thus, undergo a larger permeability reduction than low permeability zones. Fluid flow after treatment will be better divided between the originally high and low permeability zones.

Testing

Equipment: Tests were conducted in glass columns 50 cm long with an inside diameter of 4 cm. Fluids were injected by positive displacement pumps to maintain constant flow rates. Constant test temperature was maintained by placing the columns inside a heated cabinet.

Materials: Six different sands, three crude oils, and two solvents were used in testing. The polyacrylamide solution was prepared by dissolving 1.50 g Betz Hi Vis in one liter of tap water (4). The solutions were generally stirred overnight at low speed to insure that the polymer was completely dissolved and hydrated. Solution viscosity was generally about 500 cp at the time of use.

The aluminum citrate solution was prepared by dissolving 6.66 g $Al_2(SO_4)_3 \cdot 18H_2O$ and 2.10 g citric acid in one liter of tap water. A 20% NaOH solution was then added until the pH reached 6. The salt water used was prepared using laboratory chemicals and tap water to approximate the composition of injection water used at the Wilmington, CA, field.

Test Columns: Test columns were prepared in the laboratory by the free fall of sand into the vertically standing vibrating, glass columns. Porosities varied from 35 to 44% and pore volumes varied from 245 to 280 ml depending upon the sands used. Columns were prepared from fresh sands for each test. Both homogeneous and heterogeneous columns were prepared for these tests.

Heterogeneous columns were packed using a thin divider between the different sands which was continuously removed from the column as the sand level was raised The resulting pack left each sand continuous for its length of the column with no barrier to crossflow. The permeability ratio in heterogeneous columns was about 3.5 to 1. The weight ratio of sands was 4 to 1, low to high permeability. Permeabilities were not determined for individual test columns due to the heterogeneous sands. Permeabilities were determined for each of the six different test sands and ranged from 2 to 17 darcys.

After packing, the columns were placed under vacuum and saturated with salt water. Crude oil was then pumped into the columns, displacing the salt water. Oil saturations attained varied from 80 to 93%.

Procedure: All tests were conducted at 100°F with columns in a horizontal position. Injection rates were maintained to provide a frontal advance of about 5 ft/day. Test columns were first waterflooded to a minimum of 20:1 produced water:oil ratio. This generally required about one pore volume (pv) throughput. Slugs of polymer and multivalent ion were usually then injected to reduce flow through channels opened by waterflooding. Slug size was normally 0.1 pv polymer, 0.1 pv aluminum citrate, and 0.1 pv polymer. Variations occurred in the polymer slug size and aluminum citrate was omitted in some test series.

In most test series solvent injection, about 0.1 pv, followed polymer and multivalent ion injection. Solvent was followed with 0.1 pv of polymer solution to act as a mobility control. A saltwater flood then followed until a 20:1 minimum produced water:oil ratio was again reached. The entire sequence was usually repeated one or more times until most of the original oil had been produced from the test column.

Samples of produced oil were collected and analyzed on a photospectrometer to determine the quantity of solvent in the mixture.

Results

The results show that polyacrylamide thickened water, aluminum citrate solution and solvent were effective in recovering additional heavy oil from waterflooded sands.

Table 12.4 shows a summary of results obtained when polymer, aluminum citrate and solvent were all used in the recovery of different viscosity crude oils. The total oil recovered was based on averages of initial waterflood oil recovery plus averages of incremental oil recovery from two injection sequences. (The individual effects of polymer, multivalent ion and solvent injection were also measured and the effects of crude oil viscosity and sand permeability determined. Details are elaborated in the original paper.)

Table 12.4: Summary of Results

Column Type	Crude Oil Viscosity (cp)	Total Oil Recovered	Initial Waterflood Recovery	Incremental Recovery
		. . . . (%).		
Homogeneous	80	76.9	52.8	24.1
Homogeneous	1,100	67.3	35.5	31.8
Homogeneous	3,500	54.9	28.8	26.1
Heterogeneous	80	63.1	38.8	24.3
Heterogeneous	1,100	55.8	24.4	31.4
Heterogeneous	3,500	48.2	17.5	30.7

Source: SFERC/RI-76/1

Incremental oil is the amount of oil produced during the first injection sequence (including produced solvent) minus the amount of injected solvent. For example: if the original oil in place was 100 ml, and 10 ml of solvent was injected, and 30 ml of oil-solvent was produced, incremental oil would be 20%. Values listed in the table are a percent of the original oil in place.

Other results concluded from these tests were:

The optimum solvent slug size was about 8 to 10% of pore volume.

Solvent produced with incremental crude oil amounted to 40 to 80% of the solvent injected, depending upon test conditions. The overall average of solvent left behind in test columns was about 35%

No difference was seen between the effectiveness of gasoline or coker oil as solvents.

Face plugging by an accumulation of polymer gel was a problem.

Figures 12.3a and 12.3b show the effect of crude oil viscosity on oil recovery from homogeneous and heterogeneous sands. The figures are averages compiled from runs using polymer, aluminum citrate, solvent, and waterflood. Initial waterflood recovery varied greatly with oil viscosity.

Figure 12.3: Effect of Crude Oil Viscosity on Total Oil Recovery from Homogeneous Sand

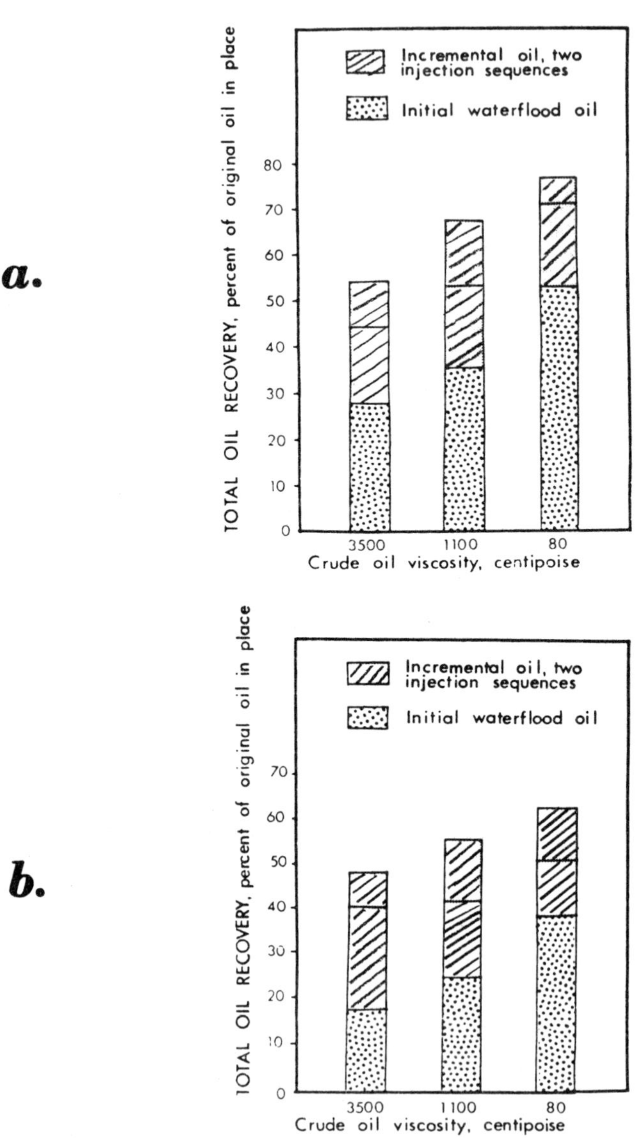

a.

b.

(a)   Homogeneous sand
(b)   Heterogeneous sands

Source: SFERC/RI-76/1

Conclusions

Sweep efficiency is inherently better in homogeneous sands, and oil recovery from homogeneous sands could be expected to show the upper limit for recovery from heterogeneous sands. When polymer, aluminum citrate, and solvent were all used, incremental recovery from heterogeneous sands equalled that from homogeneous sand.

When polymer and aluminum citrate were omitted, incremental recovery from homogeneous sand was unchanged, but incremental recovery from heterogeneous sands was substantially reduced. This demonstrated the effectiveness of polymer and aluminum citrate in controlling bypassing through the high permeability zone of heterogeneous sand columns.

The use of solvent increases oil recovery and lowers the injection pressure. For field applications the second effect may be more important than the first. Oil banks can be formed without solvent but moving the viscous crude to the producing well at an acceptable rate without fracturing the reservoir may be impractical.

The information in the following section is based on: *Evaluation of Thermal Methods for Recovery of Viscous Oils in Missouri and Kansas,* BuMines Report OFR 60-74, prepared by M.D. Arnold and A.H. Harvey, of Department of Mining, Petroleum and Geological Engineering, University of Missouri at Rolla, for U.S. Bureau of Mines, June 1974.

THERMAL RECOVERY OF HEAVY OILS IN MISSOURI AND KANSAS

An economic analysis has been made on four hypothetical heavy oil deposits considered to be typical of those found in the Missouri-Kansas border area. Oil production obtainable by various thermal recovery methods has been computed for each of the four cases.

Distribution of Heavy Oil Deposits

Shallow oil deposits, estimated to range from a few billion barrels to more than 50 billion barrels, have been reported to occur over an area of roughly 7,000 square miles in western Missouri and eastern Kansas, and a number of similar deposits are located in northeastern Oklahoma. The area of occurrence includes a substantial portion of the Forest City Basin, as well as parts of the Bourbon Arch, and the more southerly Cherokee Basin. This entire region will be termed the Forest City Basin area. Its location is illustrated by Figure 12.4 adapted from Reference (7).

The region is not underlain by any single hydrocarbon bed, but heavy oil deposits are found throughout the area. The limited data which are available suggest that both reservoir characteristics and oil viscosity are likely to be variable within individual reservoirs. Oil saturation is also variable, both from one reservoir to another and from one depth to another in the same reservoir. Heavy oil deposits have been found in this area in both sandstones and limestones ranging in age from Mississippian Osagean through the Pennsylvanian Raytown Limestone of the Kansas City Group. It is the sandstones rather than the limestones which offer a promise of economic exploitation within the foreseeable future.

Figure 12.4: Forest City Basin Area

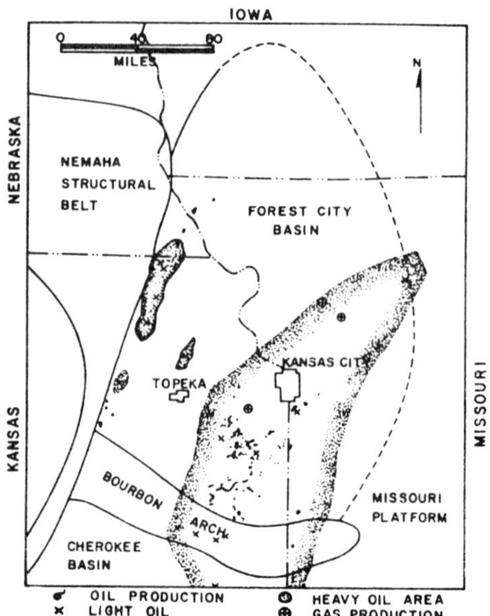

Source: OFR 60-74

Properties of Oil Reservoirs

The heavy oil deposits in the area of interest are quite low in asphalt content. Oil gravities range from <10°API to roughly 28°API, and oil viscosities range from a few hundred centipoises to >500,000 centipoises. At the maximum end of the range, the oil is a tar-like substance which is immobile under reservoir conditions.

Data on sulfur content are rather limited; one analysis reported a sulfur content of 0.75% by weight in a 10°API tar. Oil viscosity can be substantially reduced and temperatures high enough to cause thermal cracking can convert a semisolid tar into a rather low (10 cp) viscosity oil. It has been reported that some of the oil has an uncommonly high percentage of aromatic constituents. In some reservoirs the petroleum varies in composition from a live oil at the top of the zone to tar and solid hydrocarbon or gilsonite at the base.

Oil saturation is highly variable, with reported values ranging from 0 to more than 65%. Lower saturations are probably typical of outcrops and very shallow oil reservoirs, where some of the oil may have been washed away by moving ground water. Even in some deeper reservoirs, there is evidence that the oil saturation of the more permeable zones has been reduced by partial flushing.

Where this condition occurs, it will present an obstacle to commercial oil recovery. Any fluid which is injected to promote oil recovery (steam, air, solvent, etc.) will tend to enter the partially flushed zones and bypass the zones which have higher oil saturations and less effective permeability to the injected fluid.

It should also be noted that moderately high rock permeabilities do not necessarily mean adequate injectivities. Since the crude in many of the reservoirs is essentially immobile, the effective permeability of the rock to any fluid will be low until this oil has been rendered mobile by action of heat, solvent, etc.

Base Cases for Economic Study

Four base cases have been devised for detailed economic analysis. Each case is considered to be representative of a type of hydrocarbon deposit which occurs in the Forest City Basin area.

Case 1 represents a tar zone with a thickness of 15 ft. Oil viscosity is too high for forward combustion or conventional steam recovery techniques, and the reservoir is produced by reverse combustion. The oil is assumed to be sufficiently stable toward oxidation so that spontaneous ignition at the injection wells does not occur. The data used to describe this case are set forth in Table 12.5.

Table 12.5: Data Used for Case 1—Thin Tar Sand

Depth, ft	120
Net thickness, ft	15*
Porosity, %	25.5
Oil saturation, % pv	41.2
Oil in place, bbl/acre-ft	813
Initial reservoir temperature, °F	60
Oil viscosity at 60°F, cp	500,000
Oil viscosity after thermal cracking, cp	10
Effective air permeability with oil and residual water in place, md	266
Absolute air permeability with oil and residual water removed, md	814
Lease size, acres	150
Rate of development, acres/year	25
Production method	reverse combustion
Well spacing, acre/producing well	0.3306
Well pattern	direct line drive; 120 ft between like wells; 60 ft between lines
Royalty interest	⅛
Air injection rate	131.28 M cf/day/well* for bed thickness of 15 ft

*This case was also studied for net thicknesses of 25 and 40 ft. The air injection rate for these cases was increased to maintain a constant rate per foot of bed thickness.

Source: OFR 60-74

Case 2 represents a shallow viscous oil reservoir with a thickness of 26 ft. Oil saturation is rather low and water cuts will be high for any known production technique. Data describing the case are presented in Table 12.6.

Table 12.6: Data Used for Case 2—Viscous Oil Sand—29% Oil Saturation

Depth, ft	160
Net thickness, ft	26

(continued)

Table 12.6 (continued)

Porosity, %	23.5*
Oil saturation, percent pv	29
Oil in place, bbl/acre-ft	535
Initial reservoir temperature, °F	60
Oil viscosity at 60°F, cp	1,000
Air permeability, md	450
Oil gravity, °API	18
Lease size, acres	150
Rate of development, acres/year	25
Royalty interest	1/8
Foward combustion:	
Well spacing	2½-acre line drive
Oil burned/acre-ft	325 bbl
Air injection rate	1.6 MM scf/day/well
Oxygen utilization, %	70
Steam flood:	
Well spacing	2½-acre 7-spot
Injection rate	136.5 tons/day/well

*This case was also studied for porosities of 28 and 19%.

Source: OFR 60-74

Case 3 depicts a situation where the more permeable portion of the reservoirs has been partially flushed by migrating ground water. Oil saturation ranges from 53% in the upper portion of the reservoir to 37% in the middle zone and 58% in the lower zone, which is essentially a tar reservoir. Since the middle zone is more permeable than the upper and lower zones, vertical sweep efficiency will be poor for any displacement process. This case is described in Table 12.7.

Table 12.7: Data Used for Case 3—Partially Flushed Reservoir

Initial reservoir temperature, 60°F
Reservoir and fluid properties:

Depth of Productive Zone, ft	Oil Saturation (%)	Permeability (md)	Porosity (%)	Oil Viscosity at 60°F, cp
242–261	53%	125	21.6	10,000
261–276	37%*	346	24.1	10,000
276-286	58%	132	22.9	30,000

Oil gravity, °API	15
Lease size, acres	150
Rate of development, acres/year	25
Royalty interest	1/8
Well spacings:	
Forward combustion	2½-acre line drive
Steam flood	2½-acre 7-spot
Injection rates:	
Forward combustion	1.5 MM scf/day/well
Steam flood	89.25 tons/day/well

*This case was also studied with oil saturation of 27% and 47% in the interval 261–276 ft.

Source: OFR 60-74

Case 4 represents a better prospect for commercial production than the other cases. Reservoir thickness is 25 ft, oil saturation is 54.8%, and oil in place is 1,093 bbl/acre-ft. The case is described by data set forth in Table 12.8.

Table 12.8: Data Used for Case 4–Viscous Oil Sand–54.8% Oil Saturation

Depth, ft	200
Net thickness, ft	25
Porosity, %	25.6
Oil saturation, % pv	54.8
Oil in place, bbl/acre-ft	1,093
Initial reservoir temperature, °F	60
Oil viscosity at 60°F, cp	5,000
Average permeability, md	320.4
Royalty interest	$\frac{1}{8}$
Well spacings:	
Forward combustion	2½-acre line drive
Steam flood	2½-acre 7-spot
Lease size, acres	150
Rate of development, acres/year	25
Injection rates:	
Forward combustion	2.55 MM scf/day/well
Steam flood	119.35 tons/day/well
Oil burned by forward combustion	325 bbl/acre-ft

Source: OFR 60-74

Oil production from each of the four base cases was predicted. The production methods considered were forward combustion, steam flooding, cyclic steam injection, reverse combustion (for Case 1 only), and hot water flooding. The production method which appeared to be best suited to each case was selected, and an economic analysis was made for development of the reservoir by this method.

Evaluation of Case 1: The thin tar sand described as Case 1 presents a difficult problem in oil recovery. Reverse combustion was the only available thermal oil recovery technique which seemed appropriate for use in this tar sand. It was concluded that there appeared to be no economically feasible technique for producing shallow tar sands that are too low in oil content for recovery by mining and too low in permeability for recovery by reverse combustion.

Evaluation of Case 2: The viscous oil reservoir described as Case 2 presents an entirely different oil recovery problem from the tar sand which was previously discussed. Economic calculations indicate that profitability cannot be achieved for any of these three hypothetical reservoirs even at unrealistically high oil prices. The injection of either air or steam would entail a high fuel cost, and an increase in fuel cost would presumably accompany any sizeable increase in the price of crude oil. Thus higher crude prices would not necessarily lead to profitability in developing this type of low oil saturation reservoir. New production methods appear to be necessary for commercial development of this type of petroleum deposit.

Evaluation of Case 3: Oil recovery from the reservoir described by Table 12.7 was predicted for forward combustion and steam flooding. Commercial development does not appear to be feasible at any realistic oil price.

The primary difficulty in producing this type of reservoir is that most of the injected fluid (air or steam) enters the middle zone, which has a relatively high permeability and a relatively low oil saturation. The upper and lower zones contain most of the oil reserves, but most of the production will be obtained from the middle zone. In the opinion of Arnold and Harvey, this problem probably cannot be corrected by selectively injecting air or steam into the upper or lower zone. However, if an effective technique could be developed for reducing the permeability of the middle zone (at a substantial distance from the injection walls), the prospects of deriving commercial production from this type of reservoir would be greatly improved.

Evaluation of Case 4: An economic analysis of this case suggests that the steam flooding process would return the initial investment (with no profit) at an oil price of $7.09 per barrel. A 15% discounted cash flow rate of return would be realized at an oil price of $7.70 per barrel. Although field tests should be employed to verify these computations, it seems likely that this type of reservoir could be produced commercially.

Conclusions

The shallow hydrocarbon deposits in the Missouri-Kansas border area vary widely in quality. The diverse resevoirs will probably require several different oil recovery processes, causing production costs to fluctuate. These deposits will be developed with care. Low quality reservoirs containing tar and viscous oil will require improved technology to become economic, and further efforts should be directed toward developing such production techniques.

The information in this section is from: *Fifth Annual DOE Symposium on Enhanced Oil and Gas Recovery and Improved Drilling Technology, Vol. 2—Oil,* CONF-790805-P2, sponsored by U.S. Department of Energy, Tulsa, Oklahoma, August 22-24, 1979, and is based on the paper, "Field Testing of the Vapor Therm Process in the Eastburn (Cherokee) Field, Vernon County Missouri," presented by J. Sperry, F.S. Young, and R.S. Poston, of Carmel Energy, Inc.

FIELD TESTING OF VAPOR THERM PROCESS IN MISSOURI

History of Recovery Attempts

Oil production occurred in western Vernon County, Missouri before 1870. Oil was discovered in the shallow Bluejacket and Warner Sands when area farmers drilled water wells. The first serious attempt at primary production occurred between 1915 and 1920 when, during World War I a small refinery was built near Richards, Missouri. Initial primary production rates were typically 2 or 3 barrels of oil per day (bopd) but quickly fell off to less than ¼ bopd per well. It became obvious when the refinery was forced to close due to the absence of sufficient crude that economic recovery of this oil was going to require stimulation.

The first serious attempts at enhanced oil recovery from these sands occurred in the mid 1950s, when Carter Oil attempted a steam drive north of Deerfield, Missouri. Phillips, Sohio and Shell followed with similar thermal projects in the 60s. These attempts involved some form of thermal recovery either by steam injection or in situ combustion. Not one of these attempts was considered successful due to the fact that response was erratic and production rates were not sufficient to sustain commercial production.

Carmel Energy Inc. entered into an agreement in late 1977 to acquire rights to oil and gas leases covering approximately 37,000 acres in Vernon County, Missouri. In April 1978, Carmel undertook a 10-well coring program under a 12-month cost-sharing contract with the U.S. Department of Energy. From results of this coring program, Carmel selected a deposit discovered in 1915 and identified as the Eastburn (Cherokee) Field for their Vapor Therm Test.

The Vapor Therm Process

The Vapor Therm Process had previously been demonstrated to be commercial in an 870-foot Bartlesville sandstone on the Clinkenbeard Lease in Allen County, Kansas near the town of Iola, Kansas.

Carmel's Vapor Therm Process is designed to produce a mixture of superheated steam, carbon dioxide, and nitrogen which, when injected into a sandstone reservoir containing heavy oil, will stimulate oil production by: (1) reducing oil viscosity by application of heat; (2) energizing the reservoir by repressuring CO_2, N_2 gas; (3) absorption of CO_2 into the cold oil with oil-phase swelling and resulting viscosity reduction. It is the intent of the Vapor Therm Process that the stimulation mechanism take place in a way so that it is not damaging to the permeability of oil flow in the formation.

Test Site—Eastburn (Cherokee) Field

The Eastburn (Cherokee) Field test of the Vapor Therm Process was located approximately 6 miles south-southwest of Deerfield, Missouri. The Ida-Fauvergue Lease was selected as the site for the test because: (1) abandoned oil field equipment and several unplugged oil wells indicated that there had been some primary production; (2) core analyses from 8 wells in the field indicated that reservoir quality was apparently better than other locations in western Missouri; (3) Carmel was able to secure old production records which enabled them to certify the Ida-Fauvergue Lease for stripper oil prices.

Carmel cored 8 wells on the Ida-Fauvergue Lease and selected Carmel Energy, Inc. (CEI) wells No. 5, No. 6, and No. 8 to be completed as test wells. Carmel's core analysis data and that obtained from previous operators in the field illustrated that permeability in the upper Bartlesville (Bluejacket) sandstone increased with depth. Detailed analysis of sand particles, showed the sand to be medium to fine-grain, well sorted, and sub-angular, (See Table 12.9 for reservoir parameters.)

Table 12.9: Reservoir Parameters

Depth to top of Bluejacket Sand, feet	100–120
Total sand thickness, feet	
range	8–33.5
average	22
Oil sand greater than 200 md, ft	18
Aquifer	None
Oil gravity, °API	21.6
Porosity, %	
range	20.0–28.7
average	24.5

(continued)

Table 12.9: (continued)

Absolute permeability, md	
range	200–850
average	650
Oil saturation, %	
range	60–85
average	80
Initial reservoir pressure, psi	54
Initial reservoir temp, °F	70
Viscosity at 70°F, cp	710

Source: CONF-790805-P2

Carmel originally estimated oil in place on the Ida-Fauvergue Lease to be in excess of 2.5 million barrels. This figure was later reduced as more knowledge of the reservoir was obtained.

In order to certify Ida-Fauvergue Lease Production for stripper pricing, it was necessary to obtain one year's production data. This was done and the average primary production rate from 8 wells was 0.07 bopd per well.

Vapor Therm Recovery of Eastburn (Cherokee) Oil

In this process, superheated steam, CO_2, N_2, and other gases are generated for injection into heavy oil-bearing sandstone reservoirs. The approximate chemical analysis of the stream is as follows:

Component	Mol Percent
Superheated steam	68.0–70.0
Nitrogen	25–26
Carbon Dioxide	3.5–4.0
Oxygen	0–0.5
Hydrogen	Trace
Carbon Monoxide	Trace

Carmel has learned how to generate Vapor Therm gases essentially without difficulty and to inject these gases into the sandstone formations without noticeable corrosion or other adverse side effects. Carmel has, through experience, found that by operating at slightly excessive oxygen concentration, formation of coke is prevented and elemental hydrogen and carbon monoxide are generated. Hydrogen and carbon monoxide are excellent reducing agents to scavenge out oxygen downstream and to assist in corrosion control.

Carmel modified the Vapor Therm generation unit in Missouri for addition of anhydrous ammonia to the top of the steam drum for pH control which again helped in eliminating corrosion, and for assistance in preventing hydration of clays when steam condensed in the formation.

The generation unit requires approximately 15 gallons per minute of net process water which will, of course, vary with the quantity of fuel being burned in the unit, 440-volt, 3-phase power (which Carmel generated on site with a portable diesel-powered generator) and propane to fire the air compressor engine and for miscellaneous heat.

CEI No. 6 and No. 8 were drilled as 2 outside wells of a hexagonal, 7-spot pattern with one well per acre spacing. CEI No. 5 was drilled outside of the hexagonal, 7-spot pattern and was completed primarily for the purpose of determining lower limits of reservoir quality for application of Vapor Therm technology.

Vapor Therm gas after generation is distributed via insulated 2" pipelines to each of the 3 injection-production wells. Project design criteria dictated that 150 MM Btu was to be injected in 24 hr into each well or, in the event an individual injection well would not accept Vapor Therm at such high rates, then at least 4,000 lb/hr of Vapor Therm was to arrive at the well at or above 500°F.

For the duration of the field test, Carmel used fuel nozzles in the furnace that permitted the unit to burn 45 gallons of No. 2 diesel fuel per hour. This meant that it took 2 days to inject a well with 150 MM Btu of Vapor Therm. However, even at this rate, with the final exit gas temperature off the Vapor Therm unit between 650° to 700°F and only 50°F of heat loss in the transmission lines, and 50°F down hole loss, a sand face temperature of 550° to 600°F resulted. Injection pressure was limited to 180 psi at the sand face.

Problems Encountered and Their Solutions

During the first attempt at injection, in April 1978, Carmel used a diesel-fired, 750 cfm at 125 psi screw-type air compressor. This machine was designed to bring air pressure up to 125 psi and hold this pressure through action of a governor on the diesel engine. This type of compressor proved to be impossible to operate in the field.

At the first attempt to inject simultaneously into CEI No. 6 and No. 8, the wells began to take Vapor Therm; however, they quickly pressured up and air flow through the Vapor Therm generating unit was reduced. The rate of fuel injection into the unit which is regulated by the difference in system pressure and fuel injection pressure remained constant. The result was serious coking, coke plugging off the sand face with further reduction in air flow through the unit, causing an even more serious coking problem.

Operators in the field were not able to regulate the ratio of fuel to air properly with this type of compressor. Finally, the sand face at each well was totally sealed with coke and careful testing confirmed that the wells were not taking the gas.

Upon pulling the 2⅜" EUE tubing in each well, the well bore was found to be completely full of coke. Each well had to be washed out and the sand face was shot with 22 pounds of explosives to loosen the coke and then jetted and washed with 1% KCl water to clean up the sand face.

The problem of air compressor selection was solved by moving in a three-stage compressor capable of delivering 2.2 MM scfd of air at 300 psi. This compressor maintains essentially constant air throughput to the unit, and even though a well may pressure up during an injection cycle, the ratio of fuel to air is held essentially constant.

The first attempt at a sequential hydrochloric-hydrofluoric acid treatment completely destroyed the tubing and part of the wellhead on CEI No. 8. Furthermore, packing glands on each valve had to be changed to Teflon from the graphite glands originally furnished. However, much less corrosion was observed with concentrated acids, and almost no corrosion was observed in the 70% hydrochloric acid treatment. It would appear that many treatments with 70% hydrofluoric acid will be possible before appreciable corrosion.

Open hole completions apparently were the reason for the serious sand problems encountered throughout the duration of the field test. Each well, completed open hole, frequently became plugged with sand. Sand plugging could be observed by attention to daily production rates.

Frequently, the problem of sand plugging was solved by air lifting sand out of each well by intermittently injecting air down the tubing and then down the annulus and flowing loose sand up the tubing to the surface tanks. In most cases this procedure lifted out the sand and the well was quickly placed back on production, but on more than one occassion it was necessary to pull the tubing when the entire well bore became sanded up. This resulted in a very messy pulling job since the reservoir was under sufficient pressure to unload hot oil, water and sand on the pulling unit crew.

Another problem that Carmel encountered during the test was caused by extreme cold weather during the winter of 1978 to 1979. Low temperatures for the area were a one day record all time low of –24°F and two days at –20°F. A record 42 inches of snow fell during the winter. As can be seen in the production decline curves, cold weather definitely affects production. Even though the production lines were buried below expected frost depth they frequently became frozen, especially at low points in the production lines. The stock tanks were not properly insulated and had to be thawed almost every day during the winter with hand-held torches. It is recommended that any field test that is to be conducted during the winter in Kansas or Missouri should be properly designed for winter operation.

Production

Air injectivity tests were conducted on each well before firing up the Vapor Therm unit. Initial air injectivity at each well was negligible at over 200 psi. After 24 hr, however, CEI No. 6 and No. 8 began to take air at a rate of approximately 1.1 MM scfd each. No measurable air injectivity occurred at CEI No. 5.

CEI No. 6 was injected first cycle with 152.2 MM Btu of Vapor Therm, and after 25 hr of soak was placed on production. Production from the first and subsequent cycles for CEI No. 6 is shown in Figure 12.5b.

Carmel has heretofore used 150 MM Btu as a "standard" injection quantity for Vapor Therm. To get a comparison of Vapor Therm quantities, CEI No. 8 was injected first cycle with 311.8 MM Btu of Vapor Therm. Oil production from CEI No. 8 for the first and subsequent cycles is illustrated in Figure 12.5c.

Injection was attempted for 10 consecutive days into CEI No. 5. At the end of 10 days, this well had taken less than 50 MM Btu. This unacceptable rate indicated either that the reservoir quality of this well was below the minimum required for injectivity, or that severe permeability damage due to the presence of water-sensitive clays was reducing the permeability, or perhaps both of these situations existed simultaneously.

Figure 12.5: Production Rate vs Time

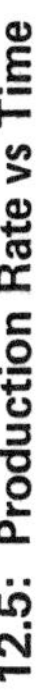

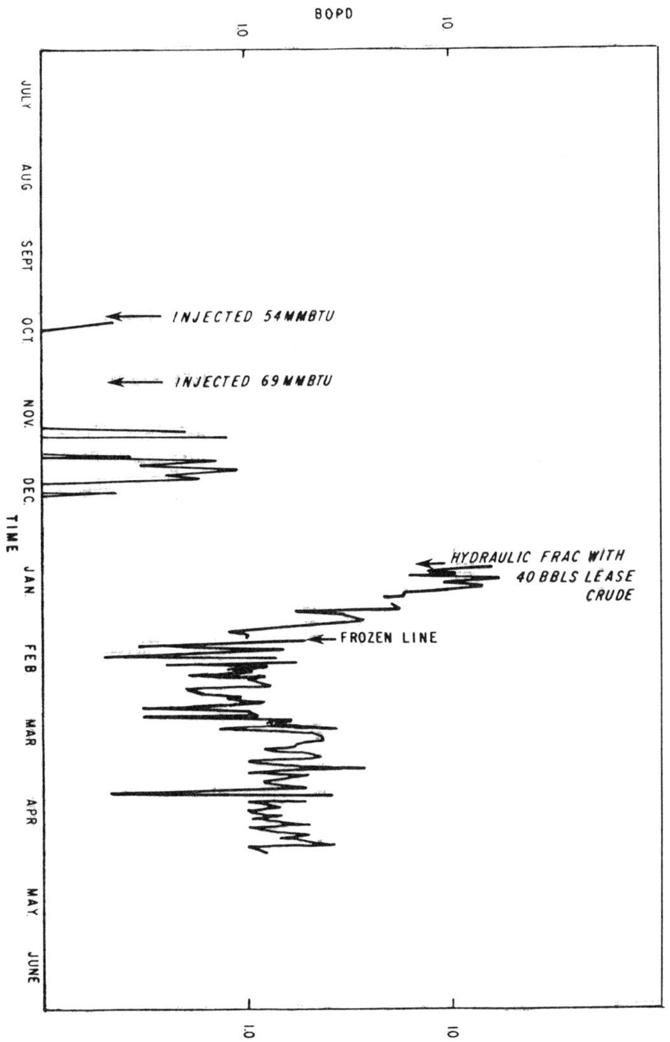

a.

(continued)

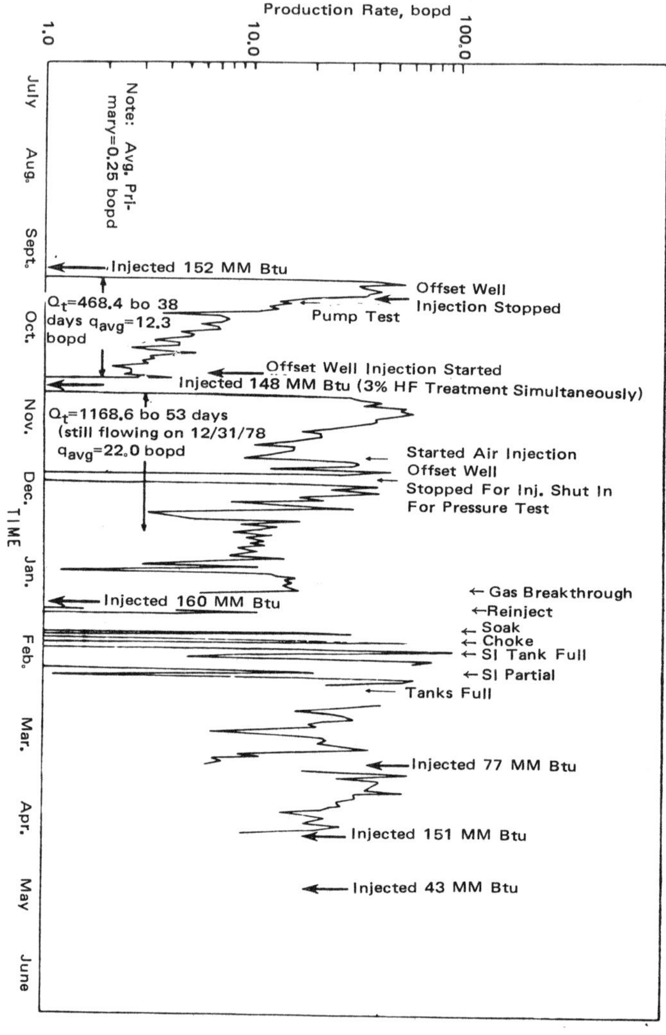

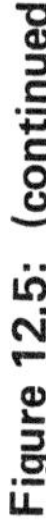

Figure 12.5: (continued)

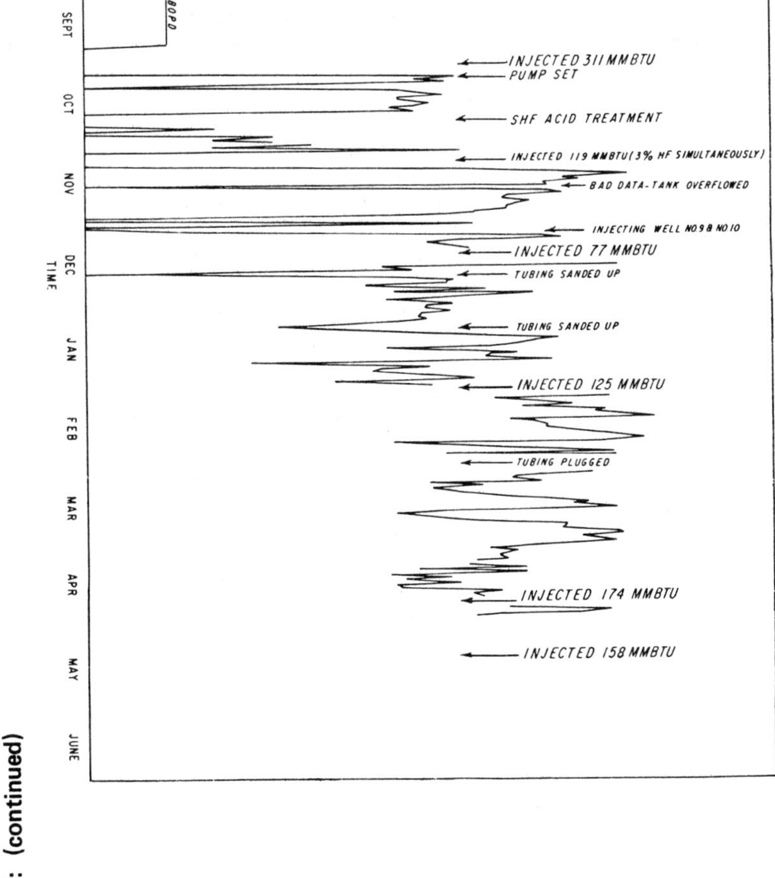

(a) CEI Well No. 5
(b) CEI Well No. 6
(c) CEI Well No. 8

c.

Figure 12.5: (continued)

Source: CONF-790805-P2

This well was later shot with 20 lb of explosives, jetted and washed, and another injection attempt was made which was also a failure. Later on, however, during the test as the reservoir was pressured-up this well produced, at a consistent rate, an excess of 2 bbl of oil per day strictly from the drive mechanism resulting from injection into wells No. 6 and No. 8. Its production is shown in Figure 12.5a.

Production response from CEI No. 6 during the first injection-production cycle was considered satisfactory. This well produced 442.5 bbl of cumulative oil during the first month of production. However, analysis of production from CEI No. 8 and confirmation of sand face temperatures indicated that this well was producing at only 21.7% of the expected rates.

Area injectivity tests on CEI No. 8 after 15 days of production were conducted and compared to the same tests conducted at the beginning of the first injection 19 days earlier. The tests confirmed that air permeability had been reduced to 19.2% of the original value. This suggested severe permeability damage due to clay and perhaps other mineral constituents in the sandstone formation. It was on this well that Carmel developed its acid treatment. A similar treatment was given CEI No. 6 at the next injection cycle.

As can be seen from the decline curves for both wells, there is no doubt that the acid treatments made a material difference in the production response to CEI No. 8 and little doubt that the acid treatment helped reduce permeability damage in CEI No. 6. In fact, peak production for CEI No. 6 increased to 97.5 bopd in this well at the beginning of the third production cycle.

Cumulative monthly production for each well is shown in Table 12.10. Cumulative injections for each well and stimulation response ratios are also presented in Table 12.10.

Table 12.10: Production Response to Stimulation

	Carmel Energy, Inc.		
Month	Well No. 5	Well No. 6	Well No. 8
	bopm		
October, 1978	0	448.2	60.2
November, 1978	0	626.2	265.9
December, 1978	6.6	613.6	195.8
January, 1979	141.1	262.5	171.3
February, 1979	33.9	576.7	362.1
March, 1979	45.5	556.4	419.9
April, 1979	40.5	848.0	233.1[*]
Cumulative bbl	267.6	3,931.6	1,708.3
Less frac oil	(40.0)		
Net total, bbl	227.6	3,931.6	1,708.3
bopd/well	1.1	18.7	8.1
Oil in place recovered, %	2.0	13.1	5.7
MM Btu injected	123	537	632
bo/MM Btu	1.8	7.3	2.7
bo/bbl steam[**]	0.8	3.1	1.2

[*]Tied into central storage and treating on 4/27/79.
[**]Cumulative field production/bbl steam injected is 2.4.

Source: CONF-790805-P2

In summary, oil production responses after Vapor Therm injection have been encouraging. The first stimulation of CEI No. 6 yielded initial rates as high as 53 bopd and a sustained average of 12.3 bopd. The second stimulation of this well was even better, with a peak rate of 58 bopd and an average of 18.2 bopd. The third stimulation, as noted above, yielded peak production rates of 97.5 bopd.

Results of oil production after stimulation of the other two wells in the three-well field research pilot test in this reservoir were obviously not as encouraging. CEI No. 5, with a k-h product of 3.4, apparently did not have the capacity to accept Vapor Therm gases and after three attempts at injecting Vapor Therm at temperatures in excess of 500°F this well was abandoned as an injector-producer well. CEI No. 8 encountered the serious difficulties noted with initial injection.

Conclusions

The Vapor Therm process will stimulate oil production from the Upper Bartlesville (Bluejacket) sandstone in western Missouri, provided the reservoir quality criteria are met on a well-by-well basis. Prospects for field-wide application of this technology in the Eastburn (Cherokee) Field on a commercial basis are excellent. In fact, since concluding the field test a commercial 30-well development drilling program has been completed. Out of 30 new wells drilled, 11 were completed as injector-producers and 10 were completed as tubingless producers only with 2⅜" casing cemented to surface. Nine wells were plugged and abandoned as not meeting a reservoir criteria of 3.5 d-ft or better.

With less than half the wells tied into production and only three new wells stimulated, peak production has been approximately 290 bopd, with an average of more than 100 bopd from 10 wells.

One may further conclude that in closed, nonbottom water reservoirs the wells may be made to flow after stimulation, completely eliminating the need for pumps and pump jacks. This is obviously due to the presence of CO_2 and N_2 gases and therefore indicates a definite technical and resulting economic advantage over straight steam alone.

Finally, this field test demonstrates the importance of attention to geologic detail as probably the most important factor in applications of Vapor Therm technology to recovery of heavy oils from "dirty" sandstone reservoirs. Successful application of Vapor Therm in a "dirty" reservoir requires a detailed knowledge of reservoir fluid composition, complete analysis of the pore space for clays and other minerals along with customary detailed core analysis. From this information, a Vapor Therm recovery system may be designed which will economically recover a high percentage of the oil in place.

REFERENCES

(1) U.S. Bureau of Mines, *Heavy Crude Oil: Resource, Reserve, and Potential Production in the United States.* IC 8352 (1967).

(2) Oil and Gas Journal, API: "Price Hikes Fail to Bolster U.S. Oil, Gas Reserves." *Oil and Gas J.,* Vol. 73, No. 14, pp. 44-46 (April 7, 1975).

(3) Bleakley, W.B., "Journal Survey Shows Recovery Projects Up." *Oil and Gas J.,* Vol. 72, No. 12, pp. 69, 76, and 78 (March 25, 1974).

(4) Sloat, B.F., Fitch, J.P. and Taylor, J.T., "How to Produce More Oil and More Profit With Polymer Treatments," Presented at 43rd Ann. Meeting, Soc. Petrol. Eng., AIME, Bakersfield, Calif., SPE Preprint 4185 (November 8-10, 1972).

(5) White, J.L., Phillips, H.M., Goddard, J.E. and Baker, B.D., "Use of Polymers to Control Water Production in Oil Wells," Presented at Improved Oil Recovery Symp., Soc. Petrol. Eng., AIME, Tulsa, Okla., SPE Preprint 3783 (April 16-19, 1972).

(6) Needham, R.B., Threlkeld, C.B. and Gall, J.W., "Control of Water Mobility Using Polymers and Multivalent Cations," Presented at Improved Oil Recovery Symp., AIME, Tulsa, Okla., SPE Preprint 4747 (April 22-24, 1974).

(7) Wells, J.S. and Anderson, K.H., "Heavy Oil in Western Missouri," *AAPG Bulletin* 52, No. 9, (September 1968).

USE OF SOLVENTS AND EXPLOSIVES TO RECOVER HEAVY OIL

The information in this chapter is based on:

Solvents and Explosives to Recover Heavy Oil, Bartlett, Kansas, BuMines, TPR 60, prepared by L.J. Heath, F.S. Johnson, and J. S. Miller, of Bartlesville (Oklahoma) Energy Research Center, for U.S. Bureau of Mines, September, 1972.

Heavy-Oil Recovery Using the SolFrac Method, CONF-7406129-1 prepared by F.S. Johnson and R.T. Johansen, of Bartlesville (Oklahoma) Energy Research Center, for U.S. Bureau of Mines and presented at Interstate Oil Compact Commission Mid-Year Meeting, Vail, CO, June 30-July 4, 1974 and published in IOCC *Bull.* 16, No. 1, 40-49, June, 1974.

Laboratory Investigation Using Solvent to Recover Heavy Oil from Fractured Reservoir, CONF-7504932, prepared by F.S. Johnson, R.A. Jones, and J.S. Miller, of Bartlesville (Oklahoma) Energy Research Center for U.S. Bureau of Mines and presented at Rocky Mountain Regional Meeting, Soc. Petrol. Eng., AIME, Denver, April 7-9, 1975, SPE Preprint 5336.

Heavy Oil Recovery Using Solvents and Explosives, BERC/ RI-77/16, prepared by L.J. Heath, F.S. Johnson, J.S. Miller, R.A. Jones and W.D. McMurtrie, of Bartlesville (Oklahoma) Energy Research Center, for U.S. Department of Energy, December, 1977.

SOLFRAC RECOVERY METHOD

The Bartlesville (Oklahoma) Energy Research Center has been developing means to recover petroleum from the many shallow, low productive, heavy oil deposits in Kansas, Oklahoma, and Missouri (1). In these states, field tests have been attempted to recover this shallow, heavy oil by steamflooding, cyclic steam injection, fireflooding, bottom hole heating, and several proprietary processes. In 1969, the U.S. Bureau of Mines reported on domestic heavy oil and tar sand recovery methods (2). That investigation brought out three factors that must be considered in any in situ recovery process for heavy oil: (1) Communication must be established between the injection and producing wells; (2) The viscosity of the oil must be lowered; and (3) Communication must be maintained. Unless all of these conditions are met in the recovery process selected, there is little chance of significant oil recovery.

The Bureau of Mines conceived the SolFrac recovery method that combines chemical-explosive fracturing with heat and solvent to extract the oil. The chemical-explosive method fractures the interwell zone and establishes communication between wells. The heated solvent dilutes the oil and lowers the viscosity. Continued injection of solvent through the fractures maintains communication. This combination, never tried before in a shallow, consolidated, heavy oil sand, met the three above criteria.

The SolFrac recovery method is being developed and evaluated through laboratory experiments and pilot field tests.

HEAVY OIL RESERVOIRS

Heavy oil reservoirs are receiving increased attention because of (a) recent increases in heavy oil production from thermal recovery operations, (b) the existence of more than 2,000 known heavy oil reservoirs, and (c) the fact that these reservoirs contain a high percentage of the oil originally in place.

Petroleum-impregnated formations have been termed at various times as tar sands; oil-, bitumen-, or asphalt-impregnated rock; and heavy oil formations. The most widely misused term is tar sands. Several rock types as well as petroleum materials are included in the term tar sands (3). In this report, the term heavy oil will be used instead of tar sand and is defined as oil having a gravity of $25°$ API or lower. Conventional secondary recovery methods used to recover heavy oil are often impaired by the unfavorable mobility ratio between the driving and driven fluids. To develop methods of producing heavy oil economically, much research has been done by the industry, including numerous field tests of thermal recovery methods.

The technology for greatly increasing the recovery of heavy crude oil by raising its temperature through steam injection is presently available. However, economics has limited general use of thermal methods of recovery to reservoirs having characteristics considered to be most favorable. Significant improvements in technology or economic conditions could result in increased profitability and use of not only thermal methods but also other methods to recover crude oil.

Many heavy oil sands are incompetent; that is, they tend to slump or cave into any voids created when the oil becomes mobile because the sand grains are not cemented to one another. Thus, when a fracture is created and heat is applied to the formation through the fracture, the incompetent formation will slump into the fracture and reduce the permeability, even when the fracture contains a conventional propping agent. Slumping can sometimes be prevented by maintaining sufficient pressure in the fracture and by regulating the fluid composition and flow rate.

Although natural fractures may exist in some heavy oil sandstones, improved fluid injectivity probably could be created by hydraulically fracturing the formation. However, if the fracture were created in a vertical plane, the fracturing would not necessarily result in a connection between wells.

The only solvents of much use in dissolving the complex hydrocarbons comprising the heavy asphaltic oils are the relatively expensive naphthenic and aromatic substances. Solvent stimulation of oil fields producing heavy oils has been practiced in California since the mid sixties and has been discussed in the previous chapter. This method has been limited to injecting a given solvent into the zone around the well bore and then placing the well back on production. The resultant stimulation is caused by cleaning up the formation and dissolving the asphaltenes, etc., adjacent to the well bore.

The production potential for oil from U.S. heavy oil deposits is poorly defined; neither industry nor government has made a comprehensive and conclusive examination of the physical extent and economic possibilities of such deposits. Bureau of Mines Monograph 12, (1) describes 546 occurrences of tar sands and 383 shallow heavy oil fields. Reserve figures, however are available for only a few deposits. Only a few of the large deposits in California and Utah have been surveyed carefully enough so that the reserve figures are meaningful; accurate volume figures for the remaining deposits are not available. Forty-two counties in Kansas, Missouri, and Oklahoma contain 40% of the recognized heavy oil deposits. Much of this is fairly shallow, less than 500 feet, and could be mined; however, because of costs and environmental considerations research should be initiated for the economical recovery of this crude by in situ methods. The geographical location of U.S. heavy oil accumulations is shown in Figure 13.1 (5).

Figure 13.1: Major Heavy Oil Deposits in the United States

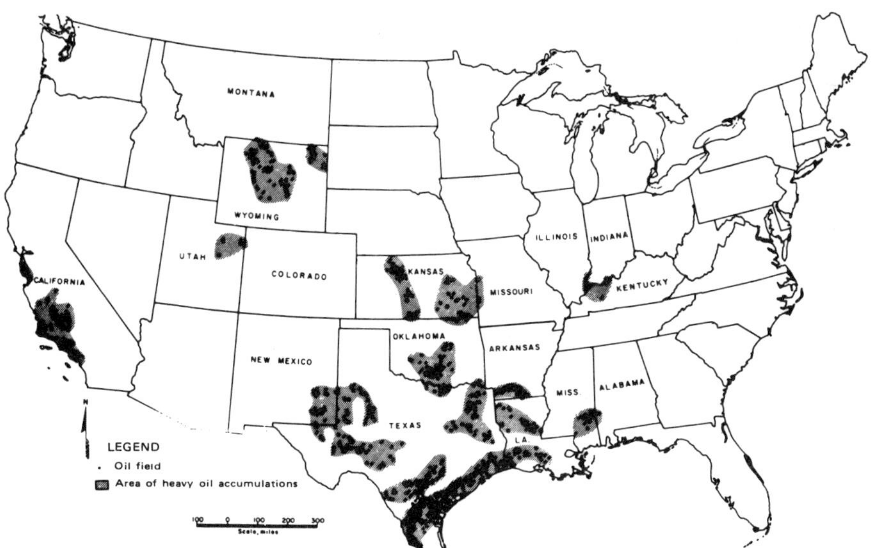

Source: BERC/RI-77/16

Potential Recovery of Heavy Oil in the United States

Statistics on cumulative oil production are available on most of the heavy oil productive reservoirs. A breakdown of heavy crude oil resources in the United States (4) indicates a proved reserve of 5.2 billion bbl with an oil resource of 106.8 billion bbl, this resource being classified at 51.3 billion bbl of 20° to 25° API gravity and 55.5 billion bbl of lower than 20° API gravity crude. This resource is also classified as to depths, at 21.3 billion bbl in reservoirs to 1,500 ft deep, 38.1 billion bbl in reservoirs from 1,500 to 3,000 feet, and 47.4 billion bbl in reservoirs deeper than 3,000 feet. These figures do not include those oil-impregnated rocks which are known but from which no crude has been recovered. It is estimated that the total amount of heavy oil in the United States would be in excess of 150 billion bbl of oil in place if deposits were included such as those in western Missouri, eastern Kansas, northeastern Oklahoma, and other reservoirs which have no reported individual oil production statistics. Considering the low primary recovery from these accumulations, the potential recovery is tremendous.

FIELD RESEARCH BY BARTLESVILLE ENERGY RESEARCH CENTER

Crude oil having a viscosity normally associated with 15° API gravity oil and under will not flow readily at ambient temperatures; therefore, heat or dilution with a suitable solvent is required to produce such a crude oil. Since the heavy oil-containing sands of Kansas, Missouri and Oklahoma generally occur at shallow depths, they are not responsive to standard fracturing or stimulating techniques.

The important difference between this oil and the shallow heavy oil in California fields is permeability. The California fields tend to have permeabilities in the range of 1,000 to 10,000 millidarcies. The Missouri deposits have permeabilities in the range of 100 to 200 millidarcies or less.

In order to solve such production problems, engineers at the Bartlesville, Oklahoma, Energy Research Center investigated the location of shallow, high-viscosity oil fields as possible site for field test. The proposed test would involve two or more wells, either in or capable of being arranged in a definite pattern. The wells would be shot with high explosives in an attempt to rubblize the interwell formation and would be cleaned out. Then an aromatic-type solvent would be circulated between the wells with varying soak periods being tested.

Test Site Selection and Treatment

A suitable reservoir was located on the H.H. Leap farm in the southeastern section of Kansas near Bartlett. Three wells were drilled and explosively fractured for use in the initial test. Although several wells had been drilled on this farm by oil-producing companies, oil production had not proved economically feasible.

Drilling Program: The first drilling was started in April 1971. Well 7 was drilled to 260 ft and cored to 375 ft. After the cores were examined the well was reamed to 375 ft, and plugged back to 360 ft. Wells 8 and 9 were drilled to 360 ft, and all three wells were completed with 7" casing to 295 ft, and cemented to the surface. A solvent injection test was performed in these wells after they were explosively fractured and before the well pattern was expanded.

The second drilling program was started in September 1972, and wells 10, 11, 12, 13, and 14 were drilled. Wells 13 and 14 were located 75 ft apart on diagonals from wells 7 and 8 for explosive fracturing wells. The wells were drilled to a depth of 360 ft. After the explosive shots were detonated, these wells were plugged in accordance with state rules and regulations. Well 12, an observation well located midway between wells 13 and 14, was drilled to 290 ft, a 6" core was obtained between 290 and 360 ft, and the well was plugged back to a casing depth of 290 ft. 5" casing was set to 290 ft and cemented to the surface, and the well was cleaned out to a total depth of 360 ft. Wells 10 and 11 were located on 50 ft spacing from wells 7 and 8 to complete the five-spot pattern. The wells were drilled to 290 ft, and 7" casing was set and cemented to the surface. The wells were then drilled to a total depth of 360 ft.

Formation Properties: Two wells were cored. Well 7 was cored prior to explosive fracturing to obtain formation properties such as permeability, porosity, and oil and water saturation. Well 12 was cored after explosively fracturing to determine postshot fragmentation of the explosively shot interval.

Well 7 was cored over the interval from 260 to 375 ft with 100% recovery. Core plugs were taken at 71 locations (273 to 375 ft) along the core and analyzed for porosity, permeability, and oil and water saturation. Information obtained from core plugs and visual observation of the core from 260 to 295 ft and 360 to 375 ft indicated little or no oil saturation in these intervals.

Fifty-five core plugs were analyzed in the zone of interest from 295 to 360 ft. Porosity ranged from 10.7 to 26.0% with a weighted average of 20.0%, permeability ranged from 0.2 to 654.6 md with a weighted average of 152.5 md, oil saturation ranged from 0 to 75.8% with a weighted average of 36.4%, and water saturation ranged from 0 to 8.4% with a weighted average of 2.1%

The interval of time from when the cores were obtained to when the core plugs were analyzed was of sufficient duration to have decreased the oil and water saturation values of the samples by evaporation. The gas saturation in this formation is negligible and was not considered in the analysis; therefore, the total fluid saturation (oil and water) should be approximately 100%. The analysis of the Leap cores indicated a weighted average fluid saturation of 38.9% The remaining 61.1% of fluid saturation is felt to be mostly a loss of water saturation with some slight loss of oil saturation due to evaporation of fluids from the cores during the time delay.

The oil in place determined through the zone of interest (295 to 360 ft) was calculated to be 37,000 bbl/acre. The cored interval was divided into 4 sections: from 273 to 295 ft, from the casing set point to the cleanout point (295 to 330 ft), from the cleanout point to the final total depth (330 to 360 ft), and the bottom 15 ft interval which was cored in well 7 (360 to 375 ft). Well 7 was plugged back to 360 ft, and the other two wells were drilled to only 360 feet. The results of weighted average calculations are shown in Table 13.1.

Well 12 was cored through the interval 290 to 357 ft, with 99% recovery. Four core plugs, taken at 295, 315, 337 and 354 ft, indicated that the permeability, porosity, oil saturation, and water saturation were within the values obtained from plugs cut from the core on well 7. Visual observation of the cores indicated that the formation characteristics correlated between the core taken in well 7 to that in well 12.

Table 13.1: Core Analysis

Depth (ft)	Permeability (md)	Porosity (%)	Oil Saturation (%)	. . . . . Oil in Place.	
				(bbl/acre)	(bbl/acre-ft)
273-295	147.1	21.9	6.6	2,464	112
295-330	132.6	18.4	33.9	16,937	484
330-360	175.6	21.9	39.4	20,069	669
360-375	13.3	15.8	4.7	868	58

Source: BERC/RI-77/16.

The preshot core, obtained from well 7, was competent through the entire cored interval, indicating that no fractures existed prior to explosive fracturing. The postshot core, obtained from well 12, indicated that fracture patterns existed in the explosively shot zone between wells 13 and 14. Near vertical to vertical type fractures were noticed between 306 and 342 ft.

Explosive Fracturing

Two explosive fracturing experiments, using a total of 15,000 lb of pelletized TNT (3,000 bbl/well) were performed in five wells on the Leap lease as part of this research program. The experiments were performed to create fragmentation and paths of communication between wells so that the heavy, high viscosity oil could be produced by the experimental solvent application.

The first explosive shot was performed in 3 wells, (7, 8, and 9) that formed part of the 50 x 50 ft 5-spot pattern. The second explosive shot was performed in 2 wells (13 and 14) that were drilled outside the pattern for explosive wells exclusively.

In both tests, all wells were bailed to total depth to gauge the rate of water influx and to determine if excess water was invading the well bores. Each well could be bailed down, some more rapidly than others, indicating a variance in water intrusion. With excessive water influx, the explosive could bridge at a height in the well bore so as not to accommodate the explosive charge. This would cause damage to the casing from the explosive detonation.

In the first explosive shot, a wireline measurement was taken to determine if a safe interval existed between the explosive and casing bottom to prevent casing damage. No wireline measurement was necessary in the second explosive shot because the shot wells did not contain casing.

Approximately 2 bbl of water was added to the shot wells in explosive shot No. 1 to fill the void space of the explosive. No water was added to fill the void space of the explosive in explosive shot No. 2 because of excessive water influx.

Each primer cap was tested with a blasting galvanometer to determine whether or not the blasting circuit was closed and in proper condition for firing. Sand was added to each well to an approximate height of 150 ft. This procedure was followed to contain the explosion in all the wells and protect casing in the cased wells. After clearance of the sites (in both explosive shots), the wells were simultaneously shot. A strong shock wave was felt in both explosive shots.

In explosive shot No. 1, well 7 blew the sand tamp to the atmosphere, and black smoke billowed for several hundred feet into the air and gradually changed to a clear gas. The wells were shut in, and after about 30 minutes the wellhead pressure had built up to 80 psig. In explosive shot No. 2, the expanding explosive gases fractured through the formation to wells 7, 8 and 9, blowing the water tamp to the surface. About 10 minutes after the shot, explosive well 13 blew its sand tamp several hundred feet into the atmosphere and continued blowing explosive gas into the air for about 45 minutes.

Airflow Tests: Preshot and postshot airflow tests were made to ascertain possible increased flow capacities between wells which would indicate greater fracture paths and fragmentation through the shot zone. In the preshot test, little pressure falloff was noted on the injection well, and no pressure buildup was noted on the production wells, which indicated negligible communication between wells.

A postshot flow test was run on wells 7, 8, and 9 after the cleanout was stopped. The flow tests were run to determine the extent of communication between wells, the flow capacities between wells, and which well to use as the solvent injection well.

Three tests were made by injecting air into wells 7, 8 and 9 in turn, with the other wells shut in until pressures in the pattern stabilized. Each of the paired wells for each test was produced at three rates through a critical flow prover for a 1 hr test period while injecting into the remaining well. The production rates were calculated by the standard critical flow prover flow equation using a 24 hr flow basis. The stabilized injection pressure (P_i) and flowing pressure (P_f) for each rate Q were recorded. Backpressure curves, plotting $P_i^2 - P_f^2$ vs Q, using well 9 as an injection and production well, are shown in Figures 13.2 and 13.3.

Figure 13.2 shows wells 7 and 8 as injection wells and well 9 as the production well.

Figure 13.3 shows well 9 as the injection well and wells 7 and 8 as production wells. Values of rate Q_{10}, (determined for the wells at a $P_i^2 - P_f^2$ of 10,000) were compared, and indicated that well 9 had better injection production relationships to the other wells, with the exception of well 7 to well 9, and consequently it was picked as the injection well.

Water-Injectivity Tests: Water-injectivity tests were run using wells 7, 8, 10 and 11 as injection wells and well 9 as the production well to determine the degree of fracture communication between wells in the five-spot pattern. Two types of tests were made: testing the four injection wells individually, and testing the four wells simultaneously.

In the individual well tests, water was injected into each well with a piston type pump for a given time. The flow rate on each well was measured by a water meter, and the pressures were obtained from pressure gauges on the wellhead. Four flow rates and corresponding pressures were recorded.

Figure 13.2: Flow Tests of Well Pairs with Well 9 as the Production Well

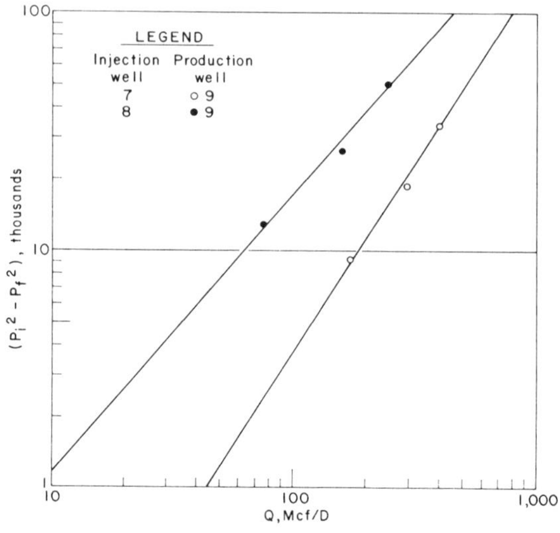

Source: BERC/RI-77/16.

Figure 13.3: Flow Tests of Well Pairs with Well 9 as the Injection Well

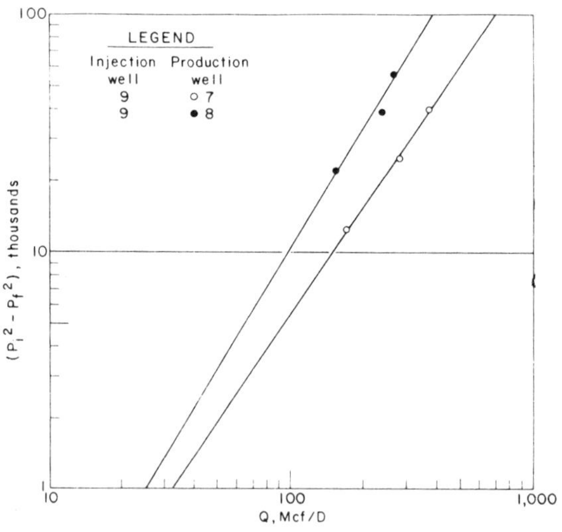

Source: BERC/RI-77/16.

The flow data for each test are given in Table 13.2 and the plotted data points
are illustrated in Figure 13.4.

Table 13.2: Q_{50} Values Obtained from Water-Injectivity Curves

Injection Well	Production Well	. . . . Q_{50} Value (bbl/day)	
		Single Well Test	Combined Well Test
7	9	72.0	39.0
8	9	71.0	53.0
10	9	17.0	13.0
11	9	5.0	31.0

Source: BERC/RI-77/16.

The plotted curves of the data points from the individual well tests (Figure 13.4) show that two of the original wells (7 and 8) have nearly identical injection profiles and that the injection profiles of the two new wells (10 and 11) are nearly identical. Table 13.2 shows that the Q_{50} rates (rate Q at 50 psia) for wells 7, 8, 10 and 11 were 72, 71, 17, and 5 bbl/day, respectively. These tests indicate that water can be injected from 4 to 12 times as fast through fracture systems in wells 7 and 8 as through fractures created in wells 10 and 11.

A combined well test was performed by injecting water simultaneously into wells 7, 8, 10 and 11. Flow rates and corresponding wellhead pressures were recorded for the 4 flow periods. The flow data for the test are given in Table 13.2 and the plotted data points are shown in Figure 13.5.

Figure 13.4: Water Injectivity Tests, Single Well

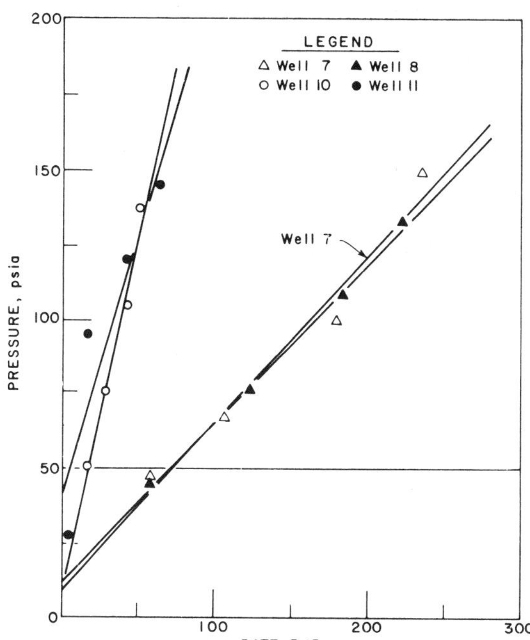

Source: BERC/RI-77/16.

Figure 13.5: Water Injectivity Tests, Combined Wells

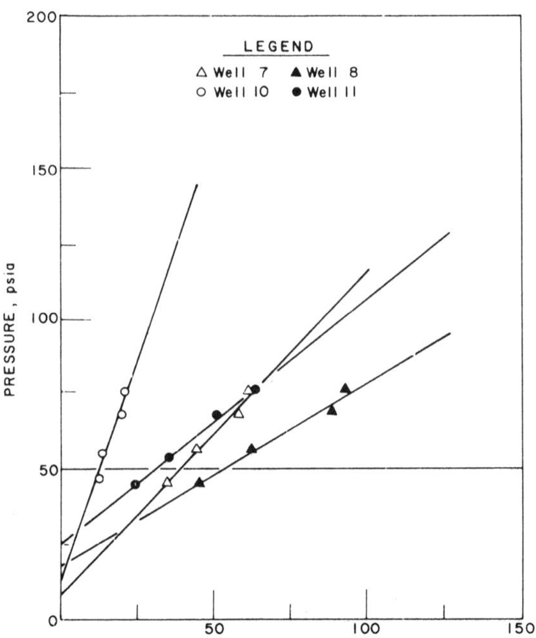

Source: BERC/RI-77/16.

To illustrate, a Q_{50} value was determined for the wells to indicate the position difference of the four curves from the previous test. The corresponding injection rates on wells 7, 8, 10 and 11 are 39, 53, 13, and 31 bbl/day, respectively.

The flow rates from this test differ from the flow rates of the previous test. The effects of simultaneous injection in the four wells are indicated in the comparison of the test results. The added injection capacity in the simultaneous test from wells 10 and 11 reduced the injection capacity of wells 7 and 8. The injection capacity of well 10 was not affected appreciably, but the injection rate of well 11 was increased, which can be attributed to flow of fluid through additional fracture systems within the pattern or leakoff of fluid through fractures off the pattern.

LABORATORY INVESTIGATION

Laboratory experiments were begun concurrently with site preparation. First the solvent was selected; then four different types of core tests were run using the solvent.

Selection of Solvent

A solvent or mixture of solvents was required that would lower the viscosity of the Bartlett oil and still retain the asphaltenes in solution. The solvent was to have an initial boiling point higher than 300°F because of plans to inject a hot solvent. An end boiling point lower than 425°F was required so that the solvent could be distilled from the oil after production.

A sample of oil from a stock tank was routinely distilled. The oil sample contained approximately 10% water, which caused considerable trouble when running the analysis. It was necessary to centrifuge the sample before the distillation could be completed.

The test used to determine the amount of precipitation was modified from ASTM D2042-66, Standard Method of Test for Solubility of Bituminous Materials in Organic Solvents. Briefly, the test consists of filtering 1 g of oil diluted with 100 ml of solvent solution through an asbestos fiber mat and determining the percentage of residual left on the mat after boiling off the solvent in an oven. The only modification was that of increasing the oven temperature to evaporate the different solvents used. In these tests, no differentiation was made between the asphaltenes and other insoluble matter. The results are shown in Table 13.3 as % precipitates.

Table 13.3: Summary of Solvents Tested

Solvent	Aromatic Content (%)	Viscosity of Mixture* (cp)	ASTM Test Precipitate (%)
Diesel fuel	32.5	183	ND
Benzene	99.0	36	1.11
2-Nitropropane	0	58	15.68
Decalin	0	148	ND
Styrene	99.9	45	0.73
n-Pentane	0	72	12.70
Kerosene	0	235	7.19
Carbon disulfide	0	17	2.16
Stoddard solvent	4.6	127	6.57
Commercial			
Mix 1	38.1	103	4.15
Mix 2	55.5	93	2.55
Mix 3	72.9	85	1.65
Mix 4	98.6	68	0.94
Mix 5	74.2	77	1.97
Mix 6	73.6	97	1.87
Mix 7	97.9	89	1.05
Mix 8	18.6	130	6.91
Mix 8 + 5% pine oil	22.7	117	4.55
Mix 8 + 10% pine oil	26.7	104	3.05
Mix 8 + 15% pine oil	30.8	118	2.15
Mix 1 + 5% pine oil	41.2	101	1.57
Mix 9	98.0	78	ND
Mix 10	74.0	73	2.17
Mix 11	0	49	12.26
Gasoline	25.0	130	4.32

Note: ND is not determined.

*75% crude oil, 25% solvent.

Source: BERC/RI-77/16

To determine the viscosity-lowering effect of each solvent on the crude oil, viscosities were determined for mixtures of 25% solvent and 75% oil at room temperature.

The literature on solvents shows that pine oil has a high kauri-butanol value, greater than 500. As this value is indicative of a solvent's power to dissolve organic matter, it was decided to see if small quantities of pine oil would accomplish the same low percentage of precipitates as did large amounts of aromatics in the solvent. Precipitate tests were run using commercial mix 8 with 5, 10, and 15% pine oil added, and on one mixture with 5% pine oil added to commercial mix 1. It was found that the pine oil content would have to be greater than 15% if mixed with commercial mix 8 before the precipitates would be lowered below 2%. Only 5% pine oil in commercial mix 1 was needed to prevent precipitation greater than 2%. However, the total cost of the mixture with even this small percentage of pine oil would exceed the cost of a higher aromatic solvent.

Both the aromatic and paraffinic solvents significantly lowered the viscosity of the oil. The aromatic solvents such as benzene, styrene, and the commercial mixes with higher than 70% aromatics gave low % precipitates and significant viscosity reduction. The solvent shown as commercial mix 10 in the table was selected for the field test. This solvent, supplied by Charter Chemicals as Espesol 1300, has a boiling range of 324° to 373°F, a gravity of 35° API, and a composition of 74.3% aromatics, 13.1% naphthenes, 12.3% paraffins, and 0.3% olefins.

Core-Solvent Testing

Fifteen experiments were conducted on cores from the lease using the solvent selected to determine the optimum operating conditions for the reservoir. Soak time, temperature, pressure, flow direction, influence of water, recycling solvent, and surface area exposed to solvent were investigated. The core experiments were classified as follows: (1) Solvent soak—The core was loosely placed in a core holder in which solvent flows around the core but was not driven through the core; (2) Solvent flood—The core was placed in a Hassler sleeve type core holder and solvent was driven through the core; (3) Solvent/water flood—Both water and solvent were driven through the core at various times; and (4) Recirculated solvent and water—Similar to solvent soak experiments, but solvent was recirculated around core.

Experimental procedures are detailed in the original report (BERC/RI-77/16).

In the solvent-soak experiments, an increase in the soaking time, the surface-to-volume ratio, and the temperature, resulted in an increase in the amount of oil recovered per pore volume produced. Soaking time appeared to be the major influence while the application of heat was not as important a factor in oil recovery as expected.

In the flow-through experiments, reversing the flow increased the rate of oil recovery, and in some cases higher recovery rates were obtained from the cores with lower oil saturations.

In the experiments where solvent and water were used, the amount of solvent required to produce the oil was much less than using solvent only, but the average recovery was lowered by about 10%. These experiments also showed that most of the solvent could be recovered by water injection.

PRODUCTION

Phase 1

The first drilling, shooting, and producing program, phase 1, involved 3 wells, part of a five-spot having a distance of 50 ft between outside wells.   All 3 wells were completed similarly.   Two strings of 1" pipe were installed through a dual-string tubing head.   One string was set at 330 ft, the other at 220 ft.   Well 9 was the injection well, and wells 7 and 8 were the producing wells.

The injection well was initially filled with solvent from a depth of 330 ft.   Subsequent solvent injection was through the 220 ft tubing.   Production was accomplished by directing compressed air down the annulus of the production well to clean out all liquids above the 220 ft level.   For the first 3 days, solvent injection was continuous.   After that a normal injection production day consisted of producing wells 7 and 8 in the morning, injecting solvent into well 9 for about 12 hr and again producing wells 7 and 8 at the end of the injection period.   This production phase was terminated on October 4, 1972.

The five-spot pattern was completed, and communication tests were made during the next few months.

The difficulties and costs encountered in cleaning out the first 3 wells after they were shot, compared with the cost of drilling a well to 360 ft, prompted the decision not to use explosives in the two additional corner wells needed to complete a five-spot pattern.   There was also a need to see whether fracture communication could be established between wells farther apart than 50 ft.   Consequently the two wells to be fractured were drilled 75 ft apart and diagonally from the 2 planned corner locations.   These are wells 13 and 14.

During the winter of 1972-73 a flash distillation unit was constructed to separate the produced oil from the solvent.   Oil and solvent mix are pumped through a 14 kW heater that heats the mixture to 360°F.   A control valve next to the 5 ft long, 8" diameter flash chamber reduces the pressure from 50 psi to atmospheric pressure.   Vaporized solvent is produced through the top of the flash chamber and condensed.   Oil containing some solvent is removed from the bottom of the flash chamber.   The distillation unit will process 18 bbl of solvent/oil mix in 24 hr.

Phase 2

Injection production operations, phase 2, were resumed on July 10, 1973, using wells 7, 8, 10 and 11 as injection wells and well 9 as the production well.   Wells 10 and 11 were completed with dual 1" tubing strings similar to the completion of the other 3 wells.   Solvent was injected down the 220 ft string, and well 9 was produced by compressing air down the annulus and displacing well fluids up the 220 ft string.   A normal injection production day consisted of producing well 9 in the morning, injecting solvent for about 3 hr, and producing well 9 at the end of the injection period.

Phase 3

In January, 1975, after a soak period of 29 weeks, phase 3 was begun.   Very little solvent was injected during this period.   Previously, production of fluid was limited to removing only the oil from the well cavity.   During phase 3, all fluids were removed from the center well; water production varied from about 60 to 95%. Fluid removal was accomplished with a sucker-rod pump installed on the production (center) well. Solvent-oil production remained about the same but the percent of oil in the oil-solvent mix increased to threefold.

Phase 4

Phase 4 (waterflood) was begun July 22, 1975, after the completion of 8 water-injection wells drilled along the perimeter of a circle having an approximate 75 ft radius from the center well of the five-spot pattern. The drill bit encountered solvent that had migrated outside the pattern in several of the wells. Water rates were determined and set to predetermined rates twice a week by use of turbine meters. Waterflooding was discontinued in November 1975, after more than 30 thousand bbl of water had been injected.

Production History

The production history is shown in Table 13.4. Soaking time, amount of daily solvent injection, and production techniques have been varied to establish the optimum ratio of oil produced to solvent injected. Soaking periods that have been tried include 1-, 2-, and 3-week periods; soaking only on weekends; and soaking on Tuesdays and Thursdays in addition to the weekends. The latter soaking period would probably give the highest percentage of oil in the solvent produced, but would give less total oil produced over a given period than soaking only on weekends. The average amount of daily solvent injection was 28, 10, 4.5 and 2.5 bbl/day during different periods. Daily production results indicate that the injection rate of 28 bbl/day caused overflushing of the fracture system between the injection wells and the producing wells and that an injection rate of 4.5 bbl/day did not displace all of the solvent in the formation each day.

The calculated fluid volume of the five-spot is 2,868 bbl with an oil content of 972 bbl. This is based on an interwell area of 0.0574 acre and includes the interval from 295 to 330 ft. Cumulative fluid produced from the zone is 2,538 bbl or 0.88 pore volume. Net crude oil production is 263 bbl, or 27% of the oil in place.

The results of the field experiments, even though less than 1 pore volume of oil-solvent mix was produced, showed more oil recovered per pore volume of solvent mixture produced than did the laboratory experiments. Intermittent injection of about 10 bbl of solvent every other day seems to be the most effective quantity to inject.

Reinjection of the produced solvent-oil mixture was done on a limited basis. Not enough additional oil was picked up during recycling of the solvent to justify this procedure.

During phase 3, the path for the flow of fluids was lowered to the depth of the pump setting. This helped remove some of the heavy oil from the lower fracture system, and the net result was an increase in the percent of crude oil in the oil-solvent mix.

The total net injection of solvent was 2,570 bbl, whereas the net solvent produced was only 2,199 bbl. The difference of 371 bbl (14% of the injected solvent) represents the amount of solvent lost to the formation.

Table 13.4: Injection and Production Data

Injection Period	. . . Solvent Injected. . . .		Solvent-Oil Mixture . . . . . .Produced.		. . .Crude Oil Produced . . .		Water Produced (bbl)	Oil in Mixture (%)	Oil in Injected Solvent (%)
	Barrels	Barrels per Day	Barrels	Barrels per Day	Barrels	Barrels per Day			
. . . . 1972–Phase 1									
Fill up	205.62	—	—	—	—	—	—	—	—
May 9–10	80.42	40.21	46.39	23.20	5.78	2.89	1.17	12.46	0
May 15–19	121.66	24.33	108.00	21.60	6.79	1.36	10.00	6.29	0
May 22–26	110.71	22.14	103.73	20.75	5.84	1.17	1.75	5.63	0
May 29–June 2 (soaked 3 weeks)	0	0	33.82	11.82	3.76	1.25	0.58	11.11	0
June 26–30	91.46	18.29	82.56	16.51	6.85	1.37	16.48	8.30	0
July 5–6 (soaked 3 weeks)	0	0	8.46	4.23	0.58	0.29	1.73	6.86	0
July 31–Aug 4	86.61	17.32	73.67	14.73	6.03	1.21	11.60	8.19	0
Sept 27–29[1]	0	0	70.60[2]	—	1.0[3]	—	ND	—	0
Oct 2–4[1]	0	0	149.90[2]	—	3.92[3]	—	ND	—	0
. . . . 1973–Phase 2									
Fill up	403.59	—	—	—	—	—	—	—	—
July 10–13	113.74	28.44	30.61	7.65	2.90	0.73	22.11	9.47	0.04
July 16–20	133.76	26.75	79.06	15.81	5.34	1.07	12.97	6.75	0.10
July 23–25 (soaked 2 weeks)	0	0	21.16	7.05	1.83	0.61	8.64	8.65	0
Aug 6–10	49.26	9.85	30.80	6.18	3.21	0.64	32.97	10.39	0.97
Aug 13–17 (soaked 2 weeks)	35.88	7.18	27.07	5.41	2.54	0.51	20.17	9.38	0.50
Sept 4–7	34.79	8.70	25.49	6.37	2.90	0.73	17.86	11.38	0.40
Sept 10–14 (soaked 3 weeks)	48.34	9.67	40.04	8.01	2.75	0.55	22.75	6.87	0.35
Oct 8–12 (soaked 3 weeks)	54.15	10.83	33.70	6.74	2.55	0.51	25.77	7.57	1.05

(continued)

Table 13.4: (continued)

	. . . Solvent Injected. . . .		Solvent-Oil Mixture . . . . . .Produced.		. . .Crude Oil Produced . . .		Water Produced	Oil in Mixture	Oil in Injected Solvent
Injection Period	Barrels	Barrels per Day	Barrels	Barrels per Day	Barrels	Barrels per Day	(bbl)	(%)	(%)
Nov 5–9	22.61	4.52	22.47	4.49	1.99	0.40	22.75	8.86	3.49
Nov 12–16[4]	12.36	2.37	15.42	3.08	1.48	0.30	25.06	9.60	1.78
(soaked 7 weeks)									
. . . . .1974.									
Jan 28–Feb 1	44.28	8.86	23.37	4.67	2.10	0.42	4.50	8.99	0.61
Feb 4–8[5]	55.25	11.05	21.33	4.27	1.58	0.32	7.55	7.41	0.58
Feb 11–15	50.46	10.09	41.22	8.24	2.75	0.55	0.15	6.67	0.46
Feb 18–22	48.01	9.68	37.59	7.52	2.36	0.47	0	6.28	0.23
Feb 25–Mar 1	47.49	9.50	38.31	7.66	2.04	0.41	0	5.32	0.06
Mar 4–8	45.48	9.10	36.86	7.37	2.07	0.41	0	5.62	0.18
Mar 11, 13, 15[6]	28.25	9.42	20.75	6.92	1.08	0.36	0	5.20	0.21
Mar 18, 20, 22	26.57	8.86	18.29	6.10	1.07	0.36	0	5.85	0.41
Mar 25, 27, 29[7]	28.16	9.39	22.71	7.57	1.41	0.47	0	6.21	13.24
April 1, 3, 5	33.82	11.27	20.46	6.82	1.66	0.55	0	8.11	8.26
April 8, 10, 12	31.64	10.55	24.38	8.13	1.93	0.64	0	7.92	6.23
April 15, 17, 19	31.20	10.40	22.20	7.40	1.66	0.55	0	7.48	6.59
April 22, 24, 26	31.06	10.35	22.06	7.35	1.67	0.56	0	7.57	6.92
April 29–May 3	32.65	10.88	21.48	7.16	1.64	0.55	0	7.62	6.89
May 6–10	34.10	11.37	25.11	8.37	1.90	0.63	0	7.56	7.33
May 13–17	30.04	10.01	21.62	7.21	1.73	0.58	0	7.99	7.62
May 20–24	29.90	9.97	19.16	6.39	1.68	0.56	0	8.79	7.66
May 27–31	26.85	8.95	16.40	5.47	1.51	0.50	0	9.20	7.49
June 3–7	30.04	10.01	18.58	6.19	1.62	0.54	0	8.74	7.92
June 10–14	40.20	13.40	27.86	9.29	2.32	0.77	0	8.33	7.99
June 17–21	30.77	10.26	23.80	7.93	2.20	0.73	0	9.23	8.38
June 24–28	35.12	11.71	25.69	8.56	2.28	0.76	0	8.87	8.60
(soaked 29 weeks)									

(continued)

Table 13.4: (continued)

Injection Period	. . . Solvent Injected . . .		Solvent-Oil Mixture . . . . . . Produced		. . . Crude Oil Produced . . .		Water Produced (bbl)	Oil in Mixture (%)	Oil in Injected Solvent (%)
	Barrels	Barrels per Day	Barrels	Barrels per Day	Barrels	Barrels per Day			
. . . . 1975—Phase 3									
Jan 20–24[8]	0	0	57.41	14.35	14.26	3.57	243.93	24.84	—
Jan 27–31	35.93	11.98	17.25	3.45	4.08	0.82	128.43	23.65	12.58
Feb 3–7	107.94	21.59	99.19	19.83	17.37	1.47	157.89	17.51	12.35
Feb 10–14	0	0	54.41	18.14	11.99	4.00	144.76	22.04	—
Feb 19–21	0	0	28.70	14.35	5.54	2.77	114.79	19.30	—
Feb 25–28	0	0	20.62	6.87	4.38	1.46	95.57	21.24	—
March 3–7	0	0	28.55	7.14	6.65	1.66	69.54	23.29	—
March 10–14	0	0	3.21	1.07	0.87	0.29	63.25	27.10	—
March 17–21	0	0	21.32	7.11	4.27	1.42	97.11	20.03	—
March 24–28	0	0	6.09	2.03	1.60	0.53	69.78	26.27	—
March 31–April 4	0	0	8.27	4.14	1.57	0.79	40.74	18.98	—
April 7–11	0	0	10.88	3.63	2.12	0.71	57.55	19.49	—
April 14–18	51.60	10.32	13.77	2.75	3.16	0.63	57.14	22.95	18.52
April 21–25	0	0	11.73	2.35	3.06	0.61	55.02	26.05	—
April 28–May 2	54.62	10.92	15.50	3.10	4.85	0.97	47.27	31.59	18.52
May 5[9]	0	0	15.89	15.89	4.49	4.49	9.93	28.25	—
July 21[10]	0	0	222.49	—	67.49	—	—	30.34	—
. . . . 1975–1976—Phase 4									
July 22–25[10]	0	0	2.32	0.58	1.22	0.31	0	52.59	—
July 26–Aug 1	0	0	1.30	0.18	0.60	0.09	2.05	46.15	—
Aug 2–8	0	0	5.94	0.85	2.17	0.31	4.58	36.53	—
Aug 9–15	0	0	2.32	0.33	1.01	0.14	5.27	43.53	—
Aug 15–22[11]	0	0	51.29	7.37	10.07	1.44	82.92	19.63	—
Aug 23–29	0	0	67.37	9.62	13.60	1.94	327.11	20.18	—
Aug 30–Sept 5	0	0	29.41	4.20	7.81	0.76	476.51	17.99	—
Sept 6–12	0	0	5.80	0.83	0.76	0.11	474.55	13.15	—

(continued)

Table 13.4: (continued)

Injection Period	. . . Solvent Injected. . . .		Solvent-Oil Mixture . . . . . .Produced.		. . .Crude Oil Produced . . .		Water Produced (bbl)	Oil in Mixture (%)	Oil in Injected Solvent (%)
	Barrels	Barrels per Day	Barrels	Barrels per Day	Barrels	Barrels per Day			
Sept 13–19	0	0	20.14	2.88	2.92	0.42	509.27	14.50	—
Sept 20–26	0	0	25.21	3.60	3.68	0.53	383.73	14.60	—
Sept 27–Oct 3	0	0	28.26	4.04	4.01	0.57	814.09	14.19	—
Oct 4–10	0	0	22.46	3.21	2.96	0.42	489.58	13.10	—
Oct 11–17	0	0	33.32	4.76	4.95	0.71	420.46	14.86	—
Oct 18–24	0	0	11.73	1.68	2.97	0.42	950.31	25.32	—
Oct 25–31	0	0	0.28	0.04	0.10	0.01	768.61	35.71	—
Nov 1–7	0	0	0.87	0.12	0.28	0.04	789.59	32.18	—
Nov 8–14[12]	0	0	1.59	0.23	0.61	0.09	582.94	40.67	—
Nov 15–Mar 9[13]	0	0	40.80	—	13.42	—	898.05	32.89	—
Total or average	2,646.39	—	2,537.87	—	339.19[14]	—	—	13.36	—

Note: ND is not determined.

[1] Emptied wells to prepare for next phase.
[2] Includes well bore, fill-up volume.
[3] Includes oil from well bore, fill-up volume.
[4] Injection into only one well each day for a 2-week period.
[5] Started flowing production well.
[6] Started injecting only 3 days per week.
[7] Started recycling solvent.
[8] Installed sucker-rod pump on center (production) well, January 6, 1975.
[9] Perforated center well from casing seat to top of formation, May 2.
[10] Waterflood begun, July 22.
[11] Cleared solvent from outside four wells, August 20.
[12] Water injection stopped, November 14.
[13] Clearing fluids from wells to March 9, 1976.
[14] Reinjected 76.63 bbl.

Source: BERC/RI-77/16

An economic analysis was not prepared, primarily because of the small size of the experimental test site. The rubblization of the formation in a normal well spacing was not considered feasible because of the present state of the art of fracturing with chemical explosives. In addition, the cost of the aromatic solvent has tripled since initiation of the solvent injection operations.

CONCLUSION

Use of explosives and solvents to stimulate shallow, low gravity, high viscosity oil reservoirs is technically feasible. Additional field testing using larger well spacing should be considered. In the final analysis, economics will govern the application of this technique.

REFERENCES

(1)   Ball Associates, Ltd., "Surface and Shallow Oil-Impregnated Rocks and Shallow Oil Fields in the United States", BuMines Monograph 12, IOCC, (1965).

(2)   Spencer, G.B., Eckard, W.E. and Johnson, F.S., *Domestic Tar Sands and Potential Recovery Methods—A Review,* IOCC Bull., 11:2, pp. 5-12; (December, 1969).

(3)   Phizackerley, P.H. and Scott, L.O., "The Major Tar Sand Deposits of the World", Proc. 7th World Petroleum Cong., Mexico, D.F., v. 3, pp. 551-571; (April 2-8, 1967).

(4)   U.S. Bureau of Mines, "Heavy Crude Oil Resource, Reserve, and Potential in the United States", IC 8352, 76 pp., (1967).

(5)   Dietzman, W.D., Garrales, M., Jr. and Jirik, C.J., "Heavy Crude Oil Reservoirs in the United States: A Survey", BuMines IC 8263, 53 pp., (1965).

CHEMICAL FLOODING—SOME PROJECTS AND RESEARCH

The information in this chapter is from:

> *Proceedings of the Fifth Annual DOE Symposium on Enhanced Oil & Gas Recovery & Improved Drilling Technology, Vol. 1—Oil,* CONF-790805-P1, sponsored by U.S. Department of Energy, Tulsa, Oklahoma, August 22–24, 1979.

The information in this section is based on the paper:

> "Commercial Scale Demonstration of the Maraflood Process: M-1 Project, Crawford County, Illinois (1978-1979)," presented by D.N. Burdge, of Marathon Oil Company.

MARAFLOOD OIL RECOVERY PROCESS

Marathon Oil Company, with Department of Energy financial assistance, is operating a commercial scale project of the Maraflood oil recovery process in the Main Consolidated (Robinson) Sand in Crawford County, Illinois. Reservoir parameters are summarized in Table 14.1.

The M-1 Project is a micellar-polymer flood of 407 acres in areal extent, including 248 acres of 2.5-acre five-spot patterns and 159 acres of 5.0-acre five-spot patterns. The project has a total of 114 injection wells and 132 production wells.

Phase II, now going on, covers the operation of the project from the start of micellar solution injection through mobility buffer and drive water injection into the 2.5-acre patterns. Micellar solution and most of the mobility buffer will have been injected into the 5.0-acre patterns when Phase II is completed in 1985. Approximately 3.3 million pounds of polymer will be utilized in the mobility buffer sequence, now being injected.

Table 14.1: Reservoir Parameters and Related Data

Parameter	. . . . Mean Value		
	2.5 Acre	5.0 Acre	Total Project
Net sand thickness, ft	29.3	25.5	27.8
Permeability, md			
Arithmetic mean	100.3	107.0	102.8
Geometric mean	75.5	79.3	76.9
Porosity, %	18.8	18.9	18.9
Area, acres	248	159	407
Reservoir pore volume, bbl	10,627,513	5,996,535	16,624,048
Total wells	181	75	256
Producers	91	41	132
Injectors			
Slug	85	29	114
Water	8	6	14
Water disposal	1	1	2

Source: CONF-790805-P1

FLUID SYSTEM DESIGN AND INJECTION

Micellar Solution

Micellar solution (slug) for the M-1 Project was manufactured by a Marathon
process which involves the direct sulfonation of crude oil. It became necessary
to reformulate the slug because of a chemical shortage.

Injection of the new slug formulation began on May 3, 1978 and proceeded
without incident until slug injection was completed simultaneously in both patterns on November 20, 1978. Both pattern spacings received 10% of a pore volume of slug (old and new formulations combined). The total slug injected was
1,663,125 barrels.

The old slug viscosity just prior to the introduction of the new slug was 27 cp.
The new slug viscosity was approximately 67 cp. To determine the effects the
higher viscosity slug would have on injection characteristics, the wells were not
adjusted for two weeks subsequent to injection of the new slug. During the initial two-week injection period, plant pressure increased from 600 psi to 740 psi
and the rate dropped from 1,522 bpd to 1,211 bpd.

Following rate stabilization, individual well rates were set to conform with a
slug injection schedule of 2,500 bpd. Two criteria were used in establishing
this schedule. Rates were desired that would: (1) assure that each well would
have as near 10% pore volume slug injected as possible, and (2) maximize the
total project's slug injection rate. Optimization work indicated that these two
goals could be readily achieved if slug injection into both patterns was completed
simultaneously. This equated to approximately 1,000 bpd in the 2.5-acre pattern and 1,500 bpd in the 5.0-acre pattern.

Mobility Buffer

The design of the mobility buffer sequence is given in Table 14.2. This design
has not been changed, the higher viscosity of the new slug formulation notwith-

standing. Laboratory core floods verified that the original mobility buffer design would be adequate for this new slug. Not only did the oil recoveries remain as high as with the old slug, suggesting no important mobility ratio problems, but the core flood pressure data revealed that the mobility relationships between the various fluid banks were quite similar to those observed with the original slug formulation.

Table 14.2: Mobility Buffer Design

Volume (% pv)	Polymer Concentration (ppm)	(lb/bbl)	Total (lb)	Completion Date* 2.5 Acre	5.0 Acre
11	1,156	0.4030	736,944	6/79	12/79
19	800	0.2789	880,925	4/80	8/81
32	625	0.2179	1,159,162	6/81	12/83
12	411	0.1433	285,867	12/81	10/84
11	200	0.0697	127,457	5/82	8/85
10	100	0.0349	58,018	8/82	3/86
10	50	0.0174	28,927	11/82	9/86
35	0	0	0	11/83	8/88

*Estimated.

Source: CONF-790805-P1

The design concentrations are based upon the use of Dow Chemical Company Pusher 700, a partially hydrolyzed polyacrylamide. At the beginning of mobility buffer injection an inventory of almost 1.4 million pounds (about 42% of the total requirement) of Pusher 700 had been accumulated.

Injection of the mobility buffer "spike" (1,156 ppm) was begun on November 21, 1978. Prior to the initiation of mobility buffer, all injection wells were equipped with a specified amount of coiled copper tubing calculated to control individual well rates at 3 bpdpf (barrels per day per foot of sand) at a plant pressure of 400 psig.

By the end of February, 1979 a plant pressure of about 670 psig was required to maintain an average rate of 2.9 bpdpf in the 2.5-acre spacing portion of the project, and 630 psig were required for an average rate of 2.7 bpdpf in the 5.0-acre portion.

As early as January, 1979 there was evidence that several of the injection wells had been formation pressure parted. Flow rate and pressure for those wells were, consequently, decreased.

On March 7, 1979, polymer and fresh polymer makeup water was found in the produced fluid of oil wells J-8 and J-12. Surrounding injection wells I-15 and K-7 are believed to have pressure parted and created a channel to these producers. As a result, injection wells I-15 and K-7 were temporarily shut in and the injection rate was decreased from 2.5 to 2.0 bpdpf for the 2.5-acre system to reduce the injection pressure.

Thirteen additional injection wells in the 2.5-acre spacing and two injection wells in the 5.0-acre spacing have also shown possible pressure parting. Pressure parting is usually indicated by a rapid rise in injection rate and a drop in injection pressure. Pressure reduction has proven to be a satisfactory technique for restoring the well to normal operating conditions.

The quality of the mobility buffer solutions is routinely monitored by Brookfield viscosity and screen factor (1). Of these, the screen factor property is considered to be the more important.

On-Site Polymerization

Five injection wells in the 2.5-acre spacing pattern (Q-3, Q-5, Q-7, O-3, and O-5) have been receiving mobility buffer prepared in pilot plant facilities for on-site polymer manufacture. Well over 20,000 pounds of partially hydrolyzed polyacrylamide have been successfully prepared in the pilot facility and injected into these five wells.

Injection Well Stimulation

Early in the project, the number of substandard injection wells was as high as 21. Attempts were made to backflow and clean out with air, but both methods were unsuccessful and hydraulic fracturing was chosen as the means of stimulation. This fracturing program, plus the gradual increase in injection rate by several wells has eliminated most substandard injection well behavior. A few wells, however, still fail to take fluids at the desired rate despite repeated attempts at stimulation. These are apparently located in an area of poor reservoir quality.

The total number of injection wells fractured before slug injection was 57. The total number of injectors which have been fractured since fluid injection was initiated is now 23.

Perimeter Water Injection Wells

Fourteen perimeter water injection wells have been activated. These wells are used to minimize off-pattern fluid migration and maintain reservoir pressure on the project perimeter. In order to monitor and regulate the individual water injection wells, a computer program was developed to calculate recommended rates for each well.

FLUID PRODUCTION

Oil Cuts

As noted in Table 14.3 the oil cut for the 2.5-acre pattern has shown no marked increase, although there has been a general trend upward. The oil cut of the 5.0-acre patterns shows no perceptible trend.

During March, 1979, 38 wells on the 2.5-acre pattern and 17 wells on the 5.0-acre pattern produced at or above a 5% oil cut. This compares to 34 and 16 wells respectively in February, 1979. In that month the 2.5-acre and 5.0-acre patterns were producing at oil cuts of 5.5% and 4.3% respectively.

It is possible that this is initial evidence of a production response in the 2.5-acre pattern. Some additional support for an impending response in this pattern is discussed under Performance Monitoring.

Table 14.3: M-1 Project Data Summary

Month	Average Injection Rate (bpd)	Cumulative Injection (% pv)	Injection/ Production Ratio	.. Injected Fluids .. Viscosity (cp)	Screen Factor	..........Produced Fluids.......... Oil Rate (bpd)	Oil Cut (%)	Sulfonate in Water* (%)	Chloride in Water* (ppm)
				2.5 Acre Spacing Pattern					
February, 1977	2,246	0.4	0.84	36.0	—	102	4.5	0.09	9,586
March, 1977	2,392	1.1	0.90	30.7	—	102	3.8	0.39	9,709
April, 1977	2,547	1.8	1.08	32.5	—	103	3.9	0.07	9,179
May, 1977	2,642	2.6	1.08	35.5	—	117	4.7	0.11	9,282
June, 1977	2,488	3.3	0.96	34.4	—	136	5.8	0.14	9,469
July, 1977	2,487	4.1	0.73	31.9	—	129	4.2	0.08	8,954
August, 1977	2,451	4.8	0.77	35.3	—	151	4.9	0.12	9,185
September, 1977	2,372	5.4	0.82	32.1	—	132	4.8	0.10	9,524
October, 1977	2,418	6.1	0.86	32.6	—	132	4.9	0.07	9,495
November, 1977	2,448	6.8	0.85	31.0	—	122	4.6	0.13	9,773
December, 1977	977	7.1	0.48	30.0	—	106	4.8	0.13	9,358
January, 1978	928	7.4	0.55	28.9	—	81	4.7	0.26	8,814
February, 1978	912	7.6	0.54	28.0	—	80	4.9	0.17	9,550
March, 1978	791	7.9	0.60	27.0	—	66	5.2	0.12	9,488
April, 1978	730	8.1	0.56	26.7	—	62	5.0	0.22	10,033
May, 1978	825	8.3	0.68	59.3**	—	58	6.8	0.21	9,604
June, 1978	994	8.6	1.18	69.6	—	65	6.9	0.24	9,534
July, 1978	1,030	8.9	1.19	77.1	—	65	7.5	0.25	9,818
August, 1978	1,017	9.2	0.88	79.4	—	67	5.7	0.29	9,764
September, 1978	1,084	9.5	1.01	78.7	—	67	5.9	0.31	10,277
October, 1978	1,114	9.8	0.97	71.4	—	66	6.0	0.41	10,185
November, 1978	3,162	10.7	2.63	54.1***	22.9	80	7.1	0.32	9,452
December, 1978	8,247	13.1	2.08	52.3	28.4	206	5.6	0.47	10,191
January, 1979	7,480	15.3	1.27	51.1	29.6	275	5.0	0.54	9,589
February, 1979	7,436	17.0	1.19	52.0	28.7	271	4.8	0.49	9,445
March, 1979	5,454	18.6	0.90	52.5	29.9	306	5.5	0.92	9,072

(continued)

Table 14.3: (continued)

Month	Average Injection Rate (bpd)	Cumulative Injection (% pv)	Injection/ Production Ratio	.. Injected Fluids .. Viscosity (cp)	Screen Factor	Oil Rate (bpd)	Oil Cut (%)	Sulfonate in Water* (%)	Chloride in Water* (ppm)
				.5.0 Acre Spacing Pattern					
February, 1977	675	0.2	0.90	36.0	—	44	6.6	0.15	8,319
March, 1977	822	0.6	0.87	30.7	—	44	3.5	0.33	9,776
April, 1977	826	1.0	0.78	32.5	—	60	3.7	0.24	9,920
May, 1977	818	1.5	0.65	35.5	—	79	4.2	0.28	9,918
June, 1977	738	1.8	0.64	34.4	—	68	4.0	0.26	8,551
July, 1977	709	2.2	0.67	31.9	—	65	4.1	0.20	9,221
August, 1977	666	2.5	0.85	35.3	—	65	5.0	0.18	9,560
September, 1977	652	2.9	0.88	32.1	—	67	5.1	0.15	9,132
October, 1977	670	3.2	0.89	32.6	—	64	4.7	0.12	10,556
November, 1977	647	3.5	0.86	31.0	—	62	4.3	0.19	8,089
December, 1977	622	3.9	0.69	30.0	—	74	4.9	0.11	9,517
January, 1978	627	4.2	0.79	28.9	—	62	4.4	0.21	9,362
February, 1978	670	4.5	0.74	28.0	—	59	4.3	0.18	9,503
March, 1978	810	4.9	0.85	27.0	—	63	4.8	0.15	10,260
April, 1978	862	5.4	0.92	26.7	—	68	5.5	0.14	11,115
May, 1978	823	5.8	1.31	59.3**	—	56	5.5	0.20	10,177
June, 1978	1,302	6.4	1.26	69.9	—	63	5.2	0.21	12,675
July, 1978	1,382	7.1	1.26	77.1	—	62	5.1	0.23	10,269
August, 1978	1,402	7.9	1.16	79.4	—	57	4.2	0.27	9,885
September, 1978	1,599	8.7	1.15	78.7	—	63	4.1	0.24	10,537
October, 1978	1,642	9.5	1.13	71.4	—	67	4.0	0.37	10,336
November, 1978	1,610	10.3	1.09	49.8***	22.9	82	4.5	0.26	9,995
December, 1978	1,978	11.3	1.37	51.7	27.2	115	5.6	0.32	9,881
January, 1979	2,007	12.4	1.04	51.3	28.2	129	4.7	0.62	10,111
February, 1979	2,024	13.2	1.11	51.8	26.7	119	4.2	0.64	10,080
March, 1979	1,927	14.2	0.99	52.5	28.8	117	4.3	0.90	9,982

*Average weighted by water production per well.
**Began injection of new slug formulation.
***Began injection of mobility buffer spike.

Source: CONF-790805-P1

Production Well Stimulation

Eighty-one oil wells have been fracture-stimulated over the life of the project. These wells produced an average 18.5 bpd total fluids prior to fracture treatment and 62.6 bpd after treatment for an average gain of 44.1 bpd.

Produced Fluid Treatment

As of March, 1979 the treatment of produced fluids to obtain pipeline quality oil remained relatively straightforward. Chemicals (Tretolite RP-578), heat, and retention were being used.

A more difficult treating problem is anticipated as sulfonate levels in the produced fluid increase. Toward this end, a trailer-mounted pilot treating unit is being used to screen treating techniques with produced fluid from more mature wells. For example, the pilot unit is presently located at oil well L-12. This well has recently shown a sulfonate-in-water content considerably above the average and its oil cut averaged 13.4% in March, 1979. As such, its produced fluid represents what will likely be experienced at some time in the future in the main treating facility.

The M-1 Project treating facility is being reconstructed to conform to a design developed during Marathon's 219-R Project (2). The use of a combination of chemical treatments, heat, and retention time is anticipated. The proper treating chemicals and operating modes for this new facility will be at least partially anticipated on a continuing basis by information derived from the pilot equipment described above.

Based upon previous experience, it is believed important to monitor treating oil quality by some technique which can be correlated with the oil's satisfactory performance in a refinery desalter. Common saleable oil analyses may be insufficient or even misleading. Also, any method of measuring the water content of the oil should be calibrated against an "absolute" analysis, such as the Karl Fisher method.

PERFORMANCE MONITORING

Sulfonate in Water Analysis

Beginning in March, 1977 produced water samples have been analyzed monthly to determine the relative concentration of sulfonate in water. Because of the complex nature of the sulfonate components, the fact that they are partitioned between oil and water, and because selective adsorption is probably occurring in the reservoir, the analytical method is not considered to be quantitative. Past experience has shown, however, that it is useful for observing trends.

Experience in the 219-R has shown that a significant increase in sulfonate in water is usually associated with an increase in oil cut. In March, 1979 there were 31 wells in the 2.5-acre pattern which showed sulfonate levels higher than the normal limit of low-level fluctuation (0.5%). This compared to 26 wells in February and 19 in January. As would be expected the weighted average of sulfonate production also increased. This latter trend is shown in Table 14.3.

As noted earlier, there has been only a slight increase in oil cut in the 2.5-acre pattern as of March, 1979. The sulfonate in water behavior for this pattern is what might be expected to precede a pattern-wide production response.

Four wells in the 5.0-acre pattern showed sulfonate levels higher than 0.5% in March, 1979. This compared to 6 wells in February and 3 in January. No significant trend is yet apparent in this larger spacing pattern.

Tritiated Water Analysis

A tritiated water tracer was injected into ten selected wells prior to slug injection. Produced water samples from every oil well are monitored monthly for the presence of this tracer.

Some 25 wells have shown a tracer content higher than background at least once, although only ten have produced tracer for six months or more. Each of these ten directly off-set a tracer injection well. Except to note that tracer has appeared sooner than theoretically predicted, no conclusions have yet been drawn from the data.

Chloride Ion in Water Analysis

Beginning in March, 1977 produced water samples having been analyzed monthly for chloride ion. Theory and past experience indicate that chloride ion concentration will decrease when portions of the fresher mobility buffer solution water are produced.

Examination of the weighted average chloride analyses in Table 14.3 shows no significant trends through March, 1979. Considering that polymer injection was only started in late November, 1978, it is probably too early to expect a chloride ion concentration decrease.

ECONOMIC MONITORING

Development Costs

The development costs for the M-1 Project, most of which were incurred prior to 1977, will total about $8.3 million. Almost 70% of all development costs are attributable to well costs. It is possible to estimate a breakdown of development costs into unit costs (i.e., costs/10 acres) for the 2.5- and 5.0-acre spacing patterns as follows: $236,756 for the 2.5-acre pattern, and $151,834 for the 5.0-acre pattern.

Slug Costs

The average micellar solution (slug) cost for the M-1 Project was $6.17/bbl. During the 22 months of slug manufacture, two different slug formulations were prepared, manufacturing rates had to be varied widely, two different methods of insuring good slug filterability were used, and other minor variations were imposed by seasonal temperature changes.

The average manufacturing rate was about one-half the design capacity of the plant. Analysis of the slug cost data shows that if the new slug formulation had been manufactured at the design plant capacity of 150,000 barrels/month, that the cost would have been about $5.53/barrel of slug.

Polymer Costs

It is estimated that the total project polymer cost (including freight and sales tax) will average $1.28/lb. This cost should not be construed as an indication of current polyacrylamide prices, since over 40% of the M-1 Project's polymer requirement was purchased in 1975-1977.

Operating Costs

Based upon the first two years of experience, an estimate of operating costs can be made. These are $329/injection well-month for 2.5-acre spacing and $444/injection well-month for 5.0-acre spacing. These estimates, unlike the numbers given above, are expressed in 1979 dollars.

The information in this section is based on:

> "Effect of Enhanced Oil Recovery Chemicals on Crude
> Refining," presented by T.P. Malloy and I. Brubaker, of
> UOP, Inc.

REFINABILITY OF CRUDE OIL PRODUCED BY EOR

UOP, Inc. is conducting a Department-of-Energy-supported study to evaluate the refining processability of crude oil produced in the presence of enhanced oil recovery chemicals. The objective of this program is to determine whether the presence of EOR chemicals in the produced crude oil affects the refining operations to which crude oil is normally subjected. The major component of a chemical-micellar flood and also the material predicted to produce the most problems in any refining operation is the surfactant, typically a petroleum sulfonate.

It is the purpose of this work to identify the level of surfactant contamination in a crude oil and to determine the extent of contamination in distillation fractions derived from atmospheric and vacuum distillation of EOR-produced crude oil.

Initial studies have, therefore, centered on preparing and collecting the various boiling point fractions normally obtained when crude oil is fractionated and examining those fractions for surfactant. The fractions found to contain surfactant will then be processed in a manner consistent with domestic refinery practice.

DETERMINING SURFACTANT CONCENTRATIONS IN EOR CRUDE OILS

The chemicals used in micellar flooding and the virgin crude oil were obtained from three DOE-supported field studies. Also received were an EOR crude oil from a producing well and an EOR crude oil from an observation well. The producing companies, together with a brief description of the crude oils, can be found in Table 14.4. The corresponding surfactant packages for the micellar floods are described in Table 14.5. The EOR-produced crude oils were analyzed for surfactant content by standard methods and results were verified through discussions with analytical staffs at the producing companies. A brief description of the general methods of analysis for the surfactants used in the micellar floods and present in the EOR or surfactant-doped crude oil is given in Table 14.6.

Table 14.4: Crude Oils Received

Company	Crude Oil	Water (%)	S* (%)
Phillips	North Burbank Unit "Virgin" crude oil (no methylene blue actives detected)	<0.1	0.21
Phillips	North Burbank Unit Tract 97 produced (EOR) crude oil (240 ppm methylene blue actives assuming eq wt = 410).	–	–
Cities Service	West cell tank "Virgin" crude oil (46 ppm methylene blue actives assuming eq wt = 430).	<0.1	0.26
Cities Service	EOR crude oil from observation well No. 131 north underground tank (3.7×10^4 ppm methylene blue actives assuming eq wt = 430).	14.1	0.48
Marathon	J.C. Carlton Battery "Virgin" crude oil (no $-SO_3NH_4$ titratables detected)	<0.1	0.26 0.27

*UOP Method 380-74 ("Leco").

Source: CONF-790805-P1

Table 14.5: Surfactants Received

Company	Surfactant	Sulfonate (wt % active)
Phillips	Phillips supplied (average eq wt = 410)	53
Cities Service	Shell surfactant package	12
	two sodium alkylaryl sulfonates	
	(eq wt 250–550; average eq wt = 430)	
	C_{12-15} alcohol (3 EO) ethoxysulfate	
	sodium salt	
	secondary butyl alcohol	
	polysaccharide	
	sodium chloride	
	water	
	Solvent stripped	77
Cities Service	Union surfactant package	–
	four sodium alkylaryl sulfonates	
	(eq wt 250–650; average eq wt = 425)	
	ethylene glycol monobutyl ether	
	crude oil	
	water	
	Solvent stripped	10
Marathon	RSO_3NH_4 (average eq wt = 475)	1.5
	oil	
	ammonium sulfate	
	Alfonic 610–50 (C_{6-10}) linear alcohol	
	(50 wt % ethoxylate)	
	Solvent stripped	64

Source: CONF-790805-P1

Table 14.6: Analysis of Surfactants

Surfactant Types	Method *
Sodium salts of alkyl aryl sulfonates or C_{12}–C_{16} alcohol ethoxysulfates	Colorimetric based on the Epton titration **; sulfonate-methylene blue complex absorbance measured at 650 nm, following extraction of the methylene blue into chloroform. Sodium lauryl sulfate used for standard calibration curve.
Ammonium sulfonates	Nonaqueous potentiometric titration using tetra-n-butylammonium hydroxide as titrant (method supplied by Marathon).

*Crude deoiled, if necessary, using procedure based on ASTM D2548-69, silica gel liquid chromatography. Sulfonates recovered by elution with isopropanol, ethanol, methanol and water.

**S.R. Epton, *Trans. Faraday Soc., 44,* 226-230 (1948).

Source: CONF-790805-P1

The surfactant concentrations in the two EOR crude oils, as can be seen from Table 14.4, differ by about three orders of magnitude (e.g., 46 ppm for the Phillips EOR crude vs 3.7×10^4 ppm for the Cities Service EOR crude produced at an observation well).

It is expected, based on EOR crude oil production data, that during the course of a chemical flood the concentration of a surfactant in the produced oil will vary and probably go through a maximum and then fall to a very low level or, in the case of a fracture, the surfactant may break through and for short periods of time be produced in high concentrations. It is, therefore, difficult to assign an average concentration of surfactant in the to-be-produced crude oil, but for this purpose a so-called worst case estimate of 10,000 ppm (1.0 wt %) has been assumed. Because the level of surfactant in the crude oils obtained to date was not always satisfactory, it was decided to prepare synthetic EOR blends using "virgin" crude oils and the corresponding surfactant supplied by the producers.

The surfactant solutions, which contained appreciable quantities of water, were stripped of low-boiling solvents prior to blending with the virgin crudes. Four blends were compounded at approximately the 1% surfactant level (Table 14.7).

In the case of the Marathon blend using the ammonium sulfonate supplied, it was observed that the resulting solution contained only 0.48 wt % ammonium sulfonates. Some of the sulfonates, even though uniformly dispersed following vigorous shaking of the desired blend mixture, separated out from the crude upon standing at room temperature.

These four blends plus the crude oil from Cities Service produced at an observation well (3.7 wt % surfactant) were then subjected to a Hempel atmospheric and vacuum distillation. The Hempel distillation simulates on a small scale (~1,700 g crude oil charged) an actual commercial-scale refinery fractionation. Six different boiling fractions were obtained from this distillation (Figure 14.1).

Table 14.7: Sulfonate Content of Crude Oils*

Crude	Sulfonate Added	Sulfonate Concentration (wt %)
Phillips North Burbank Unit "Virgin"	Phillips supplied	1.13**
Cities Service West Cell Tank "Virgin"	Union (stripped)	1.0**
Cities Service West Cell Tank "Virgin"	Shell (stripped)	0.94**
Cities Service Observation Well, No. 131 North Underground Tank	—	3.7**
Marathon J.C. Carlton Battery	Marathon (stripped)	0.48***

*Evaluated by Hempel Distillation as per UOP Methods 77-59 and 109-75T.
**As methylene blue actives.
***As $-SO_3NH_4$ titratables.

Source: CONF-790805-P1

Figure 14.1: Typical Refinery Fractionation

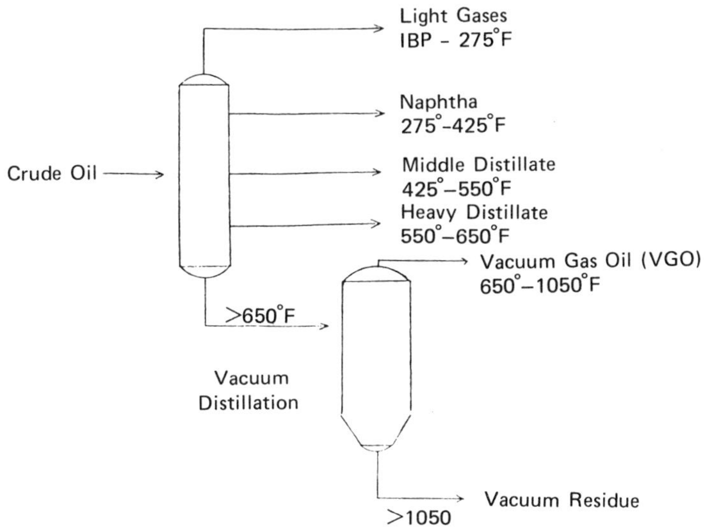

Source: CONF-790805-P1

The six individual fractions for each of the five different crude oil blends were weighed, the specific gravity and API gravity measured, and the percent sulfur and the sulfonate level as methylene blue actives or $-SO_3NH_4$ titratables determined.

These data show that in all the cases studied the surfactant does not distill over but rather remains in the vacuum bottoms. No greater than 100 ppm methylene blue actives were found in any of the fractions and that finding is probably related to some bumping resulting in mechanical carry-over during distillation.

The sulfonate recoveries are excellent (near quantitative) except for the Cities Service crude oil spiked with the Shell surfactant, in which case only 76% of the methylene blue active material was recovered. The discrepancy in this case is probably due to thermal decomposition of the $C_{12}-C_{15}$ alcohol 3(EO) ethoxysulfate component of the micellar formulation. (Paper A-3 presented at the Third ERDA Symposium on Enhanced Oil & Gas Recovery & Improved Drilling Methods, Aug. 30-Sept. 1, 1977 by Cities Service indicated the ethoxysulfate represented about 20% by weight of active surfactant.) Consistent with this observation is the high level of sulfur found in the IBP-275°F boiling fraction. The decomposition products have not yet been identified, but it is clear that there are no methylene blue actives present.

The Hempel distillations proceeded normally in all cases except for the observation well crude oil produced by Cities Service, which contained 14.1% water (by Karl Fischer) and caused some degree of bumping during the early stages of distillation. After this water was removed, the distillation proceeded smoothly. Foaming due to surfactant was also not a problem during the course of the distillations.

The information in this section is based on:

>"Characterization of Alkaline Sensitive Fraction of California Crudes," presented by T.F. Yen, R.J. Hwang, M. Chan and P.-F. Lin, of University of Southern California.

FEATURES AND MECHANISM OF ALKALINE FLOODING

Alkaline water flooding for enhanced oil recovery has been actively pursued by the oil industry and tested in selected oil reservoirs. In addition to its low cost the important feature of alkaline flooding is its simplicity in operation compared with other chemical floodings such as surfactant flooding and polymer flooding. However, alkaline flooding is considered to be a complex process as far as the operative mechanisms (6) are concerned. The large body of literature indicates disagreements and conflicts both in mechanisms and experimental results. One of the major factors behind these differences is that examination of the chemical composition of the oil phase was often neglected.

Knowledge of the identities and properties of petroleum constituents which have a high degree of interfacial activity under alkaline environment is a requisite to the understanding of the factors that may affect or control the success of flooding.

It is believed that the surfactant is formed in situ and low interfacial tension (IFT) results as alkaline solution is introduced to the reservoir for oil recovery. The lowering of the interfacial tension of the aqueous alkaline-oil system is attributed to the presence of interfacial active substances, such as carboxylic acids, phenols, or porphyrins (6)(7)(8).

Some investigators have suggested that high acid number of a given crude either by nature or due to the addition of known acid would lower its interfacial tension and would be beneficial to oil recovery. However, Cooke, Williams and Koldzie (9) found that although in situ oxidation with air further increases acid number of a given crude this artificially high-acid-number crude could not be successfully flooded with alkaline water. Apparently, acid number of crude oil is not a good index for selecting candidate crude oils for alkaline flooding and the response to alkaline flooding does not depend on a single compound but on the collective properties of a number of compounds.

Fractionation of crude oil and examination of the role of petroleum fractions during alkaline flooding is believed to allow one to eventually identify the active components and to realize its mechanism. Thus, this research effort has the following objectives:

(a) Examination of the major fractions of the crudes and characterization of their roles in alkaline flooding;

(b) Optimization of combination of petroleum fractions for the lowest interfacial tension;

(c) Control of the stability of emulsion; and

(d) Formulation of a scientific basis for screening candidate crude oils for alkaline flooding.

TESTS OF CALIFORNIA CRUDES IN ALKALINE SOLUTION

Two California crudes from Huntington Beach field, and Long Beach field, have been fractionated by vacuum distillation followed by solvent fractionation in obtaining volatile cuts and nonvolatile fractions. The ratio of volatiles to nonvolatiles is 1:2 for Huntington Beach oil and 1:2.7 for Long Beach oil under conditions of 210°C pot temperature and 0.01 torr pressure. The major fraction of each oil, isolated from nonvolatiles, is oil-resin (pentane soluble), which accounts for 52 and 68 wt % of Huntington Beach and Long Beach crudes respectively. The heavy volatile cut (~4 wt % of total crude) and oil-resin of each crude were subjected to interfacial tension and bulk viscosity studies under various concentrations of alkaline solution.

Interfacial Tension Studies

Since the lowering of interfacial tension is generally considered to be a necessary requirement for enhanced oil recovery, it is used to study the effect of alkali on petroleum samples as part of the overall work to study alkaline flooding. A spinning drop tensiometer was used in the experiments. Alkaline concentrations of 125, 250, 500, 1,500 and 3,000 ppm sodium orthosilicate, were used for the aqueous phase, which contains 7,500 ppm NaCl in all cases. Petroleum samples studied were the heavy volatile fraction, the oil-resin fraction of the nonvolatiles (note that both are defined by the above distillation and fractionation scheme) and the crude from each of the two California oil fields. For each sample, the lowest IFT reading in a time study was adopted. It usually occurred in a few minutes.

Results showed that under alkaline conditions the heavy volatile fraction of both petroleums did not exhibit any interfacial activity, i.e., did not show any noticeable drop in IFT. Both the crudes and the oil-resin fractions, however, exhibit tremendous drops in interfacial tension at sodium orthosilicate concentrations of 250 ppm and above (Figure 14.2a and Figure 14.2b). The minima in all cases occurred at 250 ppm, with IFT readings remaining low (compared to those at zero ppm alkali) for samples up through the highest alkaline concentration tested, i.e., 3,000 ppm. This observation suggested that for these two California crudes, a water alkaline concentration of at least 250 ppm sodium orthosilicate is required to give significant interfacial activity.

Figure 14.2: Effect of Sodium Orthosilicate on Interfacial Tension

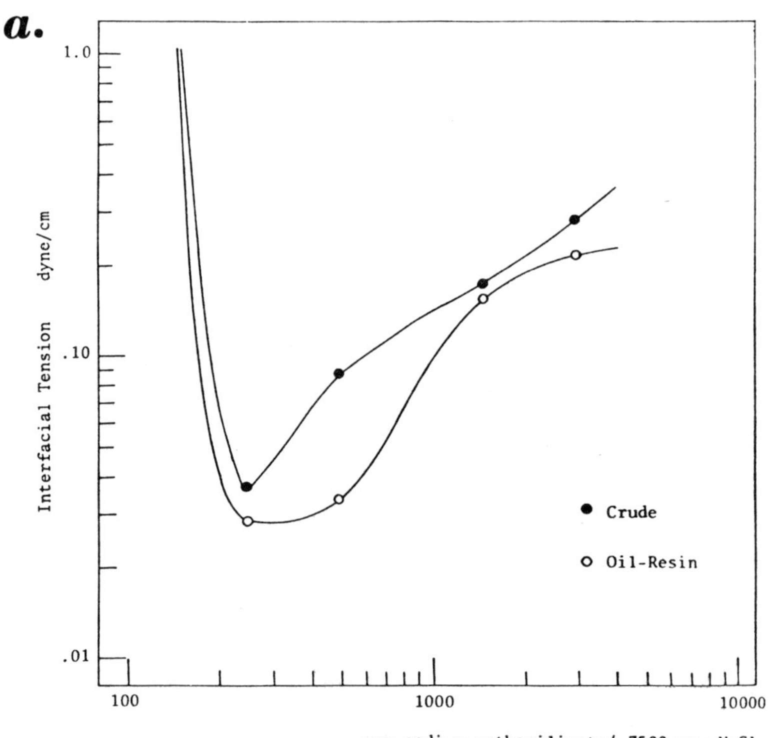

(continued)

Figure 14.2: (continued)

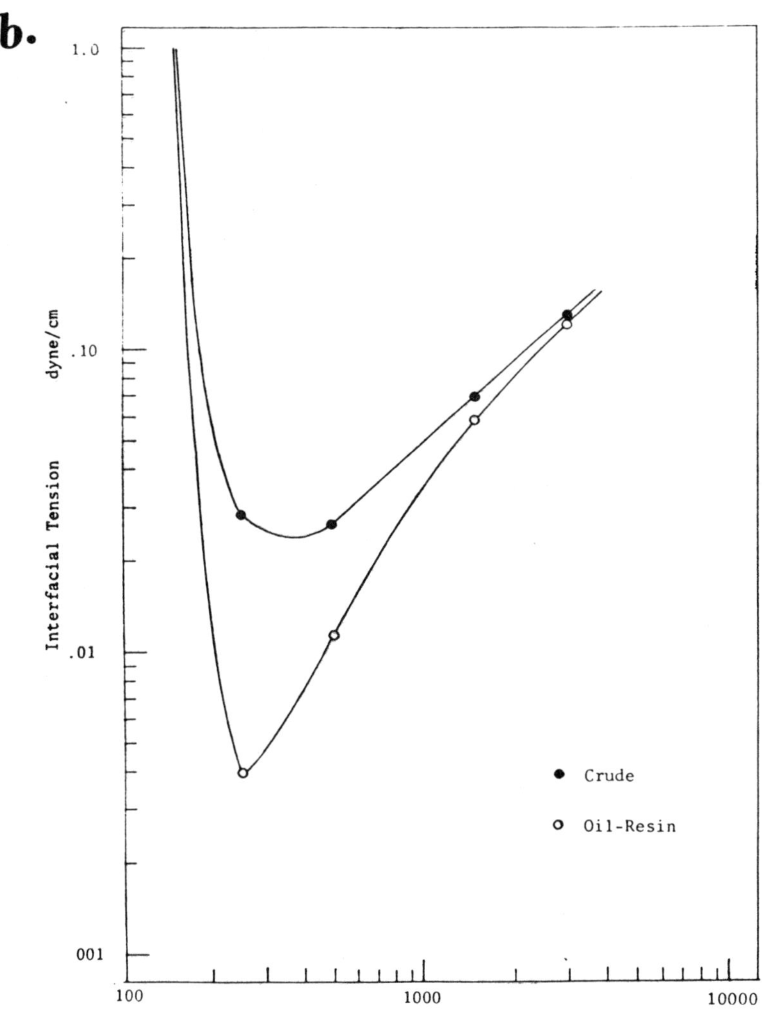

(a)   Huntington Beach (Aminoil) oil-resin
      fraction and crude
(b)   Long Beach oil-resin fraction and
      crude

Source: CONF-790805-P1

The above qualitative similarity between the IFT response of the oil-resin fraction and that of the crude suggested that the oil-resin fraction is responsible for the crude's IFT response to alkali. Also quantitatively, the greater IFT drop observed for the oil-resin than for the crude for both petroleums (especially significant in the case of Long Beach petroleum which showed a sevenfold difference between the two IFT minima), was in agreement with the analysis that the oil-resin fraction contains most of the active components which are responsible for the interfacial activity because one definitely expects a greater interfacial activity, i.e., lower IFT reading, with a sample containing a higher concentration of active components.

It is interesting to note that the IFT values drop to a minimum at 250 ppm sodium orthosilicate, and then increase gradually with increasing alkaline concentration. This is in agreement with the proposition (described under Structural Characterization) that the active ingredient of the crude cannot be a single type of functional group such as the often-proposed carboxylic group because the latter will remain in ionized form once the pH of the environment is above its pK value. A further increase in pH, therefore, could not possibly reduce the activity of the molecule. Thus, the observed reduction of activity, i.e., the increase of IFT values, at higher alkaline concentration supports the proposition that the active components are multifunctional.

A possible example is amino acid which has, on one molecule, a carboxylic group as well as an amino group. As the pH reaches a medium value, e.g., that corresponding to 250 ppm sodium orthosilicate, the amino group is still charged ($\equiv NH^+$). The pK value of the carboxylic group, however, is reached and the latter becomes charged; somehow that leads to a higher interfacial activity. As alkaline concentration increases further, the pH reaches the pK values of some of the amino groups, which lose their charge and become neutral ($\equiv N$), somehow leading to a reduction in the originally increased interfacial activity. Sequential reactions like these may well explain the observed IFT relationship with alkalinity.

Viscosity Studies

Viscosity is an important parameter for fluids because of its significant effect on the flow property of fluids. The change in viscosity of a fluid under a given circumstance reflects the changes of its intermolecular forces. To provide information on the change of molecular interaction in the sample caused by the reaction between the active ingredients of petroleum and the alkaline solution, the effect of alkaline solution on viscosity of the oil-resin fraction has been studied.

Samples for viscosity measurements were prepared by mixing various concentrations of NaOH aqueous solutions in the presence of 7,500 ppm NaCl with the oil-resin fractions in a 1:1 ratio by weight. The mixtures were vigorously shaken, and the two phases were then separated. Infrared (IR) and proton nuclear magnetic resonance (NMR) spectra of the treated oil-resins demonstrated that the water content was less than 1%. Bulk viscosities of the treated oil-resin samples from the two California crudes were measured by a Wells-Brookfield microviscometer, and results are shown in Figure 14.3.

At alkaline concentrations lower than 300 ppm, little effect on viscosities of the two oil-resins was observed. Viscosities of both oil-resins started to increase, however, as alkaline concentrations went above 300 ppm.

**Figure 14.3: Effect of Concentration of Alkaline Solution on Viscosity
of Oil-Resin Fractions***

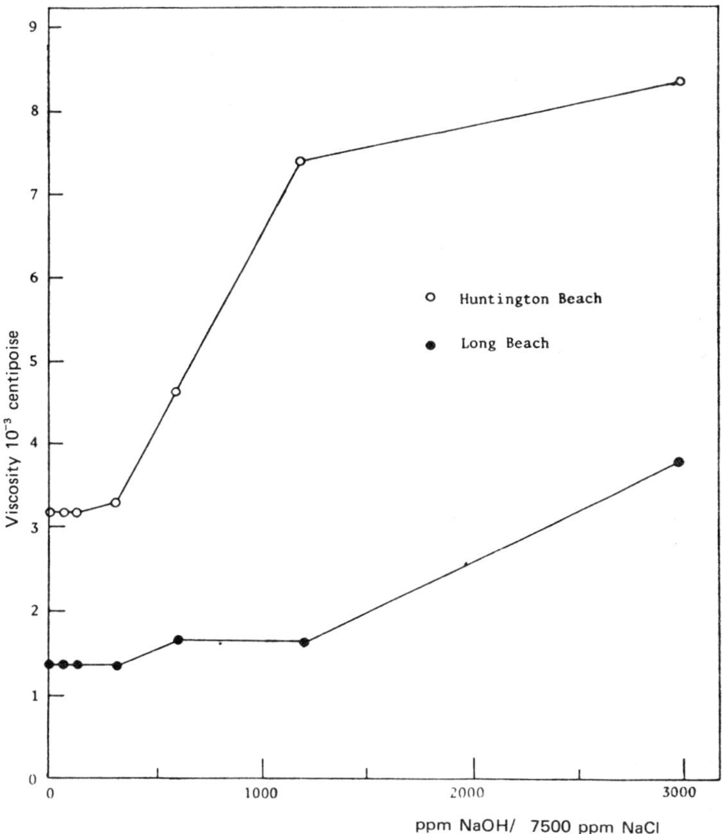

*Shear rate: 19.2 sec^{-1}. Temperature: 25°C

Source: CONF-790805-P1

The Huntington Beach sample, which has a higher viscosity than the Long Beach
sample at zero alkali environment, showed a greater increase in viscosity with
alkaline concentration than the latter. At 3,000 ppm NaOH, both viscosities
reached values almost three times those at zero alkali concentration. This increase in viscosity of oil-resin after treatment by alkaline solution may be due
to reactions between the caustic and the active ingredients of the petroleum,
such as carboxylic acids, thiols and phenols, leading to the formation of salts
of these compounds. As a result, charged species would be produced which increased interaction among petroleum molecules, thus leading to the observed
higher viscosity of the sample.

Structural Characterization

Spectroscopic techniques were used to study the average structure of petroleum fractions. Since each petroleum fraction is a very complex mixture, thorough separation and identification of each compound are not practical. Instead, structural feature and group-type analysis are the main aims of the spectroscopic study. From proton NMR data, one can derive three important structural parameters of the petroleum fractions: aromatic content (H_{Ar}), polar content (H_a) and saturated portion (H_o). While H_{Ar} and H_o are the mol fractions of protons on aromatic rings and saturated carbons respectively, H_a is the mol fraction of protons on the carbons adjacent to polar groups such as carboxyl, keto, aldehyde, ether, amino, imino, thiol, sulfide, aryl and olefin.

For both petroleum samples from Huntington Beach and Long Beach fields, aromatic and polar contents were higher in the heavy, nonvolatile fractions than in the light, volatile fractions. The opposite is true for saturated component content. NMR data of the heavy volatile cut and the oil-resin fraction of the two California crudes are given in Table 14.8.

Table 14.8: Nuclear Magnetic Resonance Data for Petroleum Fractions

Source	Fraction	H_{Ar}	H_a	H_o
Huntington Beach	heavy volatile cut	0.046	0.118	0.836
Huntington Beach	oil-resin	0.050	0.137	0.813
Long Beach	heavy volatile cut	0.033	0.149	0.818
Long Beach	oil-resin	0.050	0.158	0.792

Source: CONF-790805-P1

It is interesting to note that for the Huntington Beach samples, the H_a values of heavy volatile cut and oil-resin fraction showed noticeable difference, while the Long Beach samples showed difference primarily in the H_{Ar} values. Since IFT studies showed the oil-resin fractions of both petroleums have strong interfacial activity, the above NMR results suggested that both the aromatic and polar contents are important in determining the interfacial activity of the sample. This implication agrees with the general observation (9) that compounds, especially aromatics, containing oxygen, sulfur or nitrogen exhibit interfacial activity at the oil-water interface due to the polar nature of their hetero atoms and polarizability of aromatic rings.

Characteristics in the infrared spectra of these two petroleum fractions are shown in Table 14.9. The aromatic content of each fraction was verified by the absorptions at 860, 800 and 760 cm^{-1}. The absorption at 730 cm^{-1} is $(CH_2)_n$ rocking. The carbonyl absorption at 1,710 cm^{-1} and C–O absorption at 1,030 cm^{-1} are attributed to the presence of carboxylic acids or esters.

The oil-resin fraction has additional absorptions at 3,610 and 3,460 cm^{-1} caused by O–H stretching and amino N–H stretching respectively, indicating the presence of phenolic or alcoholic compounds and nitrogen-containing compounds such as pyrroles, carbazoles or other amines. Accordingly, combined properties of carboxylic acids, phenols and nitrogen-containing compounds, i.e., multifunctional groups, are probably responsible for the strong interfacial activity displayed by the oil-resin fractions.

Table 14.9: Characteristic Infrared Absorption Spectra of Petroleum Fractions

Frequency (cm^{-1})	Functional Group	. . .Huntington Beach . . .		Long Beach	
		Heavy Volatile Cut	Oil-Resin	Heavy Volatile Cut	Oil-Resin
3,610	$-O-H$	−	+	−	−
3,460	$\diagdown N-H$	−	+	−	+
1,710	$\diagup C=O$	+	+	+	+
1,600	$\diagup C=C\diagdown$	+	+	+	+
1,030	$-\overset{\mid}{\underset{\mid}{C}}-O-$	+	+	+	+
860 ⎫	aromatic				
800 ⎬	C–H	+	+	+	+
760 ⎭	bending				
730	$(CH_2)_n$ rocking	+	+	+	+

Note: + is present, and − is absent.

Source: CONF-790805-P1

Porphyrins, compounds with multifunctional groups and which are possible interfacial active substances (10), have been isolated from oil-resins of the two California crudes by following a modification of Erdman's procedure (11). They were identified by absorption spectra in which the four characteristic peaks in the range of 500 to 630 nm indicate the presence of demetallated porphyrins. Mass spectral analysis revealed the types of these porphyrins to be primarily deoxophylloerythroetioporphyrin (DPEP) and etioporphyrin (Etio). The ratio of DPEP- to Etio-type porphyrins in the Huntington Beach oil-resin, for example, was determined to be 1.1. Fractionation of the oil-resins using a silica gel column gave a porphyrin-rich fraction which showed absorption peaks at 550 and 510 nm, indicating that the porphyrins in these two California crude oils exist primarily as nickel complexes. The studies of the effects of porphyrins on IFT behavior of the crude oils are under way.

The information in this section is based on:

"Development of Improved Mobility Control Agents for Surfactant/Polymer Flooding," presented by F.D. Martin, of New Mexico Petroleum Recovery Research Center, and L.G. Donaruma and M.J. Hatch, of New Mexico Institute of Mining and Technology.

IMPROVING MOBILITY CONTROL AGENTS FOR SURFACTANT/POLYMER FLOODING

While projections for surfactant/polymer flooding predict a substantial contribution of this process in the area of enhanced oil recovery by 1980-1990, the process at this time is largely developmental. In order for the process to become commercial, improved process technology is necessary. Therefore, the main goal of this project will be to develop improved polymers that will increase the efficiency of the surfactant/polymer flooding process.

The research outlined in this project is applications-oriented towards this specific enhanced oil recovery process. Experiments are designed to simulate field conditions as much as possible; however, the results of the tests will be correlated with the more fundamental studies that are also being conducted at the New Mexico Petroleum Recovery Research Center. In addition, efforts will be coordinated with certain academic institutions and major oil company research laboratories where related research is being conducted.

The objectives of this project are to: (1) assess the inadequacies of mobility control agents in past and ongoing surfactant/polymer projects, (2) using Berea cores in laboratory flow tests, generate a data base on properties of commercially available products, and (3) develop improved mobility control agents for surfactant/polymer flooding.

The project has been divided into the following segments:

> Phase 1—Survey and assess applications data to outline problems
> encountered with currently available mobility control
> agents
> Phase 2—Obtain base line laboratory screening tests on commercially
> available products
> Phase 3—Generate structure–performance relationships
> Phase 4—Develop improved mobility control materials

This paper discusses the results obtained in Phase 1.

Importance of Mobility Control

In order to ensure process stability, the mobility of the surfactant slug should be equal to or less than the oil-water bank being displaced and the mobility of the polymer solution should be equal to or less than the surfactant slug (12)(13)(14). The purpose of the design is to prevent surfactant from fingering through the oil-water bank; the polymer is used to ensure a high displacement efficiency (good sweep efficiency) of the oil that is being mobilized. If adequate mobility control is not maintained, polymer or chase water will invade the surfactant slug, which could cause slug breakdown and trapping of the oil that has been mobilized.

Depending on the process design, the polymer slug may be 3 to 6 times larger than the surfactant slug. Although the polymer slug may cost only half as much as the surfactant slug, its contribution to the expense of the process is significant.

Commonly Used Mobility Control Agents

The most commonly used mobility control agents in the mobility buffer solution have been water-soluble polymers. Practical application has been limited to two polymer types: partially hydrolyzed polyacrylamides (HPAM) and a polysaccharide called xanthan gum (XG). Chemical structures of these two materials are shown in Figure 14.4 (15) and Figure 14.5 (16)(17).

Xanthan gum is a natural biopolymer produced by a microbial fermentation action of the organism *Xanthomonas campestris* on a carbohydrate. Partially hydrolyzed polyacrylamides are synthetic polymers made by hydrolysis of polyacrylamide with a caustic material (Na_2CO_3, NaOH, or KOH) or by copolymerizing acrylamide with acrylic acid or an acrylate.

Figure 14.4: Molecular Structure of Partially Hydrolyzed Polyacrylamides

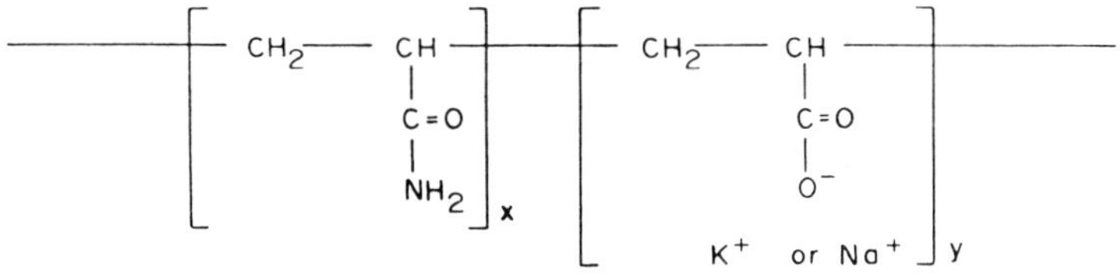

CH_2 — CH — $C=O$ — NH_2]$_x$ — [CH_2 — CH — $C=O$ — O^- — K^+ or Na^+]$_y$

Source: CONF-790805-P1

Figure 14.5: Structure of Extracellular Polysaccharide of
Xanthomonas campestris

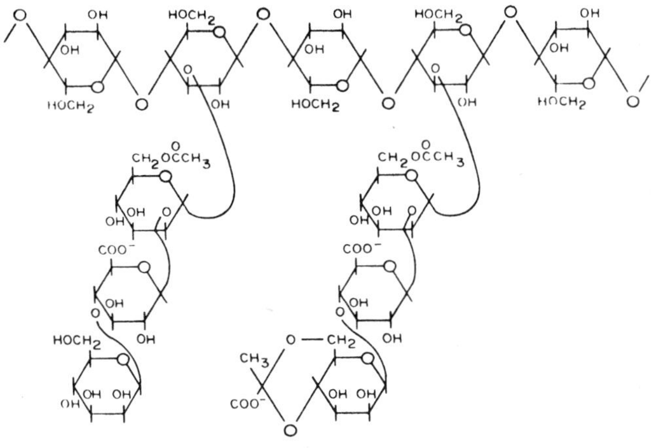

Source: CONF-790805-P1

The degree of hydrolysis of the polyacrylamide commonly ranges from 15 to 35 mol % (15). Both types of polymers are quite large macromolecules with molecular weights probably ranging from one to ten million (18).

Both types of marketed products have limitations that cause process inefficiencies or loss in cost-effectiveness. Some of the recognized inadequacies are listed in Table 14.10. Because of its lower cost, polyacrylamide is being used in a major-ity of the field applications; however, it is subject to shear degradation and lack of brine tolerance. The more expensive xanthan gum is not completely satisfactory because of the filtration or clarification requirements as well as some of the listed degradation mechanisms.

Table 14.10: Potential Inadequacies of Existing Mobility Control Agents

Inadequacy	Polyacrylamide	Xanthan Gum
Economics	moderate cost	higher cost
Shear degradation	severe	slight
Oxygen degradation	yes	yes
Viscosity loss in brine	severe	slight
Thermal degradation	$>250°F$	$>160°F$
Metal ion degradation	yes	yes
Hydrolysis reaction	yes	yes
Microbial degradation	moderate	severe
Filtration required	no	yes
Permeability limitations	>10–20 md	>5 md
Surfactant/polymer interaction	possible[*]	possible[*]

[*]Extent of inadequacy would depend on formulation of surfactant slug as well as polymer type.

Other potential inadequacies:
  Reactions with divalent ions
  Polymer retention or adsorption
  Interactions with porous media
  Reduced injectivity and wellbore impairment
  Oil emulsions
  Producing well productivity problems
  Premature breakthrough of polymer in production wells
  Handling problems
  Product quality control
  Availability for commercial-scale S/P projects.

Source: CONF-790805-P1

Questionnaire on Polymer Deficiencies

In order to assess the relative importance of deficiences of polymers used in surfactant/polymer processes, a questionnaire form was distributed to more than 200 oil industry personnel. Completed questionnaires represented the opinions of almost 100 oil industry people who are concerned with this process. Results of this input are discussed in a DOE Quarterly Progress Report (19) and are summarized in Tables 14.11 through 14.14. Data in these tables are presented in two manners: based on the total response, and based on only that response from oil company personnel with either field experience or both laboratory and field experience.

As indicated in Table 14.11, there was a preference for the hydrolyzed polyacrylamides over the xanthan gums. The primary reasons for this preference were: lower cost required to achieve the desired mobility control in the reservoir and fewer handling problems in the field.

Preferences regarding polymer form are listed in Table 14.12. If active polymer costs are comparable for the various physical forms, it is clearly obvious that solutions or liquid forms are desirable. Ease of handling was the primary consideration for the choice of solutions, emulsions, or liquid-type products.

**Table 14.11: Polymer Preference for Surfactant/Polymer Flooding
Response to Questionnaire Form**

. . . . .Total Response. . . . .

Preference	Percent
Hydrolyzed polyacrylamide	49
Xanthan gum	33.
Depends on the application	15
Neither	3

Response From Oil Company Personnel with Field Experience

Preference	Percent
Hydrolyzed polyacrylamide	41
Xanthan gum	34
Depends on the application	22
Neither	3

Source: CONF-790805-P1

**Table 14.12: Preference Regarding Polymer Form
Response to Questionnaire Form**

. . . . Total Response. . . . .

Preference	Percent
Solution	33
Emulsion	21
Powder	17
Depends on application	10
Bead	9
Liquid type	9
Solid type	1

Response From Oil Company Personnel with Field Experience

Preference	Percent
Solution	35
Depends on application	19
Emulsion	13.5
Liquid type	13.5
Powder	11
Bead	5
Solid type	3

Source: CONF-790805-P1

Instructions on the questionnaire form were to list the most severe deficiency
as No. 1, the second most serious as No. 2, the third most serious as No. 3,
etc. If a polymer was judged to be adequate in a particular aspect, that item
was marked "A" for adequate. Entries for each item were averaged so that the
most severe inadequacies are denoted by the lowest average rating. The number
of adequate ratings for each item is listed in the tables. Inadequacies of HPAM
polymers are rated in Table 14.13.

Table 14.13: Inadequacies of Partially Hydrolyzed Polyacrylamide for Surfactant/Polymer Flooding–Response to Questionnaire Form

Item	...Total Response...		Oil Company Personnel .... Response*	
	Average Numerical Ratings**	Number of Adequate Ratings	Average Numerical Ratings**	Number of Adequate Ratings
Shear degradation	2.3	2	2.3	1
Viscosity loss in brines	2.4	0	2.5	0
Reaction with divalent ions	3.4	4	3.6	0
Economics (high cost)	4.4	18	4.0	6
Oxygen degradation	4.6	7	4.8	2
Metal ion degradation	4.7	8	4.6	4
Use in low perm formations	5.3	13	5.4	8
Surfactant-polymer interactions	5.6	18	5.7	10
Thermal degradation	6.4	16	7.3	9
Wellbore plugging	6.6	18	5.8	12
Hydrolysis reaction	6.9	12	8.1	4
Microbial degradation	7.5	30	8.4	17
Filtration requirements	8.6	35	9.3	19

*With field experience.
**Instructions were to list the most severe deficiency as No. 1, the second most severe inadequacy as No. 2, the third most severe as No. 3, etc. Therefore, the most severe inadequacies are denoted by the lowest average rating.

Source: CONF-790805-P1

In the opinion of the people who participated in the survey, the most serious areas of concern with hydrolyzed acrylamides are: (1) shear degradation, (2) viscosity loss in brines, and (3) reaction with divalent ions. The least serious of the deficiencies listed was the requirement for filtration. No significant differences were observed between the total response and that response from oil company personnel with field experience. Table 14.14 indicated that for xanthan gum the most serious deficiency is (1) economics (high cost), followed by (2) wellbore plugging, (3) microbial degradation, and (4) filtration requirements. The least serious problems with XG are shear degradation and viscosity loss in brines.

Table 14.14: Inadequacies of Xanthan Gum for Surfactant/Polymer Flooding Response to Questionnaire Form

Item	...Total Response...		Oil Company Personnel .... Response*	
	Average Numerical Ratings**	Number of Adequate Ratings	Average Numerical Ratings**	Number of Adequate Ratings
Economics (high cost)	1.7	8	1.5	1
Wellbore plugging	3.2	9	3.1	4
Microbial degradation	3.3	7	3.6	2
Filtration requirements	3.6	4	3.0	2
Thermal degradation	5.0	12	5.8	8
Reaction with divalent ions	5.1	26	5.1	15
Surfactant-polymer interactions	5.4	12	4.9	7
Use in low perm formations	5.5	10	5.4	7
Metal ion degradation	5.8	16	5.9	10
Oxygen degradation	5.8	17	5.8	12
Hydrolysis reactions	6.1	27	6.9	13

(continued)

Table 14.14: (continued)

| | ...Total Response... | | Oil Company Personnel
....Response*.... | |
Item	Average Numerical Ratings**	Number of Adequate Ratings	Average Numerical Ratings**	Number of Adequate Ratings
Viscosity loss in brines	7.3	39	7.8	22
Shear degradation	8.2	44	10.8	26

*With field experience.
**Instructions were to list the most severe deficiency as No. 1, the second most severe inadequacy as No. 2, the third most severe as No. 3, etc. Therefore, the most severe inadequacies are denoted by the lowest average rating.

Source: CONF-790805-P1

It can be argued that, in some of the areas evaluated, not enough is universally known to make a fair appraisal. The results of the survey, however, presented no substantial surprises and are generally in agreement with the information obtained from the published literature. Naturally, the most significant deficiencies in polymers used for surfactant/polymer flooding will depend on the particular field application, but the major inadequacies listed above are those that are generally recognized as being problem areas.

REFERENCES

(1) Knight, B.L., "Reservoir Stability of Polymer Solutions," *Journal of Petroleum Technology,* pp 618–626, (May, 1973).

(2) Howell, J.C., McAtee, R.W., Snyder, W.O., and Tonso, K.L., "Large Scale Field Application of Micellar-Polymer Flooding," Paper SPE 7089 presented at the SPE Symposium on Improved Recovery, Tulsa, OK (April 16-19, 1978).

(3) Burdge, D.N., "Commercial Scale Demonstration of the Maraflood Process: M-1 Project, Crawford County, Illinois (1977-1978)," *Proceedings 4th DOE Symposium on Enhanced Oil and Gas Recovery,* Vol. 1, A-3/1–A-3/20 (1978).

(4) Marathon Oil Company, "Commercial Scale Demonstration, Enhanced Oil Recovery by Micellar-Polymer Flood, Annual Report for October, 1976-September, 1977," DOE Contract No. EF-77-C-02-4208, BERC/TPR-77/10, FE-4208-11.

(5) Marathon Oil Company, "Commercial Scale Demonstration, Enhanced Oil Recovery by Micellar-Polymer Flood, Annual Report, for October, 1977-September, 1978," DOE Contract No. DE-AC19-78ET13077, DOE/ET/13077-23, FE/1806-23.

(6) Johnson, E.C., Jr., *J. Petrol. Tech.,* January 85 (1975).

(7) Seifert, W.K., "Carboxylic Acids in Petroleum Sediments," *Prog. Chem. Nat. Products,* Springer-Verlag, pp. 1–49 (1975).

(8) Neumann, H.J., *Erdol Kohle,* 17, 346 (1964).

(9) Cooke, C.E., Jr., Williams, R.E. and Koldzie, P.A., *J. Petrol. Tech.,* 26 (12), 1365 (1974).

(10) Dunning, H.N., Moore, J.W., and Denekas, M.O., *Ind. Eng. Chem.,* 45, 1759 (1953).

(11) Erdman, J.G., U.S. Patent 3,190,829 (June 22, 1965).

(12) Gogarty, W.B., "Mobility Control with Polymer Solutions," *Soc. Pet. Eng. J.,* 161-173 (June, 1967).

(13) Poettman, F.H., "Microemulsion Flooding," *Secondary and Tertiary Oil Recovery Processes,* Interstate Oil Compact Comm., 67-93 (1974).

(14) Gogarty, W.B., Meaborn, H.P., and Milton, H.W., "Mobility Control Design for Miscible-Type Waterfloods Using Micellar Solutions," *J. Pet. Tech.,* 141-147 (February, 1970).

(15) Martin, F.D., and Sherwood, N.S., "The Effect of Hydrolysis of Polyacrylamide on Solution Viscosity, Polymer Retention, and Flow Resistance Properties," SPE 5339, SPE

Rocky Mountain Region Meeting (1975).

(16) Jansson, P.E., Keene, L., and Lindberg, B., "Structure of Extracellular Polysaccharide from *Xanthomonas campestris,*" *Carbohydrate Res.* (1975).

(17) Holzworth, B., "Polysaccharide from *Xanthomonas campestris*: Rheology, Solution Conformation and Flow Through Small Pores," Symposium on Advances in Petroleum Recovery, ACS Meeting, New York, NY (April 4-9, 1976).

(18) Willhite, G.P., and Dominguez, "Mechanisms of Polymer Retention in Porous Media," *Improved Oil Recovery by Surfactant and Polymer Flooding,* ed. by D.O. Shah and R.S. Schechter, Academic Press, 513 (1977).

(19) Martin, F.D., "Development of Improved Mobility Control Agents for Surfactant/Polymer Flooding," Quarterly Progress Report, September 29, 1978 to December 31, 1978, Prepared for DOE, BETC-0047-4, New Mexico Petroleum Recovery Research Center (January 17, 1979).

THERMAL RECOVERY—SOME PROJECTS AND RESEARCH

The information in this chapter is from:

*Fifth Annual DOE Symposium on Enhanced Oil and Gas
Recovery and Improved Drilling Technology, Vol. 2—Oil,*
CONF-790805-P2 sponsored by U.S. Department of Energy,
Tulsa, Oklahoma, August 22-24, 1979.

The information in this section is based on the paper:

"Bodcau In Situ Combustion Project, Bossier Parish, LA,"
presented by J. Garvey, of Cities Service Company.

SIMULTANEOUS AIR AND WATER IN SITU COMBUSTION PROCESS

The Bodcau In Situ Combustion Project is a demonstration of an efficient and economical simultaneous air and water in situ combustion process. Production in the Bellevue Field of Bossier Parish, Louisiana, which was discovered in 1921 and covers approximately 900 productive acres, is from the Upper Cretaceous Nacatoch Sand approximately 400 feet deep. Primary production by fluid expansion and gravity drainage caused high producing rates early in the life of the field but decreased very rapidly to only a few barrels of oil per day. Residual oil-in-place after primary techniques had ceased was estimated at 95%. (Estimated recoverable reserves are 700,000 bbl.) Various stimulation processes were attempted throughout the field, but none were successful in producing the heavy 19° gravity crude until an in situ combustion process was enacted.

Characteristics of Louisiana Reservoir

The Nacatoch Sand in the Bellevue Field is very unconsolidated with strings of fossiliferous lime and sandy shale. Porosity averages 33.9% with a water saturation of 27.4%. Core analysis of the formation indicated a permeability of approximately 700 md. The average thickness of the formation throughout the area was determined to be 56 feet. It was estimated that the project contained about 1,909 bbl/acre-ft or in excess of 2,000,000 barrels of oil-in-place. The reservoir temperature is 75°F and the pressure is 40 psig. Viscosity of the 19° gravity crude is 670 cp at reservoir temperature but drops rapidly with an increase in temperature (Figure 15.1).

Figure 15.1: Viscosity-Temperature Relationship (19°API, Bellevue Crude)

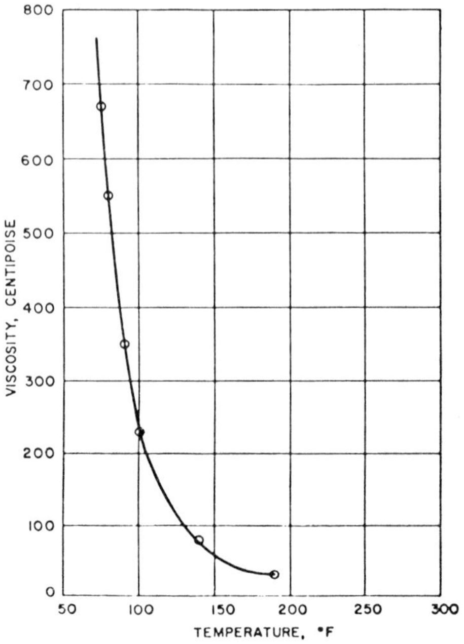

Source: CONF-790805-P2

Facilities

Thirty-three producers and five injectors were originally included in the demonstration area, with five additional producers scheduled during mid-1979. The wells were drilled to an average of 450 ft, sufficient to penetrate the entire pay interval of the Upper Nacatoch Sand, and 7" casing was set and cemented back to the surface using high-temperature resistance materials. Production wells were selectively perforated and equipped with conventional beam pumping units. A downhole water cooling system is required in each producer to combat high temperature problems (Figure 15.2).

Individual flowlines carry each well's tubing production to a centralized header system. Produced fluid is then sent to the central battery facility. Natural flow through the casing annulus is transported through an exhaust line to a concrete-lined pit. At the pits, liquids are extracted from the exhaust gas and sent to the central battery facility.

Figure 15.2: Production Well Schematic—Dry and Simultaneous Burning Phases

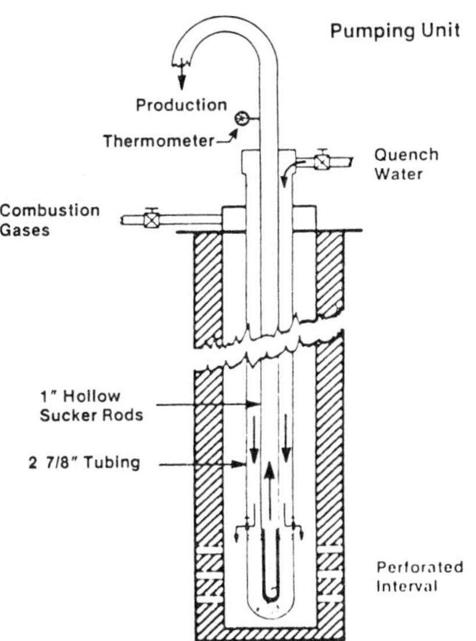

Source: CONF-790805-P2

The central battery used for the storage and treatment of the production consists of a free water knockout, emulsion treater, tankage and ACT unit. Three compressors with a combined output of 20 MMcfd at 250 psig are used to supply the air to the five injection wells. Air is delivered through a trunk system with individual metering and regulation at each injector.

Operations and Performance

Ignition of the five inverted nine-spot patterns was accomplished between Aug. 7, and Sept. 24, 1976, by injecting approximately 600°F air, generated by a 30 kW electrical immersion heater, into the formation.

To improve volumetric sweep and provide greater heat transfer ahead of the combustion zone, all five injectors were converted to simultaneous air and water injection in April, 1977. Produced water is injected down the annulus at a rate of 250 bbl/MMcf into the upper section of the zone while air is injected through tubing below a packer into the lower section of the reservoir (Figure 15.3).

Combustion front breakthrough has occurred at nineteen of the thirty-three original producing wells thus far. With the advance of the combustion front toward producing wells, the area surrounding the well becomes thermally stimulated, causing producing rates to greatly increase as the mobility of the oil increases. Commercial producing rates cannot be obtained without this stimulation. Unfortunately, the combustion front will break through, or channel, to the producer before the entire reservoir can be thermally stimulated. This breakthrough of hot gases and sands causes extreme wear of bottomhole tubulars and pumping equipment. The quench water system combats the gases and sands, but eventually the problem becomes too intense.

Figure 15.3:   Injection Well Schematic—Simultaneous Injection

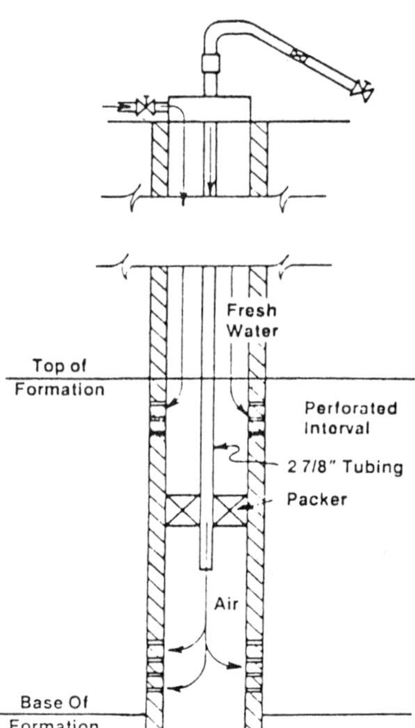

Source: CONF-790805-P2

This problem of breakthrough is eliminated by squeeze-cementing the portion of the zone that is burned. These wells continue to produce following breakthrough of a front but continual operating problems, as a result of burns, limit the withdrawal capability as the combustion progresses, hampering project economics both by increased operating costs and reducing income from crude sales.

As anticipated, problems with oil treatment exist. Treating these emulsions is difficult due to the amount of sand produced with the oil. Various treatment programs are being conducted to minimize this.

Since initiation of the project, cumulative air injection through April, 1979 is 6,381,364 Mcf. A total of 1,024,613 barrels of water have been injected during the simultaneous phase. Total recovery from the five patterns has been 401,646 barrels of the previously unrecoverable oil through April, 1979.

Monitoring and Testing

Production wells are tested monthly for oil, water and gas volumes to allocate production. Flowline temperatures are recorded daily for monitoring downhole conditions. Exhaust gas from each producer is analyzed each month for CO_2, O_2, and CO levels. Temperature profiles are run monthly in the five observation wells.

Prior to ignition, pulse tests were conducted at each injector. These tests indicated areas within the project where directional permeability was higher. This was used to determine what wells in the project would receive stimulation first.

Also, pressure falloff tests were run before injectors were ignited and again before conversion to simultaneous air-water injection. From these tests, permeabilities plus well bore damage were calculated. Average permeability, before ignition, ranged from 6 to 32 md, and after six months of dry combustion the average was 112 to 544 md.

Produced oil and water samples from each well in the project have been analyzed for composition, specific gravity and viscosity to aid in the treatment of the crude.

Twenty-five combustion tube tests have been conducted to determine burn parameters and the optimum simultaneous water and air injection ratio. It was determined that during dry combustion the fuel deposit was 2.05 lb/ft^3 with an air requirement of 17.0 MMscf/acre-ft. By switching to the simultaneous phase and injecting at a ratio of 225 barrels of water per MMscf of air the fuel deposit was reduced to 1.65 lb/ft^3 and the air requirement reduced to 10.6 MMscf/acre-ft. These tests have indicated that in the Bellevue Field a simultaneous combustion phase requires approximately 37% less injected air than a dry combustion phase project.

Economics

Development of a full-scale fireflood project is a costly adventure. One of the primary objectives of this contract is to demonstrate key economic factors that are involved with this project. Costs are kept in great detail to analyze all portions of development and operation of the project.

The total investment for wells and facilities has been $1,428,080. Total expenditures for operations and maintenance, research and production staff services, have been $2,312,016. Expenditures for Louisiana severance taxes has amounted to $676,498 and overhead expense was $545,354. The total project expenditure through April, 1979 was $4,961,948.

Operating expenses have been broken down into a cost per barrel of oil to even further demonstrate the economic factors in this project. Lease operation is $0.99, lease maintenance is $1.82, operation of compression facilities is $2.00, Louisiana severance tax is $1.92, overhead is $1.55 and depletion and depreciation is $1.97. Total operating and maintenance expense through April, 1979 is $10.25 per barrel of oil produced.

The information in this section is based on the paper:

"The '200' Sand Steamflood Demonstration Project,"

presented by W.O. Alford, of Santa Fe Energy Company—
Chanslor Division.

STEAMFLOODING A HEAVY OIL RESERVOIR UNRESPONSIVE TO CYCLIC STIMULATION

The "200" Sand Steamflood Demonstration Project is testing an enhanced steamflood technique in the Midway-Sunset Field, Kern County, California. Generally,
steamflooding on Santa Fe Energy Company (SFE) properties has been conducted in reservoirs which have responded favorably to cyclic steam stimulation
in order to accelerate cyclic production and to increase ultimate recovery. One
pool, designated the "200" Sand, made no primary production and responded
poorly to cyclic steam stimulation. With the cyclic method, estimated ultimate
recovery was virtually zero percent because of its ineffectiveness and unfavorable
economics.

The high capital and operating costs of steamflooding made a pilot desirable to
evaluate the process in this pool. Therefore, this project was initiated to demonstrate the operational, recovery, and economic aspects of steamflooding such a
typical heavy oil reservoir.

The test is being conducted in four-2.35 acre inverted seven-spot drive patterns,
which are not fully developed with producers. The project area consists of ten
producers (exclude two stepout producers), four injectors, and two observation
wells. Within the four patterns are two interior producers that will be key wells
for determining production response for expanding the project.

Reservoir and Geological Data

The formation contains approximately 50 million barrels of oil-in-place. The
crude oil is very viscous with ambient viscosities in the range of 6,500 cp. Relatively no gas is produced in conjunction with the crude.

The reservoir pressure in the "200" pool is very low, in the range of 20 to 60 psi.
This kind of pressure does not provide sufficient energy for a producing mechanism. Most of the project area is located on top of an anticlinal structure where
dips range from 5 to 15 degrees. With the shallow dips, gravity drainage does
not provide an effective mechanism to obtain economical production rates and
effectively deplete the reservoir. Herein lies the necessity of steamflooding, to
not only reduce the oil viscosity, but also increase the displacement process, so
the crude can flow to the well bore.

The reservoir is unconsolidated with sand, silt, and conglomerates. The sand is
fine to medium grained, with scattered coarse to very coarse grains and pebbles;
at some locations they are interbedded with well-cemented sandstones. Usually
the sands are clean and extremely friable with the grains being held together by
the heavy viscous crude.

Two observation wells were conventionally cored through the entire producing
sand and completed in January, 1979. They are located in the only fully developed pattern between Injector 243 and Producers 231 and 254 in order to
determine the degree of channeling and sweep efficiency of the pilot.

The wells were completed with $2\frac{7}{8}$" o.d. tubing cemented in the hole to observe the depletion growth and define the position of the heat front and steam and hot water zones.

The subject field trial is being conducted in the bottom sand interval, approximately the bottom 50 feet of the reservoir. Santa Fe Energy's experience to date indicated that intervals greater than 50 feet will permit steam overriding in the reservoir, which results in poor vertical sweep efficiency. When the bottom interval is depleted, the upper sands will be steamflooded successively from bottom to top if the technique is successful.

The average petrophysical properties for all productive sands are 2,250 md permeability, 30% porosity, and an oil saturation before steamflooding of 59%, equivalent to an average oil content of 1,373 barrels per acre-foot. Table 15.1 has a summary of reservoir data obtained from the pilot wells and surrounding core holes.

Table 15.1: Summary of Reservoir Data

Depth, feet	400–700
Areal extent, acres (est.)	250
Oil gravity, °API	12
Reservoir pressure, psi (est.)	40
Reservoir temperature, °F	90
Average formation dip, degrees	12
Average gross thickness, feet	200
Average net thickness, feet	150
Permeability to air, md	1,050–3,400
Average porosity, percent	30
Average oil saturation, percent	59
Average oil content, bbl/acre-ft	1,373
Oil viscosity at 90°F, cp	6,500
Oil viscosity at 212°F, cp	60

Source: CONF-790805-P2

Project Performance

Steam Injection: Injection of steam into the four input wells started in October, 1975. Since startup, 1.67 million barrels of steam have been injected into the four injection wells at an average injection interval temperature and quality of 350°F and 72%, respectively. Steam at the two 20 million Btu/hr steam generator outlets are 420°F with 80% quality.

Due to the lag in response time, estimated at 18 to 24 months, increased injection rates the last year have been averaging 450 bbl/day/well. Also, the project is rate sensitive; therefore, the higher the injection rate, the higher the oil production. Sensitivity can be expected in this steam drive area where the reservoir with very little dip does not allow for gravity drainage. With the low gravity oil and flat formation, the burden of displacing the crude oil to the producing wells lies primarily with the injected steam. There is an optimum injection rate, however, which results in the highest oil production per dollar invested for steam injection; and seldom do the early requirements for a steamflood operation represent an optimum steam injection rate. Realizing this, the initial objectives were getting the producing formation hot and observing maximum production rate potential, then optimizing the injection rate later in the project life.

No limits have been put on how high the injection rates can be. Considerations with respect to fracture gradient are not a problem; the only limitations now and foreseen are the generators' discharge pressures. The reservoir is thick, highly permeable, and unconsolidated, which makes fracture gradient determination virtually impossible. Tests have shown similar sands undergoing a plastic deformation rather than fracturing under high rate injection.

All ten pilot area producers (exclude two stepout producers) have temperature and gross fluid produced response. Five wells have moderate to good steam breakthrough, including the two enclosed producers that will be key wells for determining production response for expanding the pilot. All massive steam breakthrough will be avoided by reducing injection rates and restricting producing wells' drawdown (discussed in the next topic).

Production: Production from the ten pilot producers (exclude two stepout producers) have averaged 136 bbl/day oil and 276 bbl/day water the last year. Several wells were given steam stimulation treatments due to decline in well bore temperatures. Most of the response has been coming from the two enclosed producers, 254 and 255, and because they are the only producers surrounded by injectors, the two will be key wells for determining expansion.

Two different completions were used in producing wells 254 and 255. Well 254 is a 60-mesh slotted liner set across the entire sand. Peak production was 67 bbl/day oil in March, 1978 with an average production during the last year of 34 bbl/day oil and 33 bbl/day water. Well 255 reached 42 bbl/day oil in May, 1978. Production in the last year was 13 bbl/day oil and 39 bbl/day water.

It is a cased-through completion with jet perforations in the bottom pay to confine steam in the drive interval by restricting production only to the steam drive interval. Recent temperature surveys show the steam to be limited in the drive interval. The average oil response from the different completion methods indicates the slotted liner completion is the best in the steam drive. Two additional producers, 241 and 287, are showing favorable response from the drive.

One of the major limitations to the effectiveness of steam drive recovery are patterns not fully developed with producers. Fully developed patterns would have improved control of frontal movement and volumetric recovery, but capital availability did not permit additional drilling. As a result of delayed production and temperature response, the project is behind schedule. Observations from operating the pilot indicate that steam channeling from injector to producer is necessary to efficiently produce the highly viscous oil. Therefore the existing high steam injection volume will continue in the future.

Producing wells that have shown excessive steam breakthrough (channeling) have been controlled by restricting the communicating wells' drawdown. Partially shutting in the casing vents creates a back pressure which reduces the rate of frontal advance in the direction of the channel and enhances channeling to other producing wells of the patterns. Thus far, restricting responding wells' drawdown has been more advantageous than steam injection rate reduction, because the average reservoir temperature for the total pilot area is not high enough to allow a Btu input reduction.

The following section is based on the paper:

> "The Use of Chemical Additives with Steam Injection to
> Increase Oil Recovery," presented by L.L. Handy, V.M.
> Ziegler and M.O. Amabeoku, of University of Southern
> California.

STEAMFLOODING ADDITIVES TO REDUCE OIL TO STEAM RATIOS

Steam injection has been demonstrated to have considerable potential in improving oil recovery from fields containing high viscosity oils. The steam is known, however, to move through the upper portion of the reservoir while the lower portion of the reservoir is flooded only with hot or cold water. The residual oil saturation in the steam-swept portion of the reservoir has been observed to be near zero, but normal residual oil saturations are observed in the water-swept part of the reservoir. As has been suggested by others, oil:steam ratios could possibly be reduced if a suitable additive was found which would reduce the residual oil saturation in the waterflooded part of the reservoir (1)(2). The objective of research at the University of Southern California was to find chemical additives which may be suitable for this purpose.

The emphasis in this section is on results obtained in evaluating some organic surfactants which have shown promise in reducing residual oil saturations in conventional waterfloods. It was thought that some synergistic effect might be observed between temperature and detergency such that the surfactants would be more effective recovery agents in the hot water portion of a reservoir undergoing steam injection than at reservoir temperatures in a conventional waterflood. If this is the case, the surfactant must move in the heated portion of the reservoir.

Since the heat front will move more slowly than the hot water, the surfactant would have to be adsorbed to some degree if it is not to move out ahead of the temperature front. Ideally, then, a surfactant would be adsorbed adequately to remain always in the hot water zone. It must, also, be effective in reducing the residual oil saturation in this region and it must be stable at the high steam temperatures. Consequently, the project has several subsidiary aspects.

The stability of types of surfactants which have shown possible application as oil recovery agents, principally the anionic petroleum sulfonates, but also some nonionics, were evaluated. The results of these studies have been reported in a previous paper (3). Data on the effect of temperature on the interfacial tension between surfactant solutions and a San Joaquin Valley crude oil were reported at the DOE Enhanced Oil Recovery Symposium in 1978 (4). These measurements are being continued using a spinning drop method. The present report discusses the results of studies of the effect of temperature on adsorption and on surfactant floods.

Effect of Temperature on Surfactant Adsorption

The position of the surfactant front in the reservoir relative to the temperature front is determined by the adsorption of the surfactant. If the surfactant is less strongly adsorbed in the heated portion of the reservoir, the surfactant will move more rapidly at the high temperatures than at low temperatures. In that case the surfactant may have a tendency to run ahead of the hot zone but be slowed when it reaches the cold zone where the surfactant is more strongly adsorbed. This effect would, then, tend to keep the surfactant in the hot region. Consequently, measurement of the effect of temperature on the adsorption of specific surfactants would be essential in selecting surfactants for use in steamfloods.

To establish trends in the dependence of adsorption on temperature, the pure anionic sulfonate, sodium dodecylbenzene sulfonate (NaDDBS) and the nonionic surfactant, nonylphenoxypolyethanol with a molecular weight of 1,100 were selected (GAF's Igepal CO-850). Although the NaDDBS is not used as a field surfactant, it is similar to the petroleum sulfonates. Nonionics similar to Igepal CO-850 are used as cosurfactants in some micellar solution formulations.

For a porous material fired and crushed Berea sandstone, sieved to various size fractions was used. Static and dynamic tests were performed to determine adsorption. For both of these the surface areas of the matrix material were determined by the B.E.T. method using a Nelsen-Eggertsen type apparatus (5). The surface area determined for the sand used in the static adsorption test was 1,250 m^2/kg. The sands selected for the dynamic tests had specific surface areas of 340 and 460 m^2/kg.

The anionic sodium dodecylbenzene sulfonate showed a decreasing adsorption with increasing temperature but at higher concentrations and in brine solutions the sulfonate precipitated. A definitive measurement of adsorption under these circumstances was not possible, but substantial loss of surfactant during floods was observed. The nonionic Igepal CO-850 was, on the other hand, well-behaved in the adsorption experiment. It does show, however, a cloud point. The adsorption as a function of temperature is the inverse at high temperatures over that at lower temperatures. Near the cloud point adsorption increases with increasing temperatures.

Nonionic surfactants similar to Igepal CO-850 are commonly used as cosurfactants in micellar solutions (6). Their presence in sulfonate formulations has been shown to provide a synergistic effect on interfacial tension reduction. At room temperature the nonionics by themselves do not give interfacial tensions low enough to mobilize residual oil. As temperatures are increased toward the cloud point, however, ultralow values of the interfacial tension are observed (less than 10^{-2} mN/m). When temperature becomes a process variable, surface-active ethylene oxide adducts can become effective oil recovery agents in their own right. The high tolerance of nonionic surfactants to dissolved electrolytes makes their application to the steamflood process particularly attractive. The cloud point, however, places an upper limit on the surfactant concentration which can be injected.

The Effect of Temperature on Surfactant Flooding

An important aspect of adding a surfactant for steamflooding is the effect of temperature on surfactant flood performance. Preliminary studies have been completed evaluating the flood performance for two surfactants at elevated temperatures. These studies have been limited to a maximum temperature of 177°C. At temperatures greater than this the surfactants have been determined to be unstable. Even at 177°C, the experiments had to be designed to minimize the length of time the surfactants were exposed to the elevated temperature.

The porous material used for the floods was fired consolidated Berea sandstone. The cores were three inches in length and one inch in diameter. Some experimentation was required to find a satisfactory method for mounting these cores to withstand the high temperatures. The method eventually selected was to mount the cores in an Armstrong A-2 epoxy resin filled with aluminum powder. The resin was poured between the core sample and an outer aluminum jacket. The jacket was required to withstand core pressures in excess of the vapor pressure of water at 177°C.

The literature generally indicates that for a surfactant to be an effective oil recovery agent the interfacial tension between the surfactant solution and the oil must be less than 0.01 mN/m. Surfactant concentrations were chosen which would meet this interfacial tension requirement (4). For Witco TRS 10-80 the lowest interfacial tension at 177°C was observed at a surfactant concentration of 2.5 g/ℓ and a NaCl concentration of 10 g/ℓ.

Floods were conducted in the absence of salt, however, for which the interfacial tensions as measured by the pendent drop method were in excess of 0.01 mN/m. All floods were conducted at constant temperatures of 82°C, a representative reservoir temperature, or 177°C, a minimum steamflood temperature. In one experiment the waterflood was conducted at 82°C and the surfactant flood at 177°C to simulate the temperature sequence for the surfactant-steamflood process. All floods were run at a constant pressure drop across the core sample. The actual flood rates varied only between one and four feet per day. Consequently, rate was not a significant variable in the experiments.

Experimental Results: The first flood was at 82°C with a TRS 10-80 concentration of 2.5 g/ℓ and a salt concentration of 10 g/ℓ. The results of this flood are shown in Figure 15.4a.

Figure 15.4:　Oil Recovery for TRS 10-80 Surfactant Flood

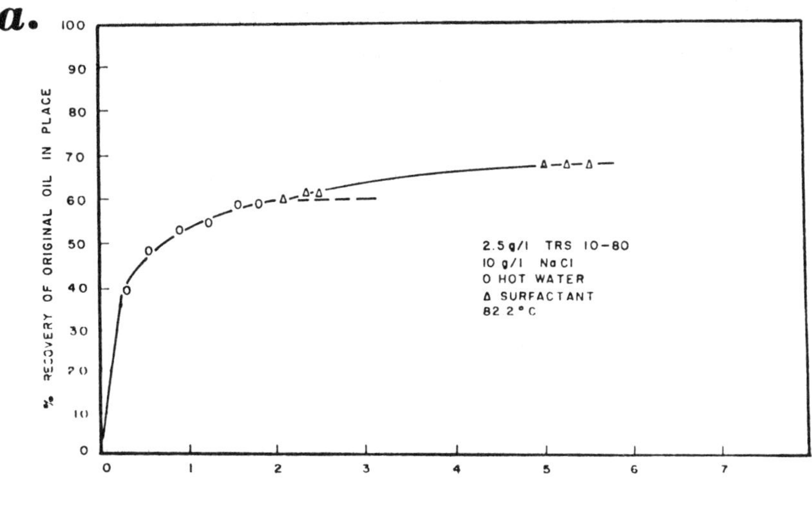

(continued)

Figure 15.4: (continued)

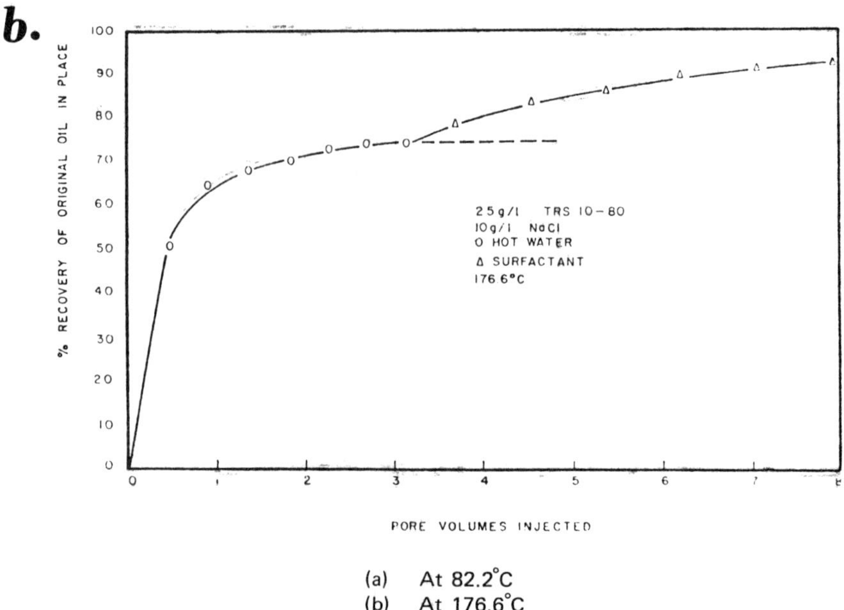

(a)   At 82.2°C
(b)   At 176.6°C

Source: CONF-790805-P2

A normal waterflood was conducted until two pore volumes of water had been injected. At that time 59% of the initial oil in-place had been recovered. The surfactant flood produced an additional 10% of the oil in-place after about three pore volumes of surfactant had been injected. These results are to be compared with a flood with the same surfactant and salt concentrations but at 177°C, as shown in Figure 15.4b.

The oil recovery after waterflooding at the higher temperature was 67% of the initial oil in-place (8% greater than at 82°C). This increase would have been expected as a result of the effect of temperature on the oil viscosity. However, an additional 20% of the initial oil in-place was recovered during the surfactant portion of the flood. This is significantly greater than that obtained at the lower temperature and the increase cannot be attributed to reduction in the oil viscosity.

The effect of a salt concentration on surfactant floods is illustrated in Figures 15.4b, 15.5a, 15.5b and 15.6. Although the oil recovery is somewhat larger with 10 g/ℓ salt (93.7% of the oil-in-place compared to 91.4%) the difference is less than might be expected, considering the pronounced effect of salt concentration on the interfacial tension. The water:oil ratio performance is also similar with or without salt as shown in Figure 15.5a and 15.5b. These results are at variance with results others have reported for low-temperature floods. If this result is confirmed in additional experiments, it will indicate that the oil recovery will not be adversely affected by the fact that, in a steamflood, the surfactant will be moving in a low salt concentration.

Figure 15.5: Water-Oil Ratio History for Surfactant Flood at 176.6°C

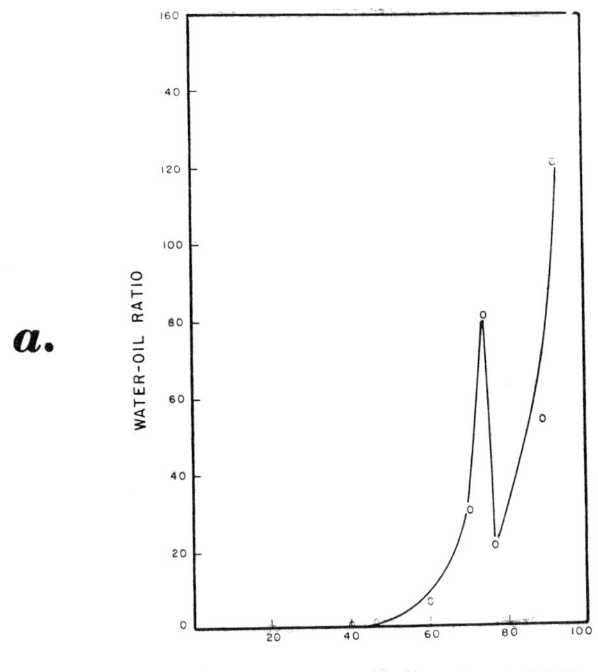

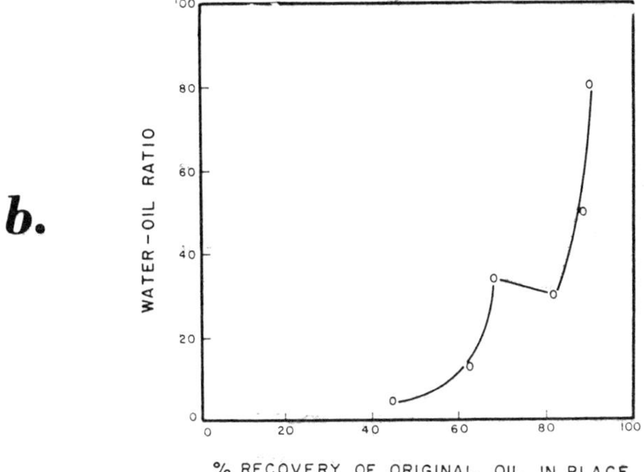

(a)   25 g/ℓ TRS 10-80 and 10 g/ℓ NaCl.
(b)   2.5 g/ℓ TRS 10-80 and 0.0 g/ℓ NaCl.

Source: CONF-790805-P2

Figure 15.6: Oil Recovery for TRS 10-80 Surfactant Flood with no Salt at 176.6°C

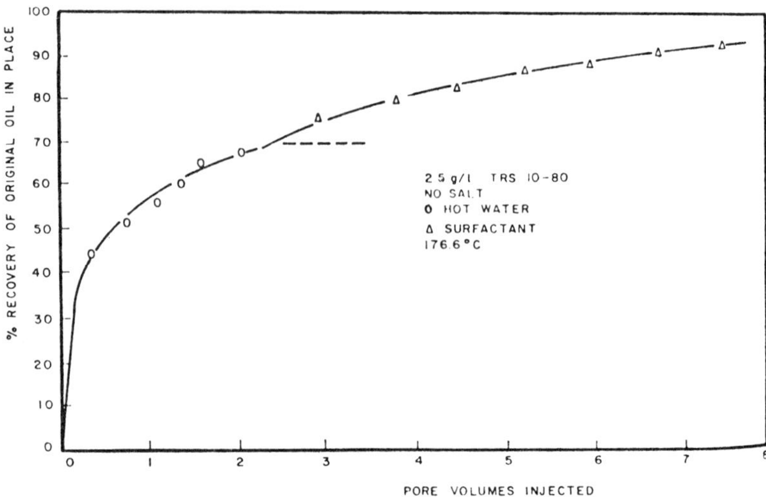

Source: CONF-790805-P2

Figure 15.7: Oil Recovery for Petrostep 465 Surfactant Flood at 176.6°C

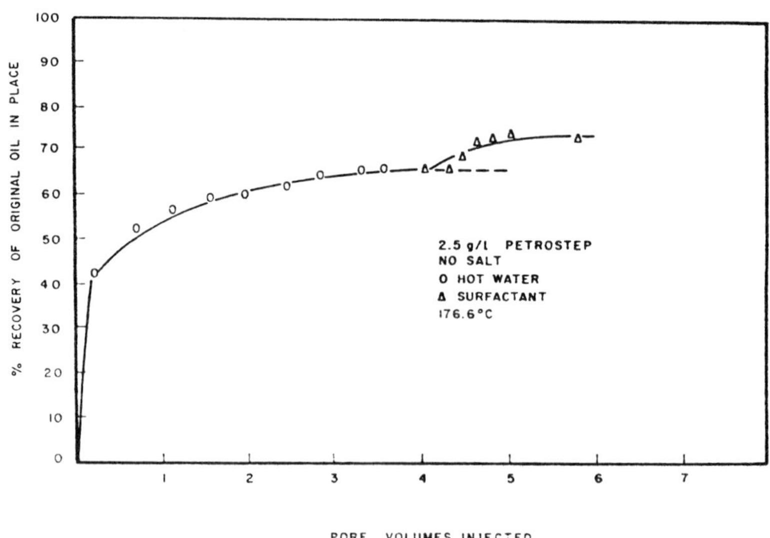

Source: CONF-790805-P2

In a flood in which the waterflood was carried out at 82°C and the surfactant flood at 177°C, no substantial difference was observed in the flood performance at two temperatures as compared to the flood at the single temperature of 82°C. The total recovery was 72.9% of the initial oil in-place when the surfactant flood was at the higher temperature as compared to 68.8% at the lower temperature. Additional floods are required to investigate the effect of variable temperature on the surfactant floods.

Petrostep 465 was the most stable at elevated temperature of all the surfactants studied (3). One flood has been completed with this surfactant. The surfactant concentration was 2.5 g/ℓ and the salt concentration was zero. The results are shown in Figure 15.7 on the previous page. The surfactant is effective in recovering additional oil at 177°C.

References

(1) Brown, A., Carlin, J.T., Fontaine, M.F., and Haynes, S., "Methods for Recovery by Hydrocarbons Utilizing Steam Injection," U.S. Patent 3,732,926, (May, 1973).

(2) Gopalakrishnan, P., Boreis, S.A., and Cambarnous, M., "An Enhanced Oil Recovery Method—Injection of Steam with Surfactant Solutions," Report of Group d'Etude IFP-IMF Sur les Milieux Poreux Toulouse, France, (1977).

(3) Handy, L.L., Amaefule, J.O., Ziegler, V.M., and Ershaghi, I., "Thermal Stability of Surfactants for Reservoir Application," SPE 7867, presented at California Regional Meeting of SPE, Ventura, CA, (April, 1979).

(4) Handy, L.L., Ershaghi, I., Amaefule, J., and Hsu, J., "The Use of Chemical Additives with Steam Injection to Increase Oil Recovery," D-9, presented at Fourth Annual DOE Symposium—Enhanced Oil and Gas Recovery and Improved Drilling Methods, Tulsa, Oklahoma, (August 29-31, 1978).

(5) Nelsen, F.M., and Eggertsen, F.T., "Determination of Surface Area Adsorption Measurements by a Continuous Flow Method," *Anal. Chem.,* Vol. 30, No. 8, 1387-1390, (1958).

(6) Hayes, M.E., Bourrel, M., El-Emary, M.M., Schechter, R.S., and Wade, W.H., "Interfacial Tension and Phase Behavior of Nonionic Surfactants," SPE 7581, presented at 53rd Annual Fall Technical Conference of Society of Petroleum Engineers, Houston, Texas, (October, 1978).

CO$_2$ FLOODING
STUDIES ON TEMPERATURE EFFECT
AND CO$_2$ RESERVES

The information in this chapter is from:

> *Fifth Annual DOE Symposium on Enhanced Oil and Gas Recovery and Improved Drilling Technology, Vol. 2—Oil*, CONF-790805-P2, Sponsored by U.S. Department of Energy, Tulsa, Oklahoma, August 22–24, 1979.

The information in this section is based on:

> "The Role of Reservoir Temperature in CO$_2$ Flooding", presented by K.I. Kamath, J.R. Comberiati and A.M. Zammerilli of U.S. DOE, Energy Technology Center, Morgantown, West Virginia.

ROLE OF RESERVOIR TEMPERATURE IN CO$_2$ FLOODING

Laboratory research on enhanced oil recovery by the CO$_2$ process has generally focused attention on oil reservoirs at temperatures well above 88°F, the critical temperature of CO$_2$, where the oil displacement is by the supercritical gas drive. However, many oil reservoirs in the U.S., including some which are currently under CO$_2$ flooding through joint DOE-industry participation, are at temperatures well below 88°F and pressures sufficiently high so that the displacement is by liquid CO$_2$.

Physicochemical considerations imply that highest oil recoveries by either process are to be expected when the reservoir temperature is as close to the critical temperature as possible. These conclusions are supported by data obtained from secondary floods made on linear Berea cores saturated with a light crude oil and water and swept by water-driven slugs of CO$_2$ at temperatures below and above 88°F under similar flooding conditions. These results appear to have special implications for the CO$_2$-flooding process for the Appalachian region of the U.S.

Phase Behavior of Hydrocarbon-CO$_2$ Systems

All pure substances are known to exist in at least three physical states: solid, liquid, and gas. Only the latter two, liquid and gas, are pertinent to this discussion. The equilibrium between the liquid and gaseous states of a pure substance, such as carbon dioxide, in a closed system is controlled by the temperature and pressure of the system. Figure 16.1 shows the saturated vapor pressure (pressure-temperature) curves for various pure substances including carbon dioxide. Points above and below the CO$_2$ curve denote liquid and vapor conditions, respectively. Points on the curve represent conditions at which liquid and vapor coexist in stable equilibrium. The lower terminus of the curve is the freezing point and the upper terminal point, its critical point (88°F and 1,070 psia). At the critical temperature of 88°F, the cohesive or surface tension forces that hold the CO$_2$ molecules together in the form of liquid are overcome by their kinetic forces. Above the critical temperature, the CO$_2$ molecules show ideal gas behavior.

Figure 16.1: Vapor Pressure of Pure Substances vs Temperature in °F

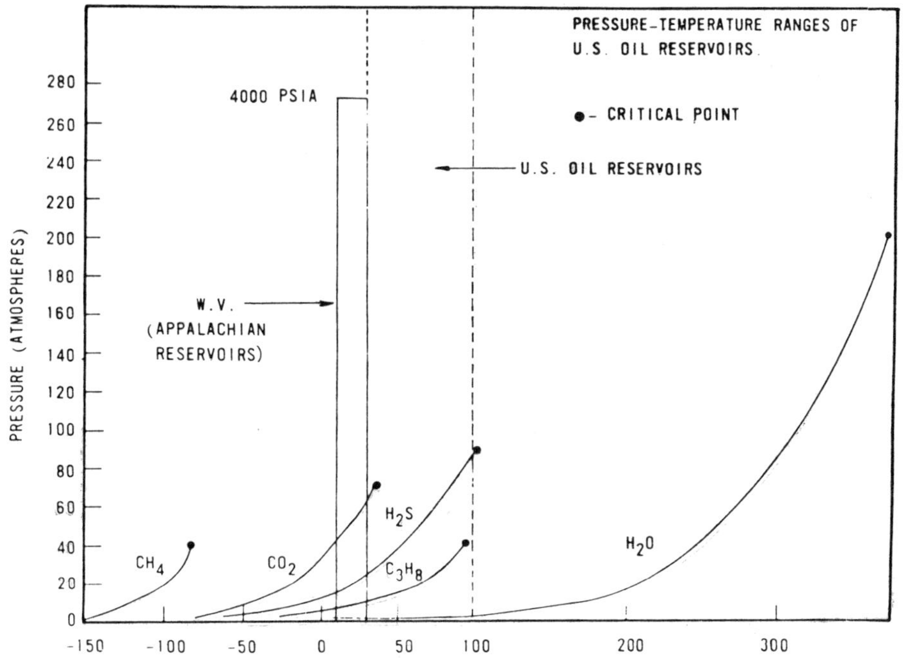

Source: CONF-790805-P2

The above discussion applies equally well to all pure substances, including hydrocarbons. It also applies, with proper modifications, to mixtures of pure substances. In general, the critical properties of pure compounds control the critical properties and phase behavior of their mixtures. Thus, the critical temperature of a binary mixture has a value generally (but not always) intermediate to that of the components. The phase behavior of multicomponent systems of hydro-carbons and CO_2 are of fundamental importance to CO_2 flooding and deserve consideration. However, an understanding of multicomponent systems can best be gained by an examination of the properties of simple binary systems as start-ing points. Three such systems that cover a wide range of natural reservoir con-ditions are represented by mixtures of carbon dioxide with methane, ethane and n-butane, with the latter three having critical temperatures of $-117°$, $90°$ and $306°F$, respectively (1).

A mixture of CO_2 and methane has a critical temperature between $-117°$ and $88°F$, depending on composition. Under laboratory conditions, both are gases and, therefore, their mixture is a homogeneous phase. On progressive isothermal compression of a methane-CO_2 mixture, a liquid condensate is formed at first. Further compression results in additional condensation and the attainment of a maximum condensate-to-gas volume ratio. Increase of pressure at this stage initiates the revaporization of the liquid phase (retrograde vaporization) and the pressure at which revaporization is complete may be considered to be the critical pressure of the mixture.

The locus of this critical pressure on the pressure-temperature diagram, Figure 16.2, is the critical locus of the methane-CO_2 system. It is characterized by a pressure maximum above which the mixture will always be a gas. The critical loci of the CO_2-CH_4, CO_2-C_2H_6 and CO_2-C_4H_{10} systems are shown in Figure 16.2.

Figure 16.2: Critical Loci for CO_2-Hydrocarbon Systems

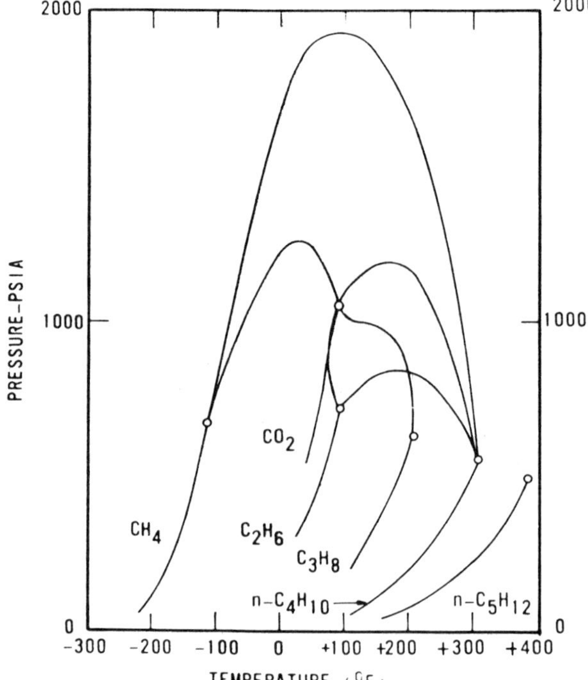

Source: CONF-790805-P2

Unlike the CO$_2$-CH$_4$ system, the CO$_2$-C$_2$H$_6$ system has no pressure maximum (miscibility pressure maximum), but shows a temperature minimum indicative of azeotrope formation (2). The CO$_2$-C$_4$H$_{10}$ system is similar to the CO$_2$-CH$_4$ system, except that the critical temperature has shifted nearly 200°F to the right to the predominantly liquid-phase region. Over the temperature range of 88° to 306°F, which encompasses nearly all reservoir temperatures, the CO$_2$-C$_4$H$_{10}$ system is far closer to the liquid state than the CO$_2$-CH$_4$ system. The CO$_2$-C$_4$H$_{10}$ system is also more representative of natural systems in the sense that dead oil, the target of enhanced oil recovery, is essentially all liquid hydrocarbons and contains little or no methane gas.

In CO$_2$ flooding, each of the crude oil hydrocarbons forms numerous binary systems (in reality, multicomponent systems) on contact with CO$_2$. It is believed that the lighter components of these systems have superior solvent properties with respect to the whole crude and are involved in an extraction process that ultimately results in an efficient miscible displacement. This concept implies that the CO$_2$ process would be more efficient for recovering heavier crudes than lighter ones. It is likely that in deep, warm reservoirs the injected supercritical CO$_2$ gas is in equilibrium with liquid oil so that the flow is essentially two-phase (discounting the aqueous phase) and overall miscibility of the CO$_2$-hydrocarbon system is never attained.

It follows from Figure 16.2 that at temperatures below 88°F, such as in Appalachian reservoirs, both the dead oil and CO$_2$ are, as a general rule, in the liquid state.

Reservoir depths and the flooded-out conditions prior to CO$_2$ injection both ensure that the fluid systems in these reservoirs are all liquid, in other words, essentially noncompressible. The effect of an increase of pressure on such systems is not likely to bring about any substantial improvement in the solubility or miscibility of the liquid CO$_2$ with the oil, and consequently, no increase in oil recovery. Hence, the use of an arbitrarily determined miscibility pressure as a screening factor for candidate reservoirs in the Appalachian region appears to be questionable. It is also significant that the usual definition of miscibility pressure as the minimum injection pressure at which CO$_2$ displacement of oil from a linear sand park such as $\frac{1}{4}$ i.d., 80 ft long, 10 darcy slim tube, would give a CO$_2$-breakthrough recovery of 80% of the original oil in place and an ultimate recovery of 94% is at best arbitrary.

Displacement Mechanisms

It was already stated earlier that CO$_2$ floods involve a supercritical or liquid CO$_2$ drive, depending on reservoir temperature. In the more general case, represented by the supercritical gas drive, the displacement mechanism is essentially a Buckley-Leverett type with mass transfer (3). In this respect, CO$_2$ gas is no different from methane or nitrogen, except that high pressures are required by the latter gases to bring about the same degree of gas-oil interaction as CO$_2$ to achieve comparable oil recoveries. Another crucial element of the recovery mechanism of particular significance to heavier crudes is believed to be the ability of CO$_2$ to vaporize the lighter hydrocarbons of the crude under supercritical conditions (retrograde vaporization) at first contact and the miscible displacement of the less volatile ends by some type of condensing drive.

In general, a CO_2 flood should not be substantially different in operation and performance from the so-called pressure maintenance or recycling operations of earlier days with ethane-rich natural gas.

The recovery mechanism with liquid CO_2 appears to be considerably more complex. CO_2 oil mixing, upon which oil recovery depends, is controlled by concentration gradients that are possibly higher than in the gas-liquid systems, but at lower diffusion rates. The unfavorable mobility ratio of the CO_2 process may serve to augment mixing by lateral diffusion resulting in a somewhat better sweep. Another possible complication that can complicate the CO_2-liquid flood is the formation of additional liquid phases and their influence on oil displacement. It has been observed during PVT tests, that under certain conditions, liquid CO_2 and oils can form a pair of partially miscible phases (4). The formation of two or more such phases at the CO_2-oil front in a CO_2 flood is not unlikely.

The flow, under such circumstances, would involve three phases, including the drive water, and may result in trapping some of the CO_2-rich oil phase or the oil-rich CO_2 phase behind the flood front. This could result in substantial losses of CO_2 to the formation that may explain material balance losses observed in CO_2 floods.

Solubility of CO_2 in Hydrocarbons

The effectiveness of the CO_2 process may be attributed to its unique chemical and physical properties. As a very mildly polar substance, CO_2 gas is several times more soluble in crude oil and hydrocarbons than it is in water (4). When dissolved in oil, it enhances oil mobility (increase of oil saturation by swelling and reduction of viscosity) in proportion to the amount dissolved. The solubility of a gas in a liquid, ideally, increases with pressure (Henry's Law) and decreases with increasing temperature. CO_2-hydrocarbon systems follow these general solution laws.

On the other hand, the mutual solubility of a pair of immiscible or partially miscible liquids generally increases with temperature and is unaffected by pressure unless the solution process is accompanied by a net change in volume. Mixing of nonpolar liquids such as hydrocarbons and CO_2, however, is usually volume-additive. Other conditions remaining the same, then, the solubility of CO_2 gas in crude oil will increase directly as the pressure and inversely as the temperature while the solubility of liquid CO_2 in crude oil will be unaffected by pressure but will increase with increasing temperature. It follows that maximum CO_2 solubility in oil would occur at the highest temperature at which it is still in the liquid state, i.e., at the critical temperature as the limit. Hence, oil recovery by CO_2 (liquid) flooding will tend to increase with reservoir temperatures up to $88°F$.

Reservoir Environment and CO_2 Flooding

Any oil recovery process that involves fluid injection is subject to certain constraints inherent in the reservoir. Among the most significant of these are reservoir depth and temperature. Reservoir depth is generally a rough measure of the maximum pressure the formation can withstand during fluid injection. This pressure, the overburden or parting pressure, on the average, is numerically equal in pounds per square inch to the reservoir depth in feet. On the other hand, the natural hydraulic pressure within the reservoir pores (pore pressure) has an average value of 0.5 psi/ft of depth. In fluid injection operation, therefore, the production engineer has a choice of injection pressures between the overburden pressure and the pore pressure of approximately 0.5 psi/ft of depth. Thus, for a formation depth of 2,000 ft, the injection pressures can be between 1,000 and 2,000 psi.

While the production engineer has a choice in injection pressure control, no choice in regard to temperature is obviously possible. It was pointed out earlier that under most existing reservoir pressures, CO$_2$ will be liquid at temperatures below 88°F and gaseous at higher temperatures, regardless of pressure. The physical state of carbon dioxide and the displacement mechanism would consequently be dependent on these reservoir conditions.

Table 16.1 lists reservoir conditions that correspond to the different types of CO$_2$ drives, supercritical gas, liquid or vapor: (a) at temperatures above 88°F, the displacement would be simple (below 1,070 psia) or supercritical (above 1,070 psia) gas drives; (b) under reservoir conditions represented by points above the pressure-temperature curve, the displacement will be by liquid CO$_2$; and (c) a vapor drive would result if the injection pressure is below the saturated vapor pressure of CO$_2$ (below the pressure-temperature curve) at reservoir temperature, e.g., the expansion of CO$_2$ from a tank of liquid CO$_2$ connected to the injection wellhead.

Table 16.1: Drive Mechanism and Reservoir Conditions

Temperature (T, °F)	Injection Well Pressure	Production Well Pressure	Drive Mechanism
	. . . . (psia)		
above 88	above 1,070	—	supercritical gas
above 88	below 1,070	—	gas
below 88	above 1,070	above 1,070	liquid
below 88	above 1,070	below Ps*	liquid and vapor
below 88	below 1,070	below Ps*	vapor

*Saturated vapor at T, °F.

Source: CONF-790805-P2

The distinction between a vapor drive and a gas drive is that the gaseous state is associated with temperatures above the critical. A vapor can exist only below the critical temperature. While a gas is subject to the normal gas laws, a vapor on compression condenses to a liquid with evolution of (latent) heat.

Experimental Work

The experimental work with Berea cores and a light Appalachian crude oil was specially oriented towards determining the disparate influence of temperature on liquid CO$_2$ and supercritical CO$_2$ gas flooding in oil recovery.

The setup for this study is shown in Figure 16.3. The pump used for brine, oil and isopropyl alcohol is a piston-type, positive displacement unit that provides a constant flow. Flow rates may be varied from 0.8 to 200 ml/hr with increments of 0.008 ml/hr in the low region and 2.0 ml/hr in the high region.

Figure 16.3: CO_2 Displacement Test Apparatus

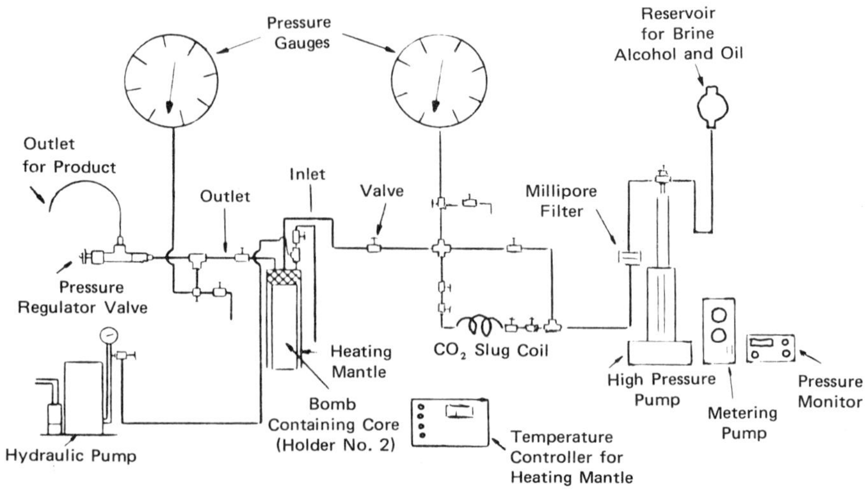

Source: CONF-790805-P2

The core is a Berea sandstone obtained from quarries in Ohio, with specifications listed in Table 16.2, along with data on materials used for the tests. The core is covered with heat-sensitive shrinkable plastic tubing and placed in a stainless steel bomb-type cylinder. Hydraulic pressure of 1,700 psi is applied inside of the bomb against the outside of the core. A pressure of 1,155 psi is used for all tests and is controlled by back pressure (pressure relief valve) at the outlet of the system.

Table 16.2: Cores and Materials

	. . . . Run No.		
	E-19	E-59	E-89
Core Characteristics			
Sandstone core	Berea	Berea	Berea
Diameter, cm	3.81	3.81	3.81
Length, cm	20.0	20.2	20.3
Pore volume, ml	50	47	46
Porosity, %	21.9	20.4	19.9
Permeability, md			
Ka (brine)	72	72	97
Ko (oil)	62	25	60
Materials			
Oil	Hilly Upland Crude Oil (3.64 cp)		
Brine	2% NaCl in distilled water		
Alcohol	commercial grade isopropanol		
CO_2	commercial grade cylinder with outlet tube reaching to bottom of inside cylinder for syphoning of liquid CO_2		

Source: CONF-790805-P2

For each test, the core was cleaned by flooding with 4 pore volumes of isopropanol, followed with 4 pore volumes of 2% brine to insure 100% water-saturation of the core. 4 pore volumes of Hilly Upland oil was then passed through the core to insure maximum oil saturation and to reach irreducible water-saturation. Flow rates for all of the above liquids was 40 ml/hr with no back pressure applied.

A sleeve-type heating jacket was placed around the core holder to heat the core to the desired temperature. A thermocouple-type probe was inserted between the heating jacket and the core holder to measure the temperature. The heating jacket and the probe were connected to an on-off type indicating temperature controller. The time necessary for the heat to pass through the core was determined by drilling a hole halfway through the center of the core and inserting a thermocouple. The core was then placed in a water bath at the desired temperature, and the time for the thermocouple to reach that temperature was measured. The temperatures (°F) and corresponding times in minutes are as follows: 86°, 6; 100°, 10; 130°, 10; 160°, 12; and 190°, 16.

A calibrated sample CO$_2$ coil (⅛" stainless steel) was weighed and filled with liquid CO$_2$ and reweighed. The difference in weights gave the weight of the CO$_2$ slug. For tests below the critical temperature of CO$_2$, density correction for temperature at the cylinder provided the volume of CO$_2$, which was then calculated for the pore volume of the slug. For tests above the critical temperature of CO$_2$, back calculations were made to determine the amount of liquid CO$_2$ that would provide the desired volume of gaseous CO$_2$. The displacement system was heated to the desired temperature and pressurized to 1,150 psig. The CO$_2$ was then permitted to flow into the core and was then followed by 1 pore volume of 2% brine at the desired flow rate (10 or 20 ml/hr). This was followed by blow-down, at which time the pressure was lowered in steps until atmospheric pressure was reached and the test terminated.

Oil Recovery and Reservoir Temperature: Table 16.3 and Figure 16.4 give oil recovery data on Berea cores containing Hilly Upland (a light WV crude) oil and a 2% NaCl brine. These floods were made in the secondary (not tertiary) mode to most effectively illustrate temperature effects. The data show the definite trend of increasing recovery with increasing temperature up to the vicinity of 88°F, and decreasing recovery thereafter, even though in the runs at higher temperatures, the throughput of gaseous CO$_2$ on a pore volume basis was from 2.15 to 3.9 times that of the liquid slugs used in the lower temperature floods on the same core.

**Table 16.3: The Effect of Temperature on Secondary
Oil Recovery by CO$_2$ Flooding**

Run No.	Flow Rate (ml/hr)	Core Temp (°F)	CO$_2$ Slug Liquid (pv)	CO$_2$ Slug Gas (pv)	So (pv)	Oil Recovery Displaced (pv)	Oil Recovery Blow-down (pv)	Oil Recovery Total (pv)	Oil-in-Place (%)
E-19-7	20	70	0.150	—	0.620	0.244	0.112	0.356	57.4
E-59-2	20	86	0.137	—	0.660	0.268	0.170	0.438	66.4
E-19-6	20	100	0.153	0.395	0.630	0.308	0.124	0.388	61.6
E-19-26	20	130	0.177	0.691	0.640	0.288	0.068	0.356	55.6

Source: CONF-790805-P2

Figure 16.4: Effect of Temperature on Secondary Oil
Recovery by CO_2

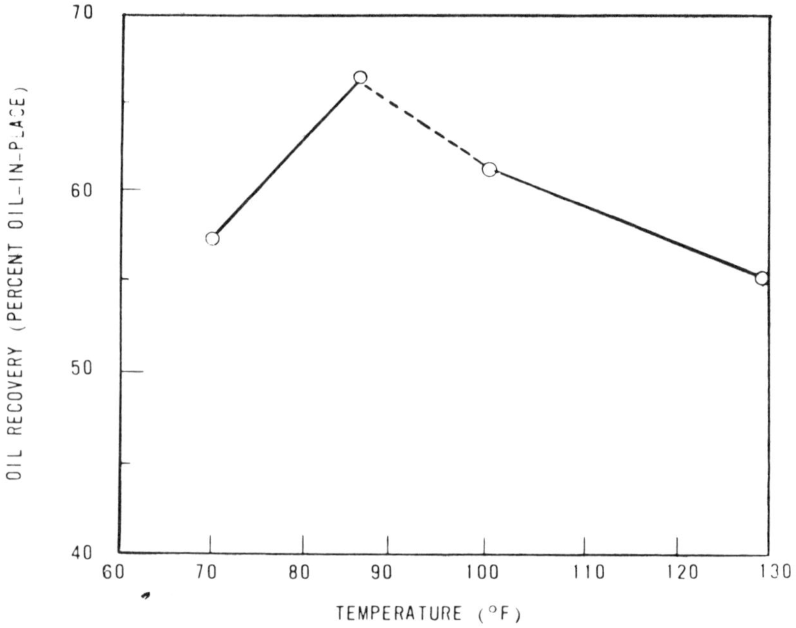

Source: CONF-790805-P2

These results imply that liquid CO_2 floods are superior to supercritical gas drives
and that higher reservoir temperatures up to 88°F show greater oil recovery potential for liquid CO_2 floods.

Oil Recovery and Back Pressure: Table 16.4 gives data on the effect of applied
back pressure on oil recovery. In the supercritical gas drive, the pressure of the
carbon dioxide in the gas phase controls the transfer rate and amount of gas
transferred into the oil phase. This has a maximum value at the injection well
where gas pressure is highest. In a liquid CO_2 flood, however, the corresponding
pressure (concentration potential) is at a maximum value as long as the CO_2 slug
is not completely dissolved in the oil.

Table 16.4: Effect of Applied Back Pressure on Oil Recovery

Run No.	Back Pressure (psig)	So (pv)	CO_2 Slug Liquid (pv)	Oil Recovery Dis-placed (pv)	Blow-down (pv)	Total (pv)	Oil-in-Place (%)
E-89-6	1,150	0.696	0.160	0.226	0.230	0.456	65.5
E-89-4	1,100	0.696	0.163	0.235	0.234	0.469	67.4
E-89-1	850	0.696	0.169	0.261	0.144	0.405	58.2
E-89-7	650	0.696	0.169	0.278	0.172	0.450	64.7

Source: CONF-790805-P2

On the other hand, if the natural sand face pressure in a production well is below the saturated vapor pressure of CO$_2$ at reservoir temperature, the liquid CO$_2$ slug would tend to vaporize at points in the reservoir at which the ambient pressure is just below the saturated vapor of CO$_2$, and a gas drive from this point on will be superimposed on the liquid CO$_2$ drive. Because such conditions may exist in field tests, a few runs were made in which the back pressure at the producing end of the core was maintained at pressures above and below 920 psia, the saturated vapor pressure carbon dioxide at the flood temperature of 75°F. These results which should be considered preliminary appear to indicate slightly better recoveries at the lower back pressures when the displacement is by CO$_2$ vapor.

Conclusions

- Experimental results indicate that oil recovery by CO$_2$ flooding under similar conditions is higher with a liquid CO$_2$ flood than with supercritical CO$_2$ gas.

- In liquid CO$_2$ floods, the oil recovery increases with increase of core temperature. In supercritical CO$_2$ floods, the opposite trend was observed.

- In the limited number of runs performed, it would appear that liquid CO$_2$ floods give higher oil recoveries than CO$_2$ vapor floods. This may mean that application of back pressure to production wells to ensure that the carbon dioxide remains a liquid during the displacement is desirable.

The information in the following section is based on:

"Naturally Occurring Carbon Dioxide Sources in the United States—A Geologic Appraisal and Economic Sensitivity Study of Drilling and Producing Carbon Dioxide for Use in Enhanced Oil Recovery", presented by F. Zimmerman of Gulf Universities Research Consortium.

AVAILABILITY AND COST OF CARBON DIOXIDE

Preliminary estimates (5) indicate 5 to 10 billion barrels of enhanced oil may be recovered by carbon dioxide flooding, which may require on the order of 20 to 50 trillion cubic feet (Tcf) of carbon dioxide. Research and field tests should resolve some of the uncertainties concerning the process so that the economics of carbon dioxide miscible flooding can be appraised more reliably. Two critical questions, however, remain to be resolved. Will the large volumes of carbon dioxide required be available and at what price will they be available?

Gulf Universities Research Consortium (GURC) was requested under its Department of Energy contract to assist Pullman-Kellogg (6) in an assessment of naturally-occurring carbon dioxide. GURC located and characterized the major source areas and estimated costs to drill and produce the carbon dioxide into a trunkline. Kellogg provided investment and operating cost estimates for compression, treatment and transportation.

GURC's overall responsibility included the evaluation of reservoir rock properties and gas composition for each potential carbon dioxide source; the estimate of costs for drilling and producing naturally occurring carbon dioxide reservoirs; and the analysis of field utilization of carbon dioxide in enhanced oil recovery (EOR) projects.

Source Delineation

GURC developed a data base available through the Office of Information Systems at the University of Oklahoma of naturally occurring carbon dioxide wells early in the study. Based in part on data originally published by the U.S. Bureau of Mines Helium Division (7) which included over 11,000 analyses of gas samples from oil and gas wells, the data base presently contains over 600 records representing occurrences of natural carbon dioxide greater than 5 mol % in 27 states. Initial delineation of areas containing potential reserves of carbon dioxide was conducted using the gas quality and reservoir characteristics data in the carbon dioxide data base.

The regions considered include West Virginia, the Jackson Dome area in Mississippi, the Delaware-Val Verde Basins in West Texas and nine subareas of the Rocky Mountains, as follows:

> Southwest Colorado, McElmo Dome area;
> South Central Colorado, Sheep Mountain area;
> North Central Colorado, North Park Basin;
> Northeast New Mexico, Sierra Grande Arch area;
> Northwest New Mexico, San Juan Basin;
> Central Utah, Farnham Dome area;
> Central Utah, Gordon Creek area;
> Southeast Utah, Paradox Basin; and
> Wyoming, Sweetwater and Uinta Counties area.

Table 16.5 briefly characterizes the gas compositions, estimated productivity, and pressures of the more significant carbon dioxide source areas.

Selection Criteria for Detailed Study

Three criteria were established to determine which of these source areas warranted detailed reservoir and economic evaluation: (1) reserves apparently sufficient for pipeline consideration; (2) proximity to miscible flood candidate reservoirs; and (3) evidence of strong industry interest.

On this basis, four areas were chosen for further geologic, reservoir and economic analysis:

> Southwest Colorado, McElmo Dome area;
> South Central Colorado, Sheep Mountain area;
> Northeast New Mexico; and
> Mississippi, Jackson Dome area.

These areas, each representing a unique geologic environment and posing varying drilling and production conditions, are reasonably representative of the range of expected naturally occurring carbon dioxide sources in the United States.

Table 16.5: Areas Reported to Have Significant CO$_2$ Deposits

State	Field/ Location	Gas Composition (mol %)	Productive Stratigraphic Units	Expected* Sustained Deliver- ability (MMcfd)	...Expected Range of...		Expected Well Depths (ft)
					Shut-In Wellhead Pressures	Initial Flowing Wellhead Pressures	
					. . . . (psig)		
Colorado (Southwest)	McElmo Dome Area, Dove Creek/ Montezuma and Dolores Cty	>96% CO$_2$, balance N$_2$	Leadville, Elbert	3-12	2,200-2,800	600-850	6,000-8,000
(South Central)	Sheep Mountain Area/ Huerfano Cty	99-99% CO$_2$, balance N$_2$, CH$_4$	Dakota, Entrada	3-12	700-1,100	200-700	3,400-7,000
Mississippi (Central)	Jackson Dome Area/ Rankin and Madison Cty	70-99% CO$_2$** many wells >98% CO$_2$	Smackover, Norphlet	4-16	4,000-6,500	2,000-4,000	12,000-17,000
New Mexico (Northeast)	Harding, Union, Mora and Colfax Cty	92-99.7% CO$_2$, balance N$_2$	Triassic-Santa Rosa, Permian-Glorieta, Tubb	0.1-2	250-700	50-500	1,500-2,800
Utah (Central)	Gordon Creek Area/ Carbon Cty	91-99% CO$_2$, balance N$_2$, CH$_4$	Sinbad, Coconino	NA	3,900-4,200	NA	11,000-12,000
(Southwest)	Farnham Dome Area/ Carbon Cty	98% CO$_2$	Navajo, Sinbad, Kaibab	NA	1,000-1,500	NA	2,800-5,000
West Virginia (South Central)	Fayette and Kanawha Cty	20-80% CO$_2$, balance N$_2$, CH$_4$	Tuscarora	0.1-2	1,800-2,900	NA	6,000-9,000
Wyoming (Southwest)	Church Buttes Area/ Uinta and Sweet- water Cty	83-86% CO$_2$, 2% H$_2$S, balance N$_2$, CH$_4$	Madison	2-6	NA	low	18,000-19,000

*Highly speculative; requires sustained production history for confirmation.
**May contain from trace to 10% H$_2$S.

Source: CONF-790805-P2

Figure 16.5: Major CO_2 Productive Areas and Oil Reservoirs Amenable to CO_2 Flooding

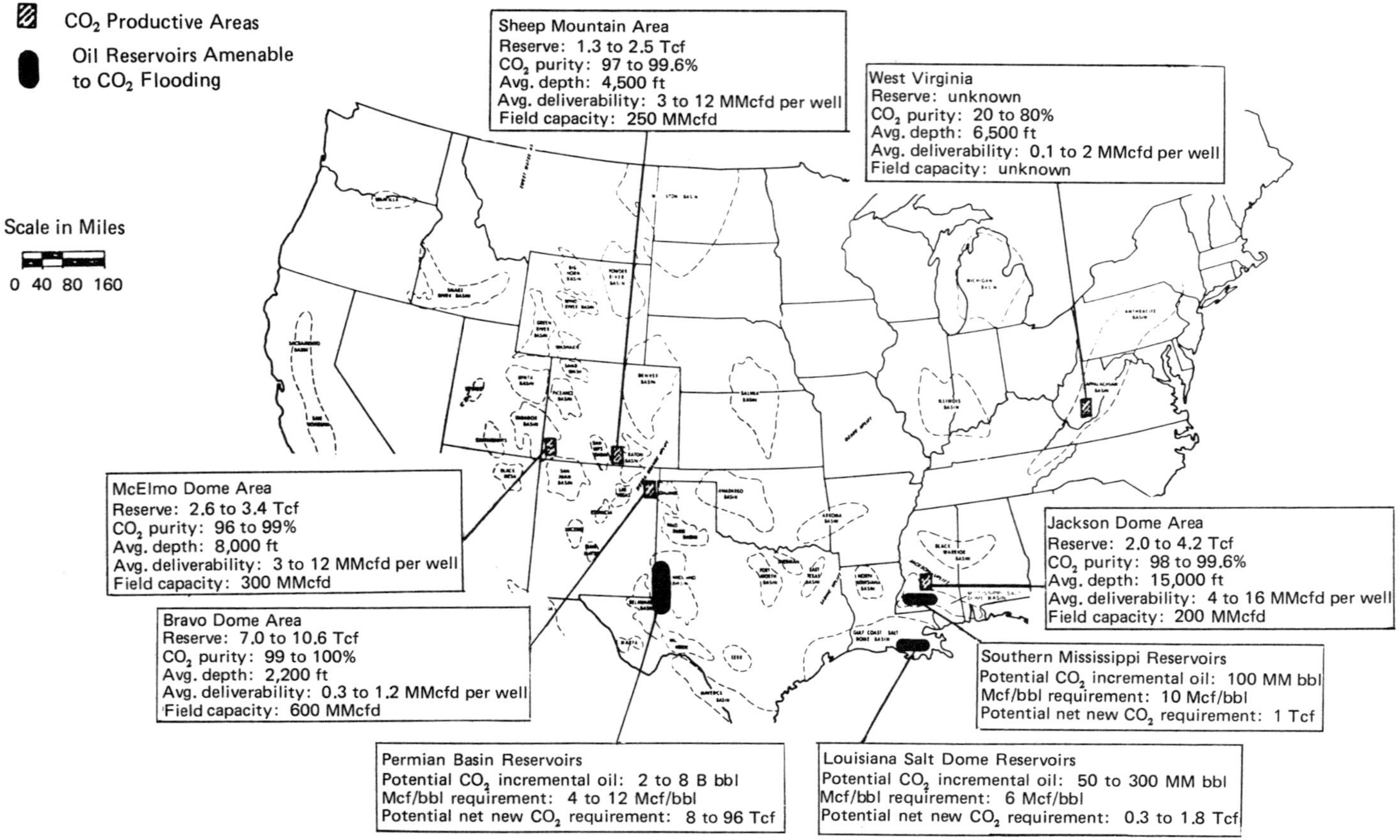

Source: CONF-790805-P2

In-Depth Analysis of Primary CO$_2$ Source Areas

The locations of several major carbon dioxide productive areas, examined in depth in GURC's report (8) on naturally occurring carbon dioxide, are identified in Figure 16.5. Brief summaries of the potential reserve, gas purity, average deliverability per well and expected field capacity are included for each of the areas. Areas containing significant oil reservoirs amenable to carbon dioxide flooding are also identified on Figure 16.5. Estimates of potential oil recoverable by carbon dioxide flooding, carbon dioxide requirement in thousand cubic feet per barrel recovered, and potential net new carbon dioxide requirements are summarized for three areas containing significant candidate reservoirs. (Net new CO$_2$ = total CO$_2$ injected – produced CO$_2$ recycled in reservoir.)

Reservoir Analysis Methodology

Data obtained from exploratory programs in each area, including well logs, core analyses and completion reports were utilized to develop estimates of porosity, gas saturation, net pay and original gas in place (OGIP) in each area.

Field test data, including flow rates, flowing tubing pressures and pressure draw-downs were then analyzed. For each area ranges of average sustained deliverabilities and flowing tubing pressures were estimated. Production was assumed to be by depletion drive for all areas investigated.

For each area, a set of well models spanning the probable variation in rock quality, OGIP and sustained productivity was created. Flowing bottom hole pressure decline curves were constructed as a function of cumulative gas produced, allowing easy correlation with varying flow rates and recovery efficiencies. Individual well economics on these representative well models were then calculated. A brief summary of four significant, naturally occurring carbon dioxide sources is presented below.

Southwestern Colorado–McElmo Dome Area: Exploration for carbon dioxide reservoirs in southwestern Colorado has focused on the McElmo Dome area in Montezuma County and the Doe Canyon–Dove Creek area of Dolores County. The two carbon dioxide productive zones in this portion of the Paradox Basin are the Leadville (Mississippian) and the Ouray (Devonian), found at depths of 6,500 to 9,000 ft subsurface. Analysis of produced gas from both formations indicates a composition of 96 to 99% carbon dioxide with 1 to 4% nitrogen.

At the time of the study, 16 wells had been drilled in the area. 9 wells were completed as potential producers, while 5 were reported as dry holes. Data were unavailable for 2 wells.

Data indicate the reservoir has an areal extent of 100 productive sections and from 75 to 100 ft net pay thickness with approximately 7% porosity. OGIP is estimated to range from 3.8 to 4.9 Tcf. Ultimate recovery (65 to 70% recovery efficiency, E_R, a fraction of OGIP) might reach from 2.6 to 3.4 Tcf. Sustained well deliverabilities are anticipated in the 3–12 MMcfd range, with flowing wellhead pressures from 600 to 850 psi. If flowing wellhead pressures in this range can be achieved, then only one stage of compression should be required for delivery into a pipeline at around 2,000 psi.

South Central Colorado–Sheep Mountain Area: The Sheep Mountain area, located in Huerfano County, Colo. is carbon dioxide productive from two stratigraphic units. The Dakota (Cretaceous), and a smaller reservoir, the Entrada (Jurassic), produce gas composed of 97 to 99.6% carbon dioxide with minor amounts of nitrogen and methane present.

Data from 15 wells supplied well control, with 10 wells completed as potential producers and 5 wells reported as dry holes. Data were available for 6 other wells in the area.

Based on available data, the productive area of the Dakota reservoir ranges from 25 to 30 sections, with 75 to 125 ft of net pay and an average porosity of 17%. OGIP in the Dakota is estimated to be 1.9 to 3.8 Tcf. Ultimate recovery (65% E_R) might range from 1.3 to 2.5 Tcf. Per well average sustained deliverability is estimated to range from 3 to 12 MMcfd. The wells are expected to be produced at sufficient flowing pressures to enable delivery to a pipeline (1,500 to 2,000 psi), with only one stage of compression.

Northeast New Mexico—Bravo Dome Area: Several areas in northeastern and north central New Mexico, including locations near Bueyeros, Wagon Mound and Turkey Mound, have been recent exploration targets for commercial carbon dioxide reserves.

The most rewarding drilling activity to date has occurred on a large structural nose on the eastern flank of the Sierra Grande Uplift. Several stratigraphic units in the area have exhibited shows of carbon dioxide. The Glorieta (Permian), Tubb (Permian), and a granite wash member appear to be carbon dioxide productive. 21 of 27 exploratory wells in the area were reportedly productive. 4 wells were reported dry and no data were available on 3 other wells.

Recent exploratory drilling in the area indicates high-grade carbon dioxide of 99.1 to 100% purity, with a maximum of 0.3% nitrogen and 0.2% hydrocarbon as contaminants.

The productive area appears to cover 880,000 acres, or roughly 38 townships, with an average of 40 to 60 ft net pay, and approximately 20% porosity. OGIP is estimated to range from 10.9 to 16.3 Tcf. The ultimate recovery (65% E_R) would then range from 7.0 to 10.6 Tcf. Anticipated sustained per well deliverabilities range from 0.3 to 1.2 MMcfd, with flowing tubing pressures at least 140 psi. With sustained flowing tubing pressures in excess of 140 psi, two stages of compression may be sufficient for delivery to the pipeline.

Central Mississippi—Jackson Dome Area: Jurassic sediments in portions of Rankin, Madison and Scott Counties, Mississippi, have tested sour gas over the past 20 years of exploration. A corridor of carbon dioxide and H_2S-rich gas is located on the east-northeast flank of the Jackson Uplift and extends in an arc from north central Madison County through central Rankin County.

Two stratigraphic units, the Smackover (Jurassic) and the Norphlet (Jurassic) have indicated strong carbon dioxide potential during exploratory drilling and testing.

Wells in the area have tested carbon dioxide concentrations ranging from 65 to 99.6% from the Jurassic section. Jurassic gas in this area is associated with up to 10% H$_2$S and varied quantities of methane. 7 of 10 wells which have tested Jurassic formations in the area are apparently productive of carbon dioxide in excess of 98% purity.

Present control indicates the combined productive area of four structural features containing gas greater than 98% carbon dioxide in a minimum of 27 sections, and total net pay of the Jurassic section is 120 to 250 ft. With 13% average porosity, minimum OGIP (CO$_2$ > 98%) would range from 3.4 to 7.0 Tcf. Minimum ultimate recovery (60% E$_R$) would then range from 2.0 to 4.2 Tcf. Sustained per well deliverabilities from both the Smackover and Norphlet, individually, are expected to range from 4 to 16 MMcfd. Initial flowing tubing pressures are expected to range from 2,000 to 4,500 psi. It seems unlikely that dual completions will be made in these two formations.

Economic Analysis

Drilling costs for carbon dioxide wells were found to be highly site-specific and regionally variant due to topographic and geologic anomalies. Completed well costs were based on those reported by operators in each area. A dry hole cost was estimated in each area, based on historical dry hole ratios and estimated intangible well costs, for both exploration and development drilling phases. Lease bonus fees and estimated probable operating costs for carbon dioxide wells were obtained from industry personnel.

Investment and operating costs for surface processing equipment, composed of compression and dehydration facilities and a gathering system, were supplied by Pullman Kellogg. Each well was assigned a pro rata expenditure for surface processing equipment based on its production contribution to a 50 MMcfd module. Actual surface processing costs may vary considerably if two-phase flow is encountered at the surface.

Discounted cash flow economic evaluations considering lease bonus, exploration costs, drilling and development expenses, gathering, treating, and compression costs, and probable time delays (as discussed below) were calculated using a commercial computer package (9).

Earning power was calculated on an after Federal income tax basis, with a tax rate of 48% and no depletion allowance for all cases. A 12.5% royalty was assumed. Economics were calculated in constant value 1978 dollars and assumed that inflation in investment and operating expenses will not exceed escalation in wellhead gas price.

As might be expected, the probable time delay between initial expenditures and initial carbon dioxide production, expected to be 5 to 6 years, had a negative impact on present value economics. Economic results must be considered speculative and depend on sustained production history for confirmation.

In each of the four areas analyzed in detail, sensitivity of earning power to variation in sustained flow rate and reserve per well was studied considering three rate cases and three reserve cases, at two gas prices, $0.25 and $0.50/Mcf. The study of the Jackson Dome area, Mississippi, is presented as an example of the economic analysis methodology.

Economics were calculated for 3 sets of model wells, which assumed average sustained producing rates of 4, 8 and 16 MMcfd. Economics for each set of the model wells were evaluated for per well reserves of 17, 51 and 152 Bcf. This range of per well reserve estimates is based on recovery efficiencies of 60% from a 640-acre drainage area and net pays of 25, 75 and 225 ft, respectively.

Capital expenditures for each well model are summarized in Table 16.6. Well operating costs and surface processing operating costs, including hydrate inhibition, drying and compressing the gas (6) are listed in Table 16.7. It is assumed that costs of purifying the carbon dioxide, i.e., removal of H_2S will be borne by the pipeline.

Table 16.6: Investment—Jackson Dome Area, Mississippi

Flow Rate (MMcfd)	Compression Required (BHP)	Compression Investment*	Dehydration Cost	Well Cost	Dry Hole Cost**	Gathering System Cost***	Total Cost†
		. . . .($)					
4	516	279,100	32,000	2,500,000	360,000	250,000	3,171,100
8	1,032	558,100	64,000	2,500,000	360,000	250,000	3,482,100
16	2,064	1,116,200	128,000	2,500,000	360,000	250,000	4,104,200

*Based on $540.8/BHP.

**Pro rata dry hole cost assigned on intangible investment for 20 dry wells per 80 producing wells.

***Gathering system based on 1 mile of 4-inch stainless steel quality pipe. Also includes $65,000 per well for hydrate control.

†Total excludes $64,000 per 640 acres lease fee.

Source: CONF-790805-P2

Table 16.7: Operating Costs—Jackson Dome Area, Mississippi

$/well/month	300
¢/Mcf for compression from 565 psi (2,000 psi when required)	7.7
¢/Mcf for dehydration	0.53
¢/Mcf for hydrate control	0.4
Total cost, $/unit	0.0863/Mcf

Source: CONF-790805-P2

The investment schedule assumes capital expenditures will occur over a 6 year period prior to actual carbon dioxide production. Investment for development drilling and surface processing equipment was scheduled over the final 2 years (years 5 and 6) prior to carbon dioxide production.

Results of discounted cash flow rate of return calculations, using the above reservoir input and gas prices of $0.25 and $0.50/Mcf, are presented in Figure 16.6. (In the figure heading, AFIT DCF-ROR is After Federal Income Tax Discounted Cash-Flow Rate of Return.)

Figure 16.6: AFIT DCF-ROR vs Per Well Reserve for Varying Flow Rates and Gas Price–Jackson Dome Area, Mississippi

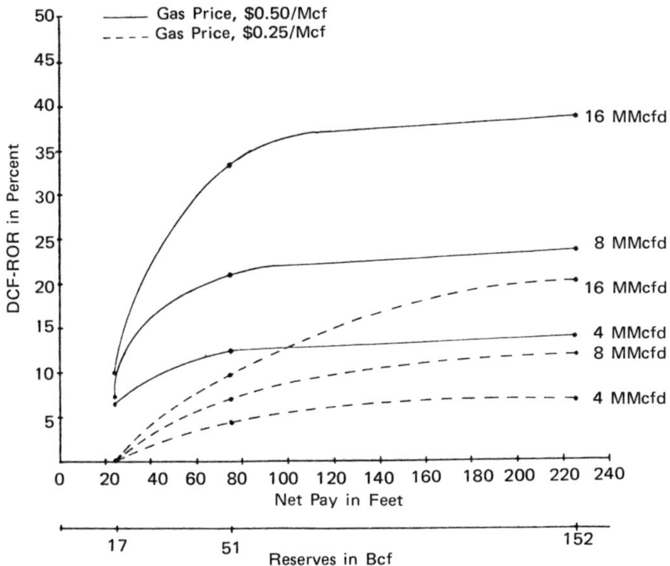

Source: CONF-790805-E2

Summary

The three areas surveyed in the Rocky Mountain province, the McElmo Dome area, Sheep Mountain area, and Bravo Dome area, potentially represent an aggregate carbon dioxide OGIP of 16.6 to 25 Tcf. Indicated carbon dioxide reserve from these areas is estimated to range from 10.9 to 16.5 Tcf. Peak pipeline delivery to west Texas oil fields will probably be on the order of 1.15 Bcfd.

Other carbon dioxide accumulations in Utah and Wyoming may eventually be developed, if demand is strong enough, and may significantly add to the indicated reserves in the Rockies.

OGIP carbon dioxide in the Jackson Dome area of Mississippi is estimated to be 3.4 to 7.0 Tcf. Ultimate recovery will probably range from 2.0 to 4.2 Tcf. Maximum transport of gas available for oil fields in southern Mississippi and Louisiana is expected to be on the order of 150–200 MMcfd.

REFERENCES

(1) Katz, D.L. et al, *Handbook of Natural Gas Engineering,* McGraw Hill, pp. 81–84 (1959).

(2) Sage, B.H. and Lacey, W.N., *Volumetric and Phase Behavior of Hydrocarbon Systems,* Stanford University Press, p. 116 (1939).

(3) Sandrea, R. and Nielson, R.F., *Dynamics of Petroleum Reservoirs Under Gas Injection,* Gulf Publishing Company, p. 115 (1974).

(4) Cales, G.L., *Role of the Aqueous Phase in Phase Behavior of the Hydrocarbon-Water-CO₂ System in Relative to Enhanced Oil Recovery by CO₂ Flooding,* M.S. Thesis, WVU (1978).

(5) Doscher, T.M. and Wise, F.A., "Enhanced Crude Oil Recovery Potential—An Estimate," SPE 5800 Symposium on Improved Recovery, Tulsa, Oklahoma, March 22–24, 1976.

(6) Hare, M., Perlich, H., Robinson, R., and Shah, M., "Sources and Delivery of Carbon Dioxide for Enhanced Oil Recovery," U.S. DOE Contract No. EX-76-C-01-2515 (1978).

(7) U.S. Bureau of Mines, *Analyses of Natural Gases, 1917–1974,* U.S. Bureau of Mines CP-76 (PB-251, 202/AS): Washington, D.C., U.S. Government Printing Office (1976).

(8) Zimmerman, F.W., "Naturally Occurring Carbon Dioxide Sources in the United States— A Geologic Appraisal and Economic Sensitivity Study of Drilling and Producing Carbon Dioxide For Use in Enhanced Oil Recovery", GURC Report 165, Bellaire, Texas (1979).

(9) "POGO (Profitability of Oil and Gas Opportunities)", PSI Energy Software, Calgary, Alberta.

THE FUTURE
OF ENHANCED OIL RECOVERY

The information in this chapter is based on:

An Investigation of Primary Factors Affecting Federal Participation in R&D Pertaining to the Accelerated Production of Crude Oil, prepared by J.M. Sharp, of Gulf Universities Research Consortium, for National Science Foundation under Contract NSF-C-942 (GURC Report 140), September 15, 1974.

A Survey of Field Tests of Enhanced Recovery Methods for Crude Oil, prepared by J.M. Sharp, of Gulf Universities Research Consortium, for National Science Foundation under Contract NSF-C-942 (GURC Report 140-S), December 27, 1974.

The Potential and Economics of Enhanced Oil Recovery, FEA Report B-76/221, prepared by Lewin and Associates, Inc., for Federal Energy Administration, April 1976.

Enhanced Oil Recovery Potential in the United States, OTA-E-59, prepared by Lewin & Associates, Inc., National Energy Law and Policy Institute and Wright Water Engineers, Inc., for U.S. Office of Technology Assessment, January 1978.

The outlook for enhanced oil recovery fluctuated radically during the previous decade. From a surge of optimism in the early 1970s, the prevailing view then held that the promise of EOR would be slow to unfold, would be prohibitively costly, and would be fraught with risk. While all of these cautions were true to some extent and still are, it is time for the pendulum of opinion to center. EOR production can, and should, contribute much to our nation's goal of domestic energy self-sufficiency. New efforts, and even the expansion of ongoing efforts to realize this goal, however, face major constraints which should be examined.

CONSTRAINTS TO EOR DEVELOPMENT

Foremost among a number of deterrents to full-scale development of EOR production are the following limitations:

(1) The unproved and thus risky nature of certain of the tertiary recovery techniques, notably surfactant and miscible flooding

(2) The large initial capital requirements

(3) The lack of a reliable long-term price of oil against which to calculate economics

(4) The time required to study reservoirs, to carry out essential laboratory tests, to design and conduct pilots, and then to expand those pilots into field-wide tertiary operations

(5) The limited available supply of certain injection materials, e.g., carbon dioxide, sulfonates, and polymers.

Uncertain Resource Availability and Process Performance

There is an uncertainty of 15 to 25% (or more) in the amount of oil remaining in reservoirs after primary and secondary recovery. In addition, there is uncertainty about the fraction of the remaining oil that can be recovered by an EOR process even after the process has been successfully pilot tested. A relatively small reduction in process performance can lead to a much larger reduction in potential EOR production; a 12 to 30% reduction in the amount of oil recovered (depending on the process) produces a 64% reduction in ultimate production at $22 per barrel, and a 163% reduction at $13.75 per barrel. ($13.75 is the January 1977 average price of $14.32 per barrel of foreign oil delivered to the East Coast, deflated to July 1, 1976, while $22.00 per barrel is the price at which the Synfuels Interagency Task Force estimated that petroleum liquids could become available from coal.) Similar reductions result for the 15 to 25% uncertainty in remaining oil. This disproportionate effect occurs because a relatively small decrease in expected production can reduce the rate of return from many reservoirs to below the 10% needed to make EOR operations an attractive investment.

Resource uncertainty represents a major technical and economic risk for any EOR project in any reservoir. Reduction of this risk would need high priority in any national program to stimulate EOR production.

Sampling a reservoir through core drilling, logging, and other well testing is an expensive, inexact, developing technology. The problem is that of finding methods which will probe outward a sufficient distance from a well bore to determine oil content in a large fraction of the region drained by the well. A further complication exists in that oil saturation variations occur both horizontally and vertically within a reservoir. Determinations at one well may not be applicable at other well sites.

A program to stimulate EOR production should contain a major effort to promote measurement of residual oil saturations in key reservoirs until confidence is gained in methods to extrapolate such data to other locations in the same reservoir and other reservoirs. Equal emphasis should be placed on the gathering of such data and on the improvement of measurement methods.

Process Mechanisms: Enhanced oil recovery processes are in various stages of technological development. Even though steam drive is in limited commercial development, the outer limits of its applicability are not well understood. Steam drive can probably be extended to light oil reservoirs but it has not been tested extensively (1). Larger gaps in knowledge exist for other processes and process modifications which are in earlier stages of development. Field tests have consistently been undertaken with incomplete knowledge of the process mechanism. Most laboratory tests of processes are done on systems of simple geometry (generally linear or one-dimensional flow), leaving the problems that occur because of the more complex flow geometry to field testing.

Extensive field testing of EOR processes will be required. As more field projects are undertaken, the tendency of industry is to shift research personnel from basic and theoretical studies to development activities. If this effort is widespread, it may limit the ability of companies to undertake fundamental EOR research. If this trend continues, additional public support for basic research applicable to EOR may become advisable.

A major industry/government coordinated effort is needed to thoroughly define the process mechanisms for each of the recognized basic EOR processes and process modifications.

Volumetric Sweep Efficiency: Recovery efficiency of all processes depends upon the fraction of the reservoir volume which can be swept by the process, i.e., sweep efficiency. Thus a strong economic incentive exists for improvement of volumetric sweep. Research in this area has been carried out for a number of years by many sectors of the oil production, oil service, and chemical industries. The importance to EOR success of improving sweep efficiency has been confirmed in a recent assessment of research needs (2).

Despite the long-term effort on this problem, success has been limited. Solutions are not available for each process. Improvements are needed for each individual process and each process variation as well as for major classes of reservoirs. Progress will be difficult and will require major field testing supported by extensive prior laboratory work.

Operating Problems: Operating problems of EOR are more severe, less predictable, and certainly less easily controlled than operating problems one faces in a plant making a new chemical product, but the industrial sector is equipped to solve such problems. For example, steam is now generated from high-salinity brines, a process that once seemed to pose serious technical problems.

Some problems exist where government assistance could be beneficial. Design of steam generators for steam-drive projects that will meet environmental pollution-control standards and use cheaper alternate fuels (heavy crudes, coal, etc.), as well as large-scale steam generation, are areas that have been recently highlighted (1). Equipment for retrofitting existing generators to permit them to meet new standards and lower their unit pollution level is also needed.

Capital Requirements

Tertiary recovery entails considerable initial capital. Not only must the operator carry the traditional field development costs but he must bear the cost of additional field equipment and injection materials associated with tertiary recovery. While capital requirements can be eased by phased start-up (over 15 years), lengthy field testing and development (from 6 to 19 years), and judicious sequencing of project starts (initiating the most profitable projects first and then moving to other projects), the capital requirements will still be staggering. According to estimates by Lewin and Associates, Inc., costs of the injection materials and fluids (surfactants, polymers, CO_2, steam, etc.) constitute 82% of EOR costs, while costs for field development and equipment comprise the remaining 18%.

The price of enhanced/tertiary oil is the single most important determinant of how much will be produced and by when. Most directly, the price determines which fields will be developed by tertiary means, which techniques can economically be used, and how long the field will produce. Less directly, price influences the level of expenditure for the research and development needed to improve the efficiency of EOR methods. Although the total capital requirements are large, however, the bulk of capital constraints should arise only during the near-term through the early 1980s. After that time, the cash flow generated by the successful tertiary projects should begin to provide sufficient capital to finance future projects.

Timing

As the nation's declining oil fields, particularly those that are already marginal under the prevailing price structure, reach their economic limit, there are strong incentives to plug the well, pull the pipe, and abandon the field. Technically and economically, a shutdown field poses the most severe challenges to tertiary recovery and, in fact, may preclude any potential oil production from that source.

Technically, for some reservoirs the enhanced process works best when used immediately after primary recovery. For other reservoirs, the optimum time may be soon after the waterflood has established the fluid transmissibility characteristics of the reservoir. Ideally, from an economic point of view, tertiary recovery should come along before primary or secondary production reaches its economic limit. This would allow the tertiary oil to share operating costs with ongoing projected production. Thus a second constraint to be overcome would be the increasing economic and technical burdens imposed by undue delay in the inception of tertiary recovery.

Injection Materials

A shortage of certain injection materials may dictate whether or not a specific EOR technique is feasible. Cost of the injection material itself or costs of transporting the material to the location where it is needed may be prohibitive. The injection material most critically in short supply is natural CO_2. Several important sources in West Texas, eastern New Mexico, southern Colorado, and Mississippi, however, have been identified. (See the previous chapter for a discussion of naturally occurring carbon dioxide sources in the United States.) Also, the possibilities of extracting CO_2 from SNG plants, fertilizer or ammonia plants, and from refinery stack gases is under consideration.

Carbon dioxide availability is, of course, central to any major expansion of CO_2 flooding. The quantity needed has been estimated to be as high as 53 trillion cubic feet, a volume almost three times the annual volume of natural gas consumed in the United States.

The economic potential of CO_2 flooding is so great that a government effort to accelerate EOR production should include not only locating natural sources of carbon dioxide but also exploring ways to produce it economically from large-scale commercial sources.

Sulfonate and polymer needs will not pose as great a constraint as will CO_2. Considering that the domestic chemical industry already has enough capacity to produce 500 million pounds of sulfonate per year (with additional major plants coming on stream), the relatively short construction lead time (3 years) and the relatively smaller initial requirements, sulfonate supply should not be a near-term constraint. Polymer requirements could pose some constraints since production of polyacrylamide and polysaccharide has been only about 30 million pounds per year (3).

Enhanced oil recovery processes will be used largely in those parts of the country that face increasing shortages of freshwater. For surfactant/polymer and polymer flooding, relatively freshwater is still needed both for the polymer and surfactant solutions and for reservoir preflushing. Even where freshwater is available for preflushing, it is often not efficient in displacing brine. Consequently, injected fluids in such reservoirs must be brine compatible. Continued laboratory and field research is needed to develop surfactants and other oil-recovery agents which are brine compatible.

Field Equipment

Just as a scarce supply of injection materials could stall tertiary recovery, so too could a shortage of field equipment. For example, a shortage of steam generators could restrict the production of reserves attributable to steam flooding. Fortunately, the considerable amount of steaming already under way (both cyclic and flood), combined with the lengthy start-up and development time, should ease the strain. Also, the shortages of drilling rigs, tubular goods, and contractor services appear to have eased. The supplies of equipment should thus not pose serious constraints.

The shift from primary to secondary recovery over the past several decades has been generally successful but has brought with it many surprises. While waterflood and pressure maintenance methods have been generally successful, the forecast production/time profile has seldom been realized. Conventional steam flooding operations have been successful in many cases, also; however, after nearly 20 years of experience, total production by this method is only about 200,000 barrels per day.

With higher prices, production by thermal methods will increase slowly. In situ combustion projects, for example, that for specific reservoirs had forecast a return of several dollars-per-dollar-invested at an oil price of only $4 per barrel, resulted in production costs of $10 per barrel, an economic disaster. This was caused by indeterminate (but not unanticipated as probabilities) factors such as channeling or extreme corrosion resulting from high temperatures and high oxygen content coupled with reservoir chemistry.

A successful EOR venture, even in terms of an equivalent increase in reservoir proved reserves, does not attract the public visibility and the attendant increase in borrowing power and stock values that a major new oil discovery attracts. All of these factors result in an industry approach to EOR applications which, although several times more active at the higher market prices than previously, must be cautious and, therefore, too slow to allow EOR to make the contribution to the nation's needs that it should in the next several decades.

Manpower

Shortages of technically trained people to operate EOR projects may exist temporarily if a major national EOR effort is undertaken. National projections of needs for technically trained people have not been highly accurate. Data are not readily available on industrial needs since many firms do not make formal, continuing, long-range personnel forecasts. The efforts of ERDA and other agencies in national manpower forecasting could be encouraged.

All EOR processes are extremely complex compared to conventional oil recovery operations. Because of this technical complexity, highly competent personnel must be directly involved in each EOR project on a continuous basis at the managerial, developmental, and field operations level. Without close monitoring by qualified technologists, the odds for success of EOR projects will be lower. There currently is a mild short-term shortage of persons to work on EOR projects. National forecasts (4) of the number of available college-age students (all disciplines) indicate a significant enrollment decline over the period of greatest potential EOR activity. The supply of technical people (engineering, science, and business) available for EOR operations will crucially depend upon the economic climate in other sectors of the economy. In a generally favorable economic climate, increasing competition for qualified personnel could develop.

NECESSITY FOR FEDERAL PARTICIPATION

It is the opinion of Gulf Universities Research Consortium that a cooperative federal/industry program of the type required to compress the time for reduction of EOR technology to practice is essential. In addition to advancing EOR technology, the program should permit the development of jointly endorsed methods for (1) assessing the risk of EOR ventures and (2) estimating the rate of return for such ventures which are based on standard terminology and definitions of economic parameters. Industry/federal communication is vital to sound energy program planning and resources management. A common economic language is also a basic and essential element in such communication.

Normalized present worth or discounted cash flow analytical procedures using jointly endorsed parameters and values for those parameters are required as the means for federal evaluation of alternatives, the evaluation of policy impacts, etc. They also are fundamental to ultimate acceptance and application of EOR technology by non-EOR-knowledgeable companies.

Federal participation in EOR field testing would accelerate production by EOR methods. While several of the large integrated companies can essentially saturate their capabilities for EOR field testing without federal cost sharing or participation of any kind, this is not generally true throughout the industry. Operating under reasonable ground rules and procedures, a federal/industry cooperative program appears to be desirable to a majority of the EOR-knowledgeable companies. It should achieve a significant compression of the time scale for EOR application and expand the total U.S. capability for its application. The establishment of firm and realistic federal policies and procedures is another essential ingredient for the level of effort and time-scale compression needed.

Even those companies having sufficient manpower, resources, and EOR technology to proceed with large-scale field testing independently of federal participation would benefit in terms of both time-scale compression and cost reduction from a cooperative program which would permit all EOR practitioners access to pertinent EOR test data and information. While mechanisms would be required for some companies which permit the developer of a successful EOR technology to realize a return on his RDT&E investment, there appear to be no other serious obstacles to effective EOR test data and information exchange as required to achieve maximum time compression for the entire industry and thus assure the widest possible application of successful technology.

INCREASING DOMESTIC OIL PRODUCTION

All reasonable opportunities for increasing domestic production and reserves of crude oil must be pursued as rapidly as possible. National demand far exceeds domestic production. With a decline in crude oil production from known domestic fields projected for this decade and imports limited, a combination of exceedingly successful conservation measures and the development of new oil supplies will be required to overcome the shortage.

A large part of the nation's energy requirements, particularly those for its transportation systems, must be satisfied by crude oil. It is a clear consensus in the energy industry that substitutes for crude oil will not be developed quickly enough to meet our energy needs.

There are few alternatives from which to choose in an attempt to increase the domestic supply of crude oil that will be of significance in this decade. Basically, the nation must find and economically exploit new domestic fields, or it must learn to extract a larger fraction of the oil in those already discovered. Within these limitations, only two alternatives require a large investment in the development and reduction to practice of new technology if they are to result in the needed maximum possible contribution to the domestic energy supply. These are:

(a)   The application of enhanced oil recovery methods to increase production and ultimate recovery from known domestic fields over that which would be realized with conventional production technology; and

(b)   The discovery and development of new oil fields in U.S. Outer Continental Shelf (OCS) areas.

Both of these alternatives are speculative insofar as quantitative estimates are concerned. Neither is an option which can be selected on the basis of established quantitative information. Neither is expected to result in cumulative or annual production during this decade which is equal to the projected decline in production from known domestic fields. Nor in combination with all alternatives for domestic production are they expected to result in a higher annual production in 1985 than the current level. Therefore, the exclusive reliance on either is dangerous.

Offshore Potential

Relative to new discoveries of crude oil in the Outer Continental Shelf areas, it is generally agreed that the discoverable recoverable quantities of crude oil in OCS areas are, indeed, significant. Credible estimates approximate domestic cumulative production to date. Industry consistently contends that the greatest potential for additional petroleum reserves lies offshore. The discovery and development of these recoverable crude oil resources can proceed at a rate which is very significant in terms of domestic production during this decade. Production from new offshore fields in 1985 is projected to amount to 80 to 90% of the decline in production from known domestic fields at that time. The offshore production industry consistently states that new offshore discovery is the quickest way to produce the most oil.

The offshore production industry and its engineering/construction and service components can and will develop the technology required to exploit these resources at a rate which is consistent with need and opportunity. The single most effective action that could be taken by the federal government to speed up discoveries and increase oil producibility would be to take steps to simplify and speed up industry exploration and development of offshore areas.

Contrasts Between OCS and EOR Development

Technology: *Status* — That technology which determines capability for offshore production and transportation is ocean engineering technology; it does not involve advanced methods of petroleum production. In contrast, EOR technology for onshore fields is not dependent upon above-ground technology but is entirely dependent upon advances in the understanding and predictability of physical/chemical processes in the reservoir. Industry is confident of forecasts of offshore technology advances. It is much less so regarding the probable success of an EOR operation in a given reservoir because of inadequate field test or production experience.

Offshore technology is generally applicable for a specified range of water depths within a particular geographic region. Within these broad limitiations, it is relatively independent of reservoir characteristics. In contrast, an EOR operation must be designed for the specific reservoir. In light of today's level of understanding, the probability of success in a given reservoir can be estimated reliably only after completion of a large-scale field test which is designed using the best information obtainable from laboratory and small-scale field tests. The results from one reservoir cannot now reliably be extrapolated to another despite similarity of many of the reservoir parameters; such interfield extrapolation is a primary objective of the cooperative federal/industry field test program recommended.

The uncertainty applicable to offshore ventures is whether oil will be found in sufficient quantities to justify the cost of production. The uncertainty applicable to EOR ventures is whether production, itself, will be successful.

Pertinent Industry Activity — EOR technology is concentrated in the large and intermediate-size producing companies. At pre-1973 prices, EOR methods were not economically attractive so that only those companies that could afford to invest in long range research (based on the probability of future shortages and higher prices) could indulge in this expensive R&D effort. For even these companies, large-scale field tests have not, in general, been justifiable risks. Also, both the highly competitive nature of the industry and legal necessity have thus far denied an effective exchange of EOR information. Therefore, capability is concentrated and competent manpower is limited. The rise in oil prices has resulted in a severalfold increase among the EOR-knowledgeable companies in development and field test activites.

Because of joint ownership in fields, some of these are multicompany efforts. Also, most of the technology is available for license; however, the unpredictability of risk has, thus far, severely limited its application. A further deterrent to a rapid build-up of activity is the availability of materials; e.g., a large-scale field test of a micellar flood method can require several million pounds of surfactants and polymers, which are not readily available in these quantities.

Conclusion — Acceleration of EOR application is directly dependent upon the rate at which well-designed and effective field tests can be carried out. In general, EOR technology would not be used in competition for field ownership; ownership of known domestic fields is already established. Therefore, except for compensation for prior R&D expenditures, there is not the level of incentive to keep EOR technology proprietary that exists for information pertaining to oil discovery.

Even the largest EOR-knowledgeable companies express a need for data and information exchange as being the most effective means for accelerating EOR technology development, beginning with cooperative field test design and extending through test interpretation. It appears that this can be accomplished most immediately and successfully if the federal government is a participant in such a cooperative effort; in fact, many companies consider an industry-wide cooperative research effort of any kind to be legally infeasible, or at least dangerous, without direct federal participation in the program.

Investment Comparisons: *Economic Analyses* — In the project-by-project selection procedures used in industry, EOR projects must be evaluated in competition with new OCS discovery and other investment possibilities. Comparative analyses of new OCS discovery with onshore EOR application have demonstrated the principal uncertainties associated with both OCS and EOR ventures and emphasized the speculative nature of each.

New discoveries in the OCS are highly sensitive to the quantity of recoverable oil that is found; this parameter, then, assumes primary importance in a Present Worth (PW) or Discounted Cash Flow (DCF) analysis. OCS new discovery costs are dominated by lease bonuses. If the venture results in finding a recoverable quantity of oil equal to that estimated for bonus bid purposes, a PW analysis of representative OCS total revenue time-profiles indicates that total capital investment is about 15%, operating costs are about 15%, taxes are about 20%, and bonus cost is up to 30% of total revenue.

On this basis, an investment of 35 to 40% of total anticipated revenue is made prior to discovery, and profits are about 20% if the discovery is as large as was estimated. The adverse effects on profits of lower-than-estimated recoverable oil discovered is apparent. The effect of a dry hole after an investment of 35 to 40% of expected total revenue on an estimated 50-million-barrel field is equally apparent. In this regard, recoverable quantities of less than about 10 million barrels are, economically, dry holes for water depths of a few hundred feet or more.

The PW of an EOR operation is most sensitive to production rate, i.e., the additional quantity produced and the time required for production. PW analysis of a completely successful chemical flood operation indicates that total capital investment is about 5 to 10%, materials (a front-end cost) and operating expenses are about 30 to 35%, and taxes are about 20 to 30% of total revenue.

Pertinent Industry Activity — Industry's investments in oil discovery and recovery continue to amount to a large percentage of corporate net income. Because of the high risk and high cost, even the largest companies are joint-bidding for OCS leases and, thereby, reducing individual project risk by participating in a larger number of projects. Industry, however, is unable to approach EOR on a multicompany shared-risk basis except as it applies to preproduction testing of a specific field and involves only the ownership of that field. Jointly sponsored RDT&E is considered by many large companies to be a definite legal risk. Thus, EOR field tests conducted for RD&T purposes must, in general, be borne by individual companies.

Conclusion — Federal cost-sharing in a large number of cooperative field tests of EOR methods, at even a 10% success probability, is a low-risk high-pay investment. A highly simplified rationale indicates that, using industry projections of probable increase in reserves by EOR methods, the average recovery attributable to a successful field test would be one billion barrels. It should be noted that this amount would result, equally, if the probability is 1:5 at one-half the incremental increase in recovery and could, therefore, be attributable to two field tests, etc. It is notable, also, that the largest incremental production operation reported as being under way is based on an expected EOR production of 1.1 billion barrels.

Lacking an experience-based success ratio, the simplified analysis appears justified for policy formulation purposes. Thus, federal investment in a number of tests, which is sufficiently large to provide high probability of one or more successful operations, is a very good one. From the severalfold increase in field test activities among some of the EOR-knowledgeable companies, it appears that even a smaller number is, in their opinion, at least a reasonable risk as compared with other investments.

Further, using the industry projection of field test costs and of 1980-1985 production as a result of those field tests, a PW analysis of the federal share of the investment in those tests indicates that it would be returned from corporate income taxes alone in 2 to 3 years following test program completion, or 7 to 8 years from program initiation.

Summary: The following comparisons of offshore new discovery and EOR technology are pertinent to the overall appraisal of these alternative potential energy sources:

(1) Offshore technology is an ocean engineering technology; EOR technology is a reservoir engineering/petroleum geology technology.

(2) Offshore technology is widely applicable, its range of application being environment/geography-dependent rather than reservoir or field-dependent; EOR technology is reservoir-specific.

(3) Offshore operations are characterized by massive structures, vehicles and equipment; EOR operations are characterized as sophisticated chemical/thermodynamic process control operations.

(4) The probability that an R&D expenditure on offshore technology will be unproductive of useful knowledge is quite low, that of an EOR field test being unproductive of valid knowledge which produces an advance in the state of the art is quite high. Effective EOR field test design requires the combination of the best knowledge available from all sources, whereas small competent groups can make very significant contributions to offshore technology.

(5) Scale models can provide a quantitatively valid approach to offshore technology problem solution and design; EOR processes cannot be scaled.

(6) The amount, complexity and range of method/reservoir/operating data pertinent to the interpretation of EOR field test data, to the selection of reservoirs as EOR production candidates, to the selection of the method(s) most effective for a specific reservoir, and to the extrapolation of test data, require a highly compressed, completely addressable data-management system; this does not appear to be true for offshore technology development and application except as required for the interpretation of massive environmental data bases pertinent to environmental regulations and control or to the development of environmental design criteria.

(7) Offshore technology requires no federal participation in order to accelerate crude oil production and increase domestic reserves from continental shelf areas; EOR technology development will require federal participation if it is to be advanced on an accelerated time scale as necessary to develop the significant production during this decade for which it has the potential.

(8) Offshore technology can be developed by individual partnerships without significant time delays in advancing the state of the art; EOR technology can be accelerated only by (a) an effective exchange of data and information which permits appraisal and correlation of a very large number of variables as obtained from a large number of organizations, or (b) a technological breakthrough, which is not considered to be an imminent possibility.

(9) Offshore technology appears to offer opportunities for the entrepreneur, and offshore production operations are possible for essentially any company which can pay the price of its fractional share of a consortium price. EOR technology requires scientific and engineering sophistication in a large number of disciplines and specialties such that, as experience has shown, only the larger companies with such manpower and the facilities required for their work can indulge. While EOR turnkey operations can be purchased for a price, the evaluation of the proposal for them would re-

quire an equally high level of sophistication and considerable disagreement among consulting specialists if they were
employed.

Applicability of EOR Processes to Offshore Production

It should also be noted that opinion has been divided on the extent of enhanced
oil recovery that can be done offshore in view of the limited number of wells
whose geology will permit such practice. In fact, the only process found to be
generally economical in the offshore reservoirs at the world oil price is CO_2 miscible. Other processes are considered to be economical in only a very few reservoirs. Application would require a highly accelerated development program and
would probably be prohibitively expensive, due to added material and equipment
transportation costs and the difficulties of operating offshore. The environmental impacts of tertiary recovery processes offshore do not appear to have been
investigated.

Until tertiary recovery becomes economical for use in land-based reservoirs, it
will not be contemplated for use offshore. It does not appear that tertiary techniques will play a significant role offshore for another 15 to 20 years (5).

INDUSTRY'S CHOICE

Energy companies, by the very nature of their procedures for judging investment
opportunities, will select those production possibilities, whether they are onshore
or offshore and whether they are primary, secondary or enhanced recovery operations, on the basis of each company's evaluation as to which will yield the most
crude oil for the least cost under conditions which pertain to that company's operations. The allocation of industry capital and other resources to the advance-
ment of technology will, therefore, be dictated as to nature, scope and rate by
such appraisals which are based on the most sophisticated means available for estimating potentials and risk.

Competition between offshore new discovery and EOR onshore possibilities, in
terms of selecting one to the exclusion of the other, is not realistic from an industry point of view. The energy industry consistently recommends that both
offshore new discoveries and the application of EOR technology be pursued at
the maximum rate consistent with opportunity and risk. They should both be
viewed as essential components in the nation's efforts to become independent of
imports for its energy supply.

Enhanced oil recovery can play an important role in seeking energy self-sufficiency. But the oil industry and its tertiary oil recovery efforts are at a cross-
roads. The industry can proceed at a modest measured pace, seeing prospects
dwindle as the domestic oil fields decline and are shut down; it can respond to
the improved economic and regulatory conditions but, given limited federal involvement, the risky capital intensive nature of the task, and the indirect, delayed
return from research, proceed at a moderate pace; or it can launch a major concerted effort, joined by creative federal leadership and research, focused on targets of high potential and on proving this new technology.

Which choice the industry will make depends heavily on near-term decisions of
the federal government. Reaching maximum production demands that sizable

resources be committed and that technological constraints be overcome. The potential capacity of the industry and the federal government appears to be sufficient to the task; however, it will need to be harnessed and carefully guided.

REFERENCES

(1) ERDA Workshops on Thermal Recovery of Crude Oil, University of Southern California (March 29–30, 1977).

(2) *Technical Plan for a Supplementary Research Program to Support Development and Field Demonstration of Enhanced Oil Recovery*, for U.S. Energy & Development Administration, Washington, D.C., GURC Report No. 154 (March 17, 1977).

(3) *Chemicals for Microemulsion Flooding in EOR*, Gulf Universities Research Consortium (March 1976).

(4) *Projection of Educational Statistics to 1985–86*, National Center for Educational Statistics, Publ. NCES 77/402, p. 32 (1977).

(5) *Tertiary Oil Recovery: Potential Application and Constraints*, PNL Report RAP-25, prepared by C.A. Geffen, of Battelle's Northwest Laboratory for U.S. Department of Energy, p. iii (June 1978).

GLOSSARY

This glossary has been adapted mainly from information in:

Planning Criteria Relative to a National RDT&E Program Directed to the Enhanced Recovery of Crude Oil and Natural Gas, ORO-4507-1, prepared by J.R. Crump and J.M. Sharp, of Gulf Universities Research Consortium, November 30, 1973.

Potential Environmental Consequences of Tertiary Oil Recovery, prepared by C. Braxton, R. Stephens, C. Muller, J. White, J. Post, J. Norton, M. Goldberg and P. Stevenson, of Energy Resources Co., Inc., for U.S. Environmental Protection Agency, under Contract 68-01-1912, July 1976.

Enhanced Oil Recovery Potential in the United States, OTA-E-59, prepared by Lewin & Associates, Inc., National Energy Law and Policy Institute, and Wright Water Engineers, Inc., for U.S. Office of Technology Assessment, January 1978.

Mining for Petroleum: Feasibility Study, BuMines Report OFR 56-79, prepared by J.S. Hutchins, E. Bond, and D.M. Bass, of Energy Development Consultants, Inc., for U.S. Bureau of Mines, July 1978.

API Gravity

The standard American Petroleum Institute (API) method for specifying the density of crude oil. It may be calculated from the following formula: $^\circ\text{API} = [141.5/(\text{specific gravity at } 60^\circ\text{F})] -131.5$.

Aquifer

A water-bearing layer of permeable rock, sand, or gravel.

Barrel

A liquid volume measure equal to 42 U.S. gallons.

Brine

Water saturated with or containing a high concentration of sodium chloride and other salts.

Centipoise

A unit of viscosity equal to 0.01 poise. A poise equals 1 dyne-second per square centimeter. The viscosity of water at 20°C is 1.005 centipoises.

Connate Water

Water that was laid down and entrapped with sedimentary deposits, as distinguished from migratory waters that have flowed into deposits after they were laid down.

Core

A sample of material taken from a well by means of a hollow drilling bit. Cores are analyzed to determine their water and oil content, porosity, permeability, etc.

Created Fractures

Fractures induced by means of hydraulic or mechanical pressure exerted on the formation.

Crude Oil

A mixture of hydrocarbons that exists in the liquid phase in natural underground reservoirs and remains liquid at atmospheric pressure after passing through processing facilities that separate out some components. The reported volumes of crude oil also contain a very minor amount of other liquids which is statistically insignificant but technologically either difficult to separate at one place or desirable to include at another. Crude oil is reported in units of stock tank barrels of 42 U.S. gallons at atmospheric pressure corrected in volume to 60°F.

Cyclic Steam Injection

The injection of steam down a producing well for a period of days or weeks, followed by soaking for days, to heat up the area around the well bore and increase recovery of oil immediately adjacent to the well; also called steam stimulation, steam soak or huff and puff.

Darcy

A unit of permeability. A porous medium has a permeability of 1 darcy when a pressure of 1 atm on a sample 1 cm long and 1 cm^2 in cross section will force a liquid of 1-cp viscosity through the sample at the rate of 1 cm^3/sec.

Emulsion

A suspension of one finely divided liquid phase in another.

Enhanced Oil Recovery Process

A technique for recovering additional oil from a petroleum reservoir beyond that economically recoverable by conventional primary and secondary recovery. It may be a:

Thermal Recovery: Injection of heat into a petroleum reservoir in the form of steam or propagation of a combustion zone through a reservoir by air injection into the reservoir.

Miscible Flooding: Injection into a reservoir of a material such as carbon dioxide that is miscible, or nearly so, with the oil in the reservoir.

Chemical Flooding: Injection of water with added chemicals into a petroleum reservoir. The two chemical types employ (a) surfactants and (b) polymers.

EOR

Enhanced oil recovery.

Fireflooding

A synonym for in situ combustion, q.v.

Flip-Flop Process (Open Surface Flotation)

An in situ method of mining heavy petroleum which is close to the surface by the use of hot brine saturated with surfactants to reduce oil viscosity, causing the oil to flow up and out while the brine flows down into the formation, effectively interchanging (Flip-Flopping) position of the brine and oil.

Forward Combustion

A thermal oil recovery process in which air is injected, and ignition is obtained at the well bore in an injection well. Continued injection of air drives the combustion front toward producing wells.

Fracture

Any kind of discontinuity in a body of rock if produced by mechanical failure, whether by shear stress or tensile stress. Fractures include faults, shears, joints, and planes of fracture cleavage.

Gravity Drainage

An in situ method of mining light crude petroleum by mining underneath a reservoir and producing from below by controlled gravity flow.

Huff and Puff Method

Another name for cyclic steam injection, q.v.

Incompetent

Term applied to heavy oil sand formations which tend to slump or cave into any voids created where the oil becomes mobile, because their sand grains are not cemented to one another.

Injection Well

A well used for moving fluids from an oil field into a reservoir.

In Situ Combustion

A process which heats oil in the reservoir to increase its mobility by decreasing its viscosity. Heat is applied by igniting the oil sand or tar sand and keeping the combustion zone active by the injection of air.

In Situ Mining of Petroleum

A method which uses mining technology to get close to a hydrocarbon reservoir and then petroleum technology to extract the hydrocarbon, leaving the reservoir rock in place.

Interfacial Tension

The contractile force of an interface between two phases.

Micelle (and Micellar Fluid)

A molecular cluster or aggregate, generally of molecules that have an oil-seeking and a water-seeking end. It dissolves other fluids into its center, swelling to accommodate the solubilized liquids. An oriented layer of such molecules on the surface of a colloidal droplet stabilizes oil-in-water or water-in-oil emulsions, making oil and water quasi-miscible.

Miscible

Able to mix together; refers to two or more substances. Liquids that are not miscible separate into layers according to their specific gravities.

Miscible Agent

A substance miscible with each of two immiscible liquids which is used to reduce the interfacial tension between the two immiscible liquids; in oil recovery, a substance that promotes miscibility between water and oil, such as natural gas, hydrocarbon gas enriched with liquefied petroleum gas, etc.

Miscible Displacement

Displacement of oil by a fluid with which it is miscible. When such a fluid contacts the oil, the two liquids dissolve each into the other and form a single phase. There is no interface between the fluids and hence there are no capillary forces active.

Miscible Displacement Recovery

The use of various solvents to increase the flow of crude oil through reservoir rock.

Mobility

A measure of the ease with which a fluid moves through reservoir rock, expressed as the ratio of rock permeability to fluid viscosity.

Oil Saturation

The extent to which the voids in rock contain oil, usually expressed in percent related to total void.

Original Oil in Place (OOIP)

Calculated volume of crude oil in known reservoirs prior to any production. Known reservoirs include:

(1) those that are currently productive;
(2) those to which proved reserves have been credited but from which there has been no production; and
(3) those that have been depleted.

The volume of original oil in place is based on calculations using volumetric or material balance methods when sufficient factual data are available concerning reservoir rock, fluid properties, reservoir limits, and production performance. Where such data are not available, or are seriously incomplete, the volume is estimated on the basis of information and performance characteristics from reservoirs believed to be comparable.

Permeability

Capacity of rock for transmitting a fluid. Degree of permeability depends upon the size and shape of the pores and the size, shape and extent of the interconnections. The unit of permeability is the darcy.

Polyacrylamide

Synthetic combination of carbon, hydrogen and oxygen formed into a long chain molecule, which can be used to increase the efficiency of waterflooding in petroleum recovery.

Polymer

A type of organic compound characterized by a large-chain molecule formed by thousands of repeating blocks called monomers; it is added to water for polymer flooding.

Polymer-Augmented Waterflooding

A type of enhanced oil recovery in which high molecular weight chemicals (polymers) are added to water used in waterflooding to thicken it, thus reducing the tendency of water to bypass oil in less permeable portions of the reservoir.

Polysaccharide

A carbohydrate polymer commonly used in polymer-augmented water-flooding.

Pore Volume

The volume of void space in an oil reservoir which may contain petroleum, gas and/or brine.

Porosity

The fraction of the total volume of a material that is made up of empty space, or pore space. It is the volume of pore space expressed as a percentage of the total volume of the rock mass; measures the absorbent capacity of the material or the volume of liquid held by the pores.

Potential Additional Recovery

That volume of remaining oil in place that is estimated to become recoverable by improved methods not yet available or not yet sufficiently tested for commercial application, the future development of which is judged to be feasible under future conditions of technology.

Pressure Maintenance

An artificial secondary means of maintaining pressure in an oil formation by injecting natural gas into or near the reservoir.

Primary Oil Recovery

Removal of petroleum from a reservoir using only naturally occurring forces or mechanical or physical pumping methods.

Remaining Oil in Place

That volume of the original oil in place that would remain in presently discovered reservoirs after production of ultimate recovery and indicated additional recovery if no improved recovery methods nor changed economic conditions come into existence to increase directly or indirectly the recovery efficiency of present-day operations.

Reserves

The amount of a mineral expected to be recovered by present-day methods and under present economic conditions.

Proved Reserves: As of December 31 of any given year, the estimated quantities of all liquids statistically defined as crude oil, which geological and engineering data demonstrated with reasonable certainty to be recoverable in future years from known reservoirs under existing economic and operating conditions. Reservoirs are considered proved if economic producibility is supported by either actual production or conclusive formation tests. The area of an oil reservoir considered proved includes: (1) that portion delineated by drilling and defined by gas-oil or oil-water contacts, if any; and (2) the immediately adjoining portions not yet drilled but which can be reasonably judged as economically productive on the basis of available geological and engineering data. In the absence of information on fluid contacts, the lowest known structural occurrence of hydrocarbons controls the lower proved limit of the reservoir.

Indicated Additional Reserves: Crude oil potentially available from known productive reservoirs in existing fields expected to respond to improved recovery techniques such as fluid injection where (1) an improved recovery technique has been installed but its effect cannot yet be fully evaluated; or (2) an improved technique has not been installed but knowledge of reservoir characteristics and the results of a known technique installed in a similar situation are available for use in the estimating procedure. The economic recoverability of these reserves is not established with sufficient conclusiveness to allow them to be included in Proved Reserves. If and when improved recovery techniques are successfully applied to known reservoirs, the corresponding Indicated Additional Reserves will be reclassified and added to the inventory of Proved Reserves. Indicated Additional Reserves do not include reserves associated with acreage that may be added to the area of a proved reservoir as the result of future drilling.

Reservoir

A discrete section of porous rock containing an accumulation of oil or gas, either separately or as a mixture.

Reservoir Fluids

Fluids contained within the reservoir under conditions of reservoir pressure.

Residual Oil

The amount of liquid petroleum remaining in the formation at the end of a specified production process.

Reverse Combustion

An in situ type of thermal oil recovery in which the formation is ignited at the producing well and the combustion zone moves countercurrent to the injected air and reservoir fluid stream. Because the oil flows into a zone already heated, there is no tendency for it to congeal and decrease permeability.

Secondary Recovery

The injection of water or natural gas into a petroleum reservoir in the second stage of recovery to augment the natural energy of the reservoir and thus increase production.

SolFrac Recovery

A method of enhanced oil recovery utilizing chemical-explosive fracturing to enable treatment with solvent injection and improve mobility.

Steam Displacement

A synonym for steam drive, q.v.

Steam Drive (Steam Displacement or Steamflooding)

The injection of steam into a group of outlying wells to push oil toward the production wells. The heat is pushed into the reservoirs to displace oil and reduce viscosity.

Steamflooding

A synonym for steam drive, q.v.

Steam Soaking (Steam Stimulation)

A synonym for cyclic steam injection, q.v.

Sulfonates

A surfactant consisting of an organic chain with an SO_3H attached; used in oil recovery because the polar SO_3H end is attracted to water, while the organic end mixes with oil.

Surface Tension

The tension forces existing in the extreme surface film of an exposed liquid surface due to unbalanced cohesive forces within the body of the liquid.

Surfactant

A material which affects the interfacial tension between two liquids. In oil recovery, surfactants are used to reduce the interfacial tension between oil and water. Each surfactant molecule has a polar end, which is attracted to water, and an organic chain, which is attracted to oil.

Sweep Efficiency

The ratio of the volume of rock contacted by the displacing fluid to the total volume of rock subject to invasion by the displacing fluid.

Tertiary Recovery

The addition of energy to an oil reservoir through the use of fluids or solvents under pressure to increase production following primary and secondary recovery.

Thermal Recovery

A petroleum recovery process that utilizes heat to thin viscous oil in an underground formation, causing it to flow more readily towards producing wells.

Ultimate Recovery

Volume of crude oil which has been produced from a reservoir, and is expected to be produced in the future if there are no substantial changes in present economic relationships and known production technology. Accordingly, the current estimate of ultimate recovery is the sum of Cumulative Production to date plus the current estimate of Proved Reserves. Ultimate Recovery may also be expressed as the percentage of Original Oil in Place which is expected to be eventually produced. This percentage will vary from one reservoir to another in accordance with the reservoir fluid, rock characteristics, and the producing mechanism or drive which is present.

Viscosity

The internal resistance offered by a fluid to flow. Attributable to the attraction between molecules of a liquid, this phenomenon is a measure of the combined effects of adhesion and cohesion between suspended particles and the liquid environment.

Waterflooding

A secondary recovery operation in which water is injected into a petroleum reservoir to create a water drive to increase production.

Wettability

As applied to petroleum reservoirs, the relative affinity of the coexisting oil and water phases to adhere to the surface of a rock. If the rock is predominantly in contact with water, it is said to be preferentially water-wet. Similarly, if the rock is predominantly covered with oil, it is referred to as oil-wet.

Wetting

The adhesion of a liquid to the surface of a solid.

Wetting Agent

A substance or composition which, when added to a liquid, increases the spreading of the liquid on a surface or the penetration of the liquid into a material.

WOR

Water to oil ratio.

SOURCES UTILIZED

BERC/RI-77/16
: *Heavy Oil Recovery Using Solvents and Explosives,* by L.J. Heath, F.S. Johnson, J.S. Miller, R.A. Jones and W.D. McMurtrie, of Bartlesville Energy Research Center, for U.S. Department of Energy, December 1977.

BuMines OFR 4-75
: *The Benefits/Costs of Tertiary Oil Recovery,* by Arthur D. Little, Inc. for U.S. Bureau of Mines, December 1974.

BuMines OFR 55-79
: *Oil Mining. A Technical and Economic Feasibility Study of Oil Production by Mining Methods,* by A. Edey, B.A. Kennedy and L.A. Readdy, of Golder Associates, Inc., for U.S. Bureau of Mines, October 1978.

BuMines OFR 56-79
: *Mining for Petroleum: Feasibility Study,* by J.S. Hutchins, E. Bond and D.M. Bass, of Energy Development Consultants, Inc. for U.S. Bureau of Mines, July 1978.

BuMines OFR 60-74
: *Evaluation of Thermal Methods for Recovery of Viscous Oils in Missouri and Kansas,* by M.D. Arnold and A.H. Harvey, of Department of Mining, Petroleum and Geological Engineering, University of Missouri at Rolla, for U.S. Bureau of Mines, June 1974.

BuMines RI-7978
: *Solvent Stimulation Tests in Two California Oilfields,* by H.J. Lechtenberg, G.L. Gates, W.H. Caraway and O.C. Baptist, of San Francisco Energy Research Laboratory, for U.S. Bureau of Mines, 1974.

BuMines TPR 60
: *Solvents and Explosives to Recover Heavy Oil, Bartlett, Kansas,* by L.J. Heath, F.S. Johnson and J.S. Miller, of Bartlesville Energy Research Center, for U.S. Bureau of Mines, September 1972.

CONF-7406129-1
: *Heavy Oil Recovery Using the SolFrac Method,* by F.S. Johnson and R.T. Johansen, of Bartlesville Energy Research Center, for U.S. Bureau of

Mines and presented at Interstate Oil Compact Commission Mid-Year Meeting, Vail, Colorado, June 30-July 4, 1974, and published in IOCC Bull. 16, No. 1, 40-49, June 1974.

CONF-7504932
Laboratory Investigation Using Solvent to Recover Heavy Oil from a Fractured Reservoir, by F.S. Johnson, R.A. Jones and J.S. Miller, of Bartlesville Energy Research Center, for U.S. Bureau of Mines and presented at Rocky Mountain Regional Meeting, Soc. Petrol. Eng., AIME, Denver, April 7-9, 1975. SPE Preprint 5336.

CONF-790805-P1
Fifth Annual DOE Symposium on Enhanced Oil & Gas Recovery & Improved Drilling Technology, Vol. 1—Oil, sponsored by U.S. Department of Energy, Tulsa, Oklahoma, August 22-24, 1979.

CONF-790805-P2
Fifth Annual DOE Symposium on Enhanced Oil & Gas Recovery & Improved Drilling Technology, Vol. 2—Oil, sponsored by U.S. Department of Energy, Tulsa, Oklahoma, August 22-24, 1979.

EPA-68-01-1912
Potential Environmental Consequences of Tertiary Oil Recovery, by C. Braxton, R. Stephens, C. Muller, J. White, J. Post, J. Norton, M. Goldberg and P. Stevenson, of Energy Resources Co., Inc., for U.S. Environmental Protection Agency, July 1976.

EPA-68-01-2445
The Estimated Recovery Potential of Conventional Source Domestic Crude Oil, by J.W. Devanney III, R. Ciliano and R.J. Stewart, of Mathematica, Inc., for U.S. Environmental Protection Agency, May 1975.

ERDA-tr-163
"Treatment of Injection Waters," by M. Peinado, *Rev. Inst. Fr. Petr. 20:* No. 2, 280–290, 1965, translation by U.S. Joint Publications Research Service for Energy Research and Development Administration, 1976.

FEA/B-76/221
The Potential and Economics of Enhanced Oil Recovery, by Lewin and Associates, Inc., for Federal Energy Administration, April 1976.

FEA/G-75/482
Review of Secondary and Tertiary Recovery of Crude Oil, by Lewin and Associates, Inc., for Federal Energy Administration, June 1975.

JPRS 59039
"Oil Recovery by Nuclear Explosion," *Nauka i Zhizn* (U.S.S.R.), No. 2, 1973, by L. Shadrin, translation by U.S. Joint Publications Research Service, May 17, 1973.

NSF-C-942 (GURC 140)
An Investigation of Primary Factors Affecting Federal Participation in R&D Pertaining to the Accelerated Production of Crude Oil, by J.M. Sharp, of Gulf Universities Research Consortium, for National Science Foundation, September 15, 1974.

NSF-C-942 (GURC 140-S)
A Survey of Field Tests of Enhanced Recovery Methods for Crude Oil, by J.M. Sharp, of Gulf Universities Research Consortium, for National Science Foundation, December 27, 1974.

NSF-GP-44165

Basic Research Needs for Tertiary Oil Recovery, by R.S. Schecter and W.H. Wade, of the University of Texas at Austin, for National Science Foundation Workshop, June 26-27, 1974.

NSF-RA-N-74-251

Assessment of Enhanced Recovery Technology as a Means for Increasing Crude Oil Recovery in Texas, by E.A. Lohse, of Texas Governor's Advisory Council, for National Science Foundation, January 6, 1975.

OTA-E-59

Enhanced Oil Recovery Potential in the United States, by Lewin & Associates, Inc., National Energy Law and Policy Institute, and Wright Water Engineers, Inc., for U.S. Office of Technology Assessment, January 1978.

PNL-RAP-25

Tertiary Oil Recovery: Potential Application and Constraints, by C.A. Geffen, of Battelle's Pacific Northwest Laboratory for U.S. Department of Energy, June 1978.

SFERC/RI-76-1

Laboratory Study of Heavy Oil Recovery Using Polymer and Solvent, by G.D. Peterson, H.J. Lechtenberg, W.H. Caraway and G.L. Gates, of San Francisco Energy Research Center, for U.S. Energy Research and Development Administration, February 1976.

UNCONVENTIONAL NATURAL GAS
Resources, Potential and Technology 1980
Edited by M. Satriana

Energy Technology Review No. 56

Oil imports could be reduced and domestic gas production increased if additional gas production were obtained from four unconventional sources—tight gas basins, Devonian shale, coalbeds, and geopressured aquifers. Gas produced from these resources could help maintain overall production levels as supplies from conventional sources gradually decline. The Federal Energy Regulatory Commission's partial decontrol of high-cost gas sources such as these will provide the incentive for development of unconventional natural gas potential, also referred to as enhanced gas recovery (EGR).

The purpose of this book is to describe in detail unconventional natural gas resources, and their potential, and to discuss the current state of the art of production technology, with special emphasis on tight gas basin and Devonian shale technology, particularly the various fracturing techniques which can be employed.

A partial and condensed table of contents follows here. **Chapter headings and important subtitles** are given.

1. OVERVIEW OF THE RESOURCES
Tight Gas Basins
Devonian Shale Gas
Methane from Coal Seams
Methane from Geopressured Aquifers
The Resource Base
Industry Perspective
Government Actions to Spur
   Development
Future Production Possibilities

2. TIGHT GAS BASINS
Types & Properties
Stimulation Technology
Massive Hydraulic Fracturing (MHF)
Economics
Resource Potential
The Problem & the Potential

3. DEVONIAN SHALE
The Resource
General Reservoir Characteristics
Alternate Economic Analysis

4. METHANE FROM COAL SEAMS
Historical Research Efforts
Industry Criteria for Commercialization
Technology

Identification of Problems
Legal Concerns
Environmental Considerations
R&D Program Summary

5. METHANE FROM GEOPRESSURED AQUIFERS
Resource Evaluation & Assessment
   Technique
Recoverability of Dissolved Gas
Reservoir Classifications
Environmental Aspects
Economic Feasibility
Texas Gulf Coast Aquifers
Louisiana Gulf Coast Aquifers

6. FRACTURE FLUID-RESERVOIR INTERACTIONS IN TIGHT GAS FORMATIONS—LITERATURE REVIEW
Well Stimulation
Well Performance
Well Testing & Analysis
Fracturing Fluids
Gelling Agents
pH Control Agents
Fluid-Loss Additives
Friction-Loss Reducers
Fracture Proppants
Reservoir Matrix Characteristics

7. TECHNOLOGY STUDIES—TIGHT GAS BASINS
Piceance Basin—MHF
Cotton Valley Lime Matrix—MHF
Natural Buttes Unit—MHF
Fracture Mapping
Surface Electrical Potential System
Borehole Seismic System
Borehole Stacked Hydrophone System
Fracturing Fluid Interactions—Cotton
   Valley Limestone Formation
Fracture Containment Analysis
Fracturing Fluid Damage
Productivity Assessment
Foam Fracture

8. TECHNOLOGY STUDIES—DEVONIAN SHALES
Hydraulic Fracturing
Chemical Explosive Fracturing
Core Analysis
Stimulation Technology Evaluation
   Conventional Techniques
   Novel Techniques

ISBN 0-8155-0808-5

358 pages

HEALTH HAZARDS AND POLLUTION CONTROL IN SYNTHETIC LIQUID FUEL CONVERSION 1980

Edited by Perry Nowacki

Pollution Technology Review No. 68
Environmental Health Review No. 2
Chemical Technology Review No. 165
Energy Technology Review No. 57

This review addresses the environmental, health and pollution control aspects of coal liquefaction, oil shale processing, and tar sands conversion. The emphasis throughout is on well-documented processes and projects in the pilot scale or early commercial phases. After a preliminary discussion of composition and structural features peculiar to coal, oil shale, and tar sands, the major processing options, proprietary implementations, and associated pollution sources are outlined. Toxicity, carcinogenicity, synergetics, interactions, and residual effects of involved elements, trace elements, particulates, or compounds are treated at length.

The book will be of interest to engineers and biomedical and environmental scientists involved in the assessment and control of pollution hazards in synthetic liquid fuel conversion. A condensed table of contents is listed below with **chapter headings and important subtitles**.

ISBN 0-8155-0810-7

511 pages

SYNTHETIC OILS AND ADDITIVES FOR LUBRICANTS 1980
Advances Since 1977

Edited by M. William Ranney

Chemical Technology Review No. 145

Fully synthetic lubricants are highly resistant to oxidation and sludge formation having anti-wear and water tolerance not equalled by straight petroleum oils.

Synthesis and production of pure synthetics are still costly and complex however. Moreover it is claimed that fossil oils have certain lubricant qualities that could be superior to those of the synthetics.

A solution to the dilemma is to take the much less expensive natural hydrocarbon oil stock and make it resistant to deterioration by addition of synthetics without sacrificing any inherent lubricant power. This is the essence of this book.

Our prolific chemical industry has made available so many diverse fluids and additives intended to upgrade lubricating oil stock that almost any desired property can be emphasized. One need only to examine the following partial table of contents which gives **examples of some chapter headings and subtitles.** Numbers in ( ) indicate numbers of processes per topic.

ISBN 0-8155-0781-X

409 pages

FUNCTIONAL FLUIDS
FOR INDUSTRY, TRANSPORTATION AND AEROSPACE 1980

Edited by M. William Ranney

Chemical Technology Review No. 155

This review covers a variety of developments in functional fluid technology and formulation techniques. The major change, now taking place, is away from petroleum-based fluids. This tendency, while readily evident in the aircraft hydraulic fluid area, also forms the basis of much advanced research in related applications.

Many different types of liquids can be employed in many different types of machinery. Such fluids are used as electric and electronic coolants, diffusion pump fluids, pressure lubricants, damping fluids, force transmission fluids (hydraulic fluids), heat transfer fluids, in extrusion processes and in air-conditioning systems.

Composition, manufacture and applications of these fluids are accurately described in this patent-based book.

The partial table of contents below lists **numbered sections, chapter headings, examples of some subtitles** and, in parentheses, the number of processes per topic.

ISBN 0-8155-0789-5

364 pages

OFFSHORE OIL TECHNOLOGY

Recent Developments 1979

by M.William Ranney

Energy Technology Review No. 38
Ocean Technology Review No. 8

From the time the first submersible drilling structure was used in the Gulf of Mexico in the late 1940s, technology for offshore drilling developed rapidly to service the needs originated with this new era in oil and gas exploitation.

Today engineers are continuing to design new apparatus and systems to improve the world's supply of these resources. This text keeps the reader abreast of these latest developments.

The first chapter reviews the activity in offshore drilling to date, discussing the mobile and fixed structures and equipment employed. Subsequent chapters detail approximately 170 recent processes, many of which are even now being implemented throughout the industry. These processes include the key areas of platform erection, drilling operations, wellhead construction, and storage facilities. Over 80 figures and illustrations accompany the process descriptions.

The partial, condensed table of contents presented below with chapter headings and some subtitles demonstrates the comprehensive range of this text. The numbers in parentheses signify the total number of processes covered in each chapter.

1. **OFFSHORE DRILLING OVERVIEW**
   Background—History and Geology
   Survey Vessels
   Exploratory Drilling Rigs
   Development and Production Systems
   Transshipment and Storage
   Onshore Support Needs

2. **PLATFORM CONSTRUCTION (56)**
   Telescopically Mated Stanchions
   Hollow-Based Five-Column Tower
   Vertically Movable Legs
   Cellular Foam for Jacket Legs
   Structure for Earthquake-Prone Regions
   Anchoring System for Soft Substrates
   Self-Elevating Platforms
   Jacking Units for Template Legs
   Platform Self-Mobile on Land
   Vertically Moored Platform
   Grouting Systems
   Reinforced Elastomeric Seal
   Sealing Concentric Tubular Inserts
   Hollow Underwater Anchoring Apparatus
   Multiple Wells Through Single Caisson
   Parallelepipedic Box Deck Structure

Multiple Platform Crane
Interconnected Prefabricated Units

3. **MARINE RISERS AND WELL CONDUCTORS (24)**
   Hydraulic Heave Compensator
   Pressure-Compensated Dual Riser
   Risers for Platform Structures
   Well Conductors
   Dog-Leg Segment for Directional Drilling
   Well Conductor Pipe in Soft Floor
   Cutting and Recovery of Surface Casing

4. **DRILLING PROCESSES (16)**
   Fluid-Actuated Counterbalancing Means
   Drilling Fluid Diverter System
   Drilling Tool to Curb Debris Buildup
   Running and Retrieving Tools

5. **WELLHEAD APPARATUS (34)**
   Flow Control System
   Atmospheric Pressure Operation
   Reentry Guidance Apparatus
   Portable Atmospheric Cellar
   Christmas Tree Junction Housing
   Replacing Broken Guidelines
   Flow Lines and Pipe Connectors
   Using Two Flow Line Mating Vessels
   Tension Control in Flow Line Connection
   Dry Isolated Chambers in Base
   Guiding Technique for Diving Bell
   Self-Propelled Trenching Apparatus
   Multipressure, Single-Line Supply Means

6. **ARCTIC REGION OPERATIONS (18)**
   Movably Mounted Ice Cutter
   Fill Material on Natural Ice
   Ice Platform
   Ice Anchor
   Bottom-Supported Vessel
   Monopod Platform on Arctic Floor
   Casting Concrete Tanks Under Ice

7. **STORAGE AND PROCESSING (20)**
   Marine Pier with Submerged Storage
   Storage Tank Submarine
   Caisson-Defined Storage Area
   Mobile Processing Plant
   Underwater Sphere for Processing
   Oil Recovery from Underwater Fissures
   Automated Subsea Control System
   Work Arm System in Submerged Chamber
   Platform to Lower Blow-Out Preventer

ISBN 0-8155-0741-0   **399 pages**

CRUDE OIL DRILLING FLUIDS
1979

by Maurice William Ranney

Chemical Technology Review No. 121
Energy Technology Review No. 35

The drilling of oil wells into subterranean reservoirs presents many problems concerning the choice of a drilling fluid. Rotary drilling rigs are used nowadays almost universally throughout the world's oil regions. Perhaps the greatest advantages of rotary drilling over other methods is that the well bore is kept full of liquid during drilling. A weighted fluid (also called drilling mud) in the bore hole serves two important purposes: by its hydrostatic pressure it blocks the entry of formation fluids into the well, thus preventing blowouts and gushers. In addition, the drilling mud carries the crushed rock to the surface, so that the drilling is continuous until the bit wears out.

This apparent simplicity of the process is complicated by many factors, such as sealing the productive formation with a casing, which must have openings to allow the oil to enter the well under controlled conditions.

Many new, potentially oil-producing areas are being explored under oceans, and in arctic and desert regions. Increasing emphasis is also placed on secondary recovery techniques. Research and development activity therefore centers on improvements in drilling fluids, stimulation methods and secondary recovery processes involving water and polymer flooding techniques.

This book, focusing on the U.S. patent literature since 1974, encompasses over 250 processes. A partial and condensed table of contents follows here. Chapter headings and important subtitles are given. Numbers in parentheses indicate the number of processes per topic.

Note: These recent processes provide hundreds of formulations, techniques and new developments which will be used to efficiently recover crude oil from subterranean oil reservoirs in the coming years.

ISBN 0-8155-0732-1

348 pages

REPROCESSING AND DISPOSAL OF WASTE PETROLEUM OILS 1979

by L.Y. Hess

Pollution Technology Review No. 64
Chemical Technology Review No. 140

Recovery and judicious disposal of waste oil have become financially rewarding practices by reason of resource conservation and environmental protection.

Dirty and contaminated lubricating oil still has high energy values. It can be re-refined or used as a feedstock for making other petroleum products. Industrially, it can be reclaimed to nearly original quality by simple equipment.

Sufficiently worthwhile disposal practices include road oiling, combining with fuel oil, use as auxiliary fuel in municipal incinerators, and landspreading with decomposition by suitable microorganisms.

This book is based on the latest government research reports and U.S. patents which contain practical technological process information. A condensed table of contents follows here.

1. SOURCES AND CHARACTERISTICS
Generation and Disposal Figures
Automotive Service Centers
Commercial Truck Fleets
Railroad Service Centers
Aviation Service Centers

2. RECOVERY PROCESSES
Acid/Clay Process
Extraction Processes
Distillation

3. PROPRIETARY PROCESSES (118)
Treatment with Chemicals
Solvent Extraction
Adsorption
Distillation & Hydrogenation
Filtration
Other Separation Techniques
Application of Electric Fields

4. RESEARCH PROJECTS
Innovative Techniques
Ion Exchange Percolation
Chelation
Comprehensive Waste Oil Facility

5. DISPOSAL PRACTICES
Incineration and Use as Fuel
Environmental Aspects

6. DISPOSAL OF RECOVERY RESIDUES
Waste Products Generated
Acid Sludge
Caustic Sludge
Spent Clay
Distillation Bottoms
Characterization of Wastewaters
Marine Waste Oil Processing Facility
The Re-Refining Facility
Environmental Assessment

7. DISPOSAL AND RECYCLING OF AIR FORCE WASTE OILS
Generation of Waste POLs
Kelly AFB Survey
Andrews AFB Survey
Available Disposal/Recycle Techniques

8. COMBUSTION STUDIES OF WASTE POLs
Air Force Experimental Studies
Combustion Test Equipment
Energy Recovery
Waste Oil Burn-Off in Coast Guard
    Power Plants
Waste Lube Oil Added to Diesel Fuel

9. MUNICIPAL INCINERATOR FUEL
Physical and Combustion Properties of
    Waste Oil
Heating Requirements
Phenomena Occurring in a Refuse Bed
Transferring Heat Flux
Mixing Waste Oil Directly into Refuse
Utilizing Auxiliary Burners
Impact on Air Quality

10. DISPOSAL BY LANDSPREADING
Land Disposal Practices
Microbiological Studies
Shell Project Description and Objectives
Oil Decomposition Rate and
    Effect on Fertilizer
Metals Contents of Soils
Microbial Action
Rainfall Runoff
Union Carbide Study
Combined Oil/Nitrate Application
Summary

ISBN 0-8155-0775-5

322 pages